CLINICAL
ORBITAL ANATOMY

CLINICAL ORBITAL ANATOMY

Marcos T. Doxanas, M.D.

Division of Oculoplastic Surgery
Department of Ophthalmology
Greater Baltimore Medical Center
Baltimore, Maryland

Lecturer in Ophthalmology
Wilmer Institute
Johns Hopkins Hospital
Baltimore, Maryland

Richard L. Anderson, M.D.

Professor of Ophthalmology
Director of Oculoplastic, Orbital and Oncology Services
University of Utah
Salt Lake City, Utah
Formerly Professor of Ophthalmology
Director of Oculoplastics, Orbital and Oncology Services
University of Iowa
Iowa City, Iowa

WILLIAMS & WILKINS
Baltimore/London

Library of Congress Cataloging in Publication Data

Main entry under title:

Doxanas, Marcos T.
  Clinical orbital anatomy.

  Bibliography: p.
  Includes index.
  1. Eye-sockets—Anatomy. I. Anderson, Richard L. (Richard Lee), 1945–
II. Title. [DNLM: 1. Orbit—Anatomy and histology. WW 202 D752c]
QM511.D69   1984   611′.84   83-19803
ISBN 0-683-02650-X

Composed and printed at the
Waverly Press, Inc.

Preface

The last comprehensive books devoted to orbital and eyelid anatomy were written many years ago. While anatomy itself has not changed, many important discoveries have occurred in the interim. Our interpretations and utilization of anatomic principles and procedures have also advanced greatly. In the past decade, there has been a renewed interest in eyelid and orbital anatomy with the development of many new procedures based on anatomical principles. A detailed working knowledge of the anatomy of the eyelids and orbit is essential for performing these procedures. This book is intended not only as a comprehensive presentation of basic eyelid and orbital anatomy but also to relate the clinical importance of these orbital and eyelid structures. The practice of ophthalmology, ophthalmic plastic surgery, and orbital surgery is based on the orbital and eyelid anatomy described in this text.

With the belief that learning orbital anatomy does not have to be a distasteful task, we began our book. The reason that learning basic orbital anatomy is such an arduous task, and therefore neglected by most, is that the clinical applications of basic orbital and eyelid anatomy are not well recognized and seem so far removed from textbook descriptions or laboratory dissections of anatomical structures. We have taken what we feel is a refreshingly new approach of attempting to present basic and clinical orbital anatomy and their applications simultaneously. We have borrowed from the classical anatomical works of such anatomists as Whitnall, Bassett, Wolfe, Jones, Beard, Quickert, Koornneef, etc., and have incorporated anatomical and clinical findings and their applications by recent authors such as ourselves. The elegant color plates in the dissection chapter, Chapter 10, are taken from the meticulous orbital dissections of Dr. David Bassett. We elected to use these color plates as we doubt that they will ever be excelled in detail or clarity. They greatly facilitate and augment the learning of orbital anatomy as well as simplify orbital dissection. These elegant color plates not only are applicable to the orbital dissection described in Chapter 10 but also should be used frequently as a reference for the anatomical structures discussed in detail in the other chapters. While a large amount of orbital and eyelid anatomy can be learned from reading this book and studying the numerous illustrations and color plates, orbital dissections are required to completely understand and appreciate orbital and eyelid anatomy

We recommend reading Chapters 1 through 9 before performing the orbital dissection described in Chapter 10. For detailed discussions on the structures dissected in Chapter 10, references to the information provided in the prior chapters is suggested. In this way, some repetition (which, incidentally, is a very good way to learn orbital and eyelid anatomy) is incorporated into the learning process.

Collectively, the ten chapters of this book cover nearly all aspects of eyelid and orbital anatomy, while each chapter covers its own subject matter independently.

While neither of the authors is an expert in embryology, "in the beginning" seems to have been a good place to start. Chapter 1 is included for completeness. It presents an overview of orbital embryology, and should the reader desire an in-depth study of this area, other references are suggested. Chapter 2 is a clinically oriented presentation of osteology which is the framework of orbital anatomy. The sphenoid bone, which is the foundation of the skull and key to understanding orbital osteology, is presented in detail. The only chapter written by an outside author is Chapter 3. We feel this excellent chapter on radiographic evaluation of the orbit contributed by Dr. Jonathan Dutton represents one of the finest descriptions of its kind on clinical orbital radiography. The anatomy of the eyebrows, eyelids, and anterior orbit, which constitutes the basis for the practice of oculoplastic surgery is comprehensively presented in meticulous detail in Chapter 4. The anatomy and physiology of the lacrimal system as well as techniques for its clinical evaluation are presented in Chapter 5. Chapter 6 presents a review of the connective tissue planes of the orbit, based on Koornneef's extensive studies of the subject. Chapters 7, 8, and 9 present comprehensive reviews of the extraocular muscles, nerves, and vascular supply of the orbit, respectively. The clinical importance of these specific anatomical structures of the orbit is stressed. The concluding chapter, Chapter 10, which by itself could represent a very useful book on orbital dissection, presents the invaluable and almost artistic color plates of Dr. Bassett's orbital dissections. A considerable amount of time is warranted for studying these color plates which are also very useful as a cross-reference for the remainder of the book.

This book is recommended as a text for learning basic and clinical orbital anatomy for ophthalmology residents in training as well as residents of other specialties and subspecialties who work around the eyelids or orbit. It is also recommended for practicing ophthalmologists, oculoplastic surgeons, orbit surgeons, otolaryngologists, plastic surgeons, neurosurgeons, or dermatologists who perform surgery in this area. We hope this book will stimulate further interest in anatomical investigations and utilization of anatomical principles and procedures. We also hope the reader will learn as much from studying this book as the authors have from its writing.

Acknowledgments

We thank Neil Miller for his many appropriate suggestions and corrections on the chapter on neuro-opthalmology, Sohan Hayreh for similar suggestions on the vascular chapter, and Elise Torczynski for her review and suggestions on the embryology chapter. We are grateful to Patty Culotta, Jackie Gordon, and Jane Duwa for the many times they typed and retyped the manuscript, and Jean Prybil for organizing photographs. We appreciate the assistance of Michael Houck, in obtaining reference material. We also thank our teachers in this field, Richard Green, Richard Hoover, Robert Dryden, Fred Blodi, Crowell Beard, Orkan Stasior, and Marvin Quickert, who stimulated our interest and helped provide us with the tools to write this book. Most of all we thank our wives, Bo and Karen, and our children, Tom and Stacy and Mark and Erin, for their selflessness and understanding in giving up many hours of family time during the writing of this book.

Contents

Embryology of the Eyelids, Lacrimal System and Orbit

A basic knowledge of the embryology of the eye and its adnexa greatly facilitates the study of eyelid and orbital anatomy. Therefore, the embryologic derivation of these structures will be reviewed. More detailed evaluation of embryonic developmental abnormalities of the globe, however, will be found in texts devoted exclusively to this topic (Mann, 1957, 1964; Duke-Elder, 1963; Ozanic and Jakobiec, 1982).

This chapter is divided into four sections: embryology of the eyelids, the lacrimal system, the orbit, and the extraocular muscles. Brief discussions of developmental abnormalities are included at the end of each section.

Classically, the embryologic derivation of the eye and orbital structures has been postulated to occur from ectodermal, mesodermal, and endodermal precursors (Mann, 1957, 1976; Duke-Elder, 1963). However, no mesodermal somites are located in the head and neck region. In addition, experimental embryologists, utilizing electron microscopy, have documented cranial neural crest origin of many elements of the connective tissues of the eye and its adnexal structures. It is now believed that mesoderm contributes only to the extraocular muscles and vascular endothelium. As such, the basic concepts of the embryologic derivation of ocular structures has changed dramatically with the results of recent experimental analysis (Le Lievre and Le Douarin, 1975; Jakobiec and Tannenbaum, 1975; Johnston, et al, 1979; O'Rahilly, 1965; Torczynski, et al, 1977; Ozanics and Jakobiec, 1982; Noden, 1982).

Before detailed descriptions of the embryology of the eyelids, lacrimal system, orbit, and extraocular muscles are undertaken, one must first consider the globe. In comparison to other sensory organs, an important distinguishing feature in ocular development is the concept that the eye develops directly from an outpouching of the embryonic brain (neural tube) (Fig. 1.1). The retina is an extension of the neural plate, first manifesting itself as the optic vesicle. Since the retina is formed directly from the brain, the optic nerve is not a typical nerve, but is rather a fiber tract, which develops along a stalk of brain tissue.

At 3½ weeks gestation, the primary optic sulcus first appears (embryo of 7–9 somites) as a depression which occurs in the lateral portion of the neural plate (Fig. 1.2). The optic sulcus invaginates, becoming the optic vesicle, which broadens. The cranial neural crest cells are situated on either side of the invaginating neural folds. These cells undergo extensive migrations to envelop the optic primordium and form the mesenchyme of the globe and orbital tissues. By the fifth week of gestation, the dorsal portion of the vesicle flattens, then invaginates, to form the optic cup. The optic cup invaginates eccentrically toward its ventral margin, and a gap in the continuity of the optic cup is created, forming the embryonic (choroidal) fissure. The embryonic fissure permits mesenchymal tissue to enter the globe, which forms its vascular supply.

While invaginating, the layers of the optic cup rapidly differentiate, forming an outer pigmented layer (the future retinal pigment epithelium) and an inner sensory layer of the retina (the future sensory retina). The lens is formed by a proliferation of surface epithelium (lens placode) in the central portion of the optic cup. As the optic cup invaginates, the lens placode also invaginates forming the lens vesicle, which then separates from the surface epithelium. At this point, the globe is covered by surface epithelium which will form the epithelium of the cornea. The cranial neural crest cells migrate to form the mesenchyme of the globe and orbital tissues. The development of the ocular structures are summarized in Table 1.1.

EYELIDS

After the lens placode invaginates forming the lens vesicles, the eye is initially covered by surface epithelium. The embryologic derivation of the eyelids can be envisioned to occur in these basic stages. The first stage is the development of the eyelid folds. The second stage occurs following the fusion of the lids at the ninth week of gestation. At this time differentiation of the eyelid margins occurs. The third stage of eyelid development is the differentiation of tissues of the eyelid.

The eyelids first appear in the seventh week of gestation as definite folds forming a circular palpebral aperture (Fig. 1.3). The mesenchymal portion of the upper eyelid develops from a medial and a lateral portion of the frontonasal process. The lower eyelid develops from an upgrowth from the maxillary process. The mesenchymal component of the eyelids rapidly proliferates to fuse at the ninth week of gestation (Fig. 1.4).

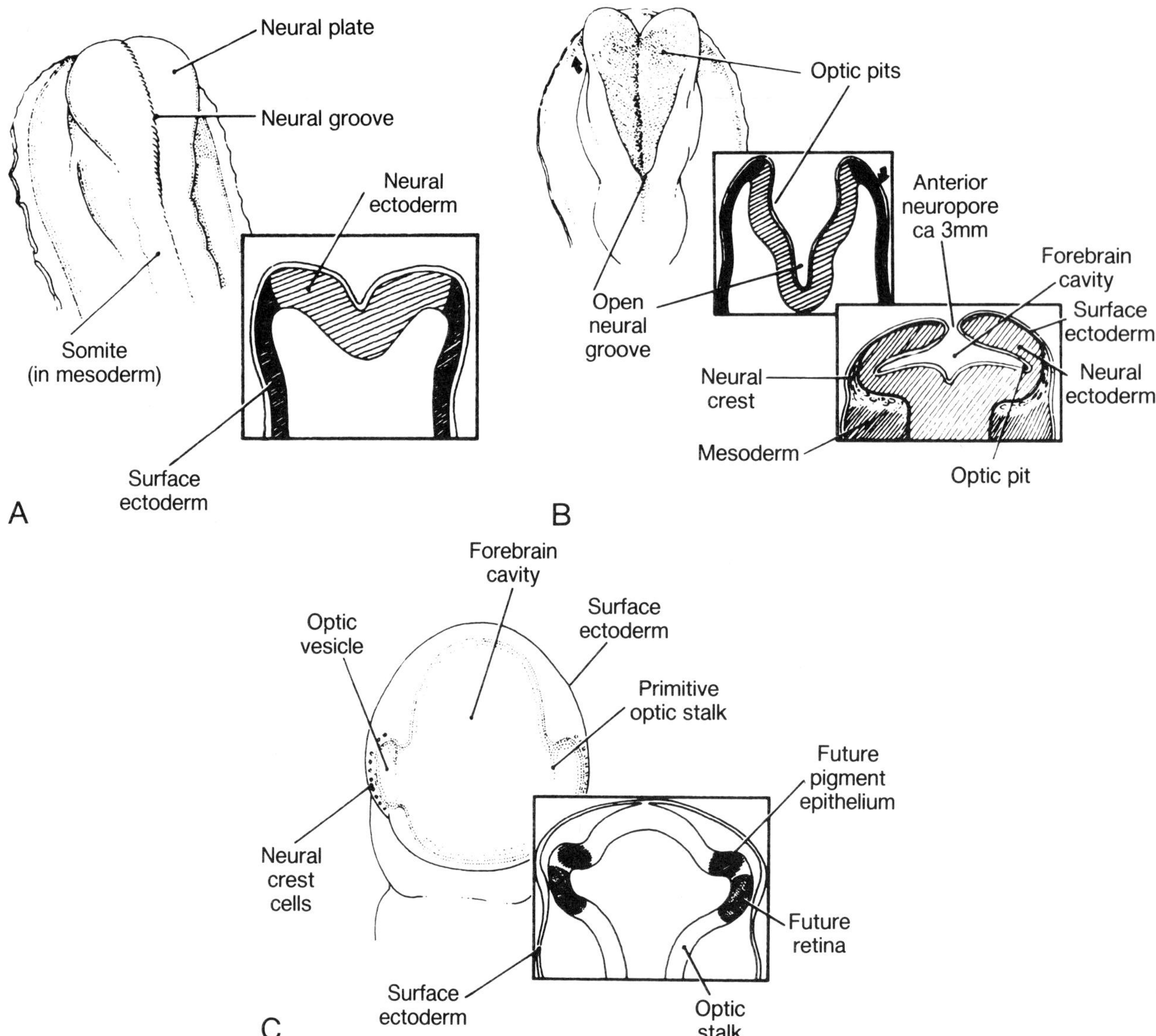

Figure 1.1. Schema of early stages in optic vesicle formation. (*A*) At 20 days gestation, the neural plate has a central depression in the neural groove. (*B*) By the 24th day, the deepened neural groove has closed. However, anteriorly the groove remains open, forming the optic pits laterally. The bottom sketch demonstrates the tissue derivation of the optic pit. (*C*) By 26 days, the neural tube is closed, and the optic pits deepen to form vesicles. These arise from the ventrolateral aspect of the brain and are connected by primitive optic stalks. (Reproduced from Ozanics V, Jakobiec FA: Prenatal development of the eye and its adnexa. In Duane TD, Jaeger EA (eds): *Biomedical Foundations of Ophthalmology*. Volume 1, Chapter 2. © 1982, Philadelphia, Harper and Row Publishers. Redrawn from the film, "Embryology of the Eye," 1950, By permission of the American Academy of Ophthalmology.)

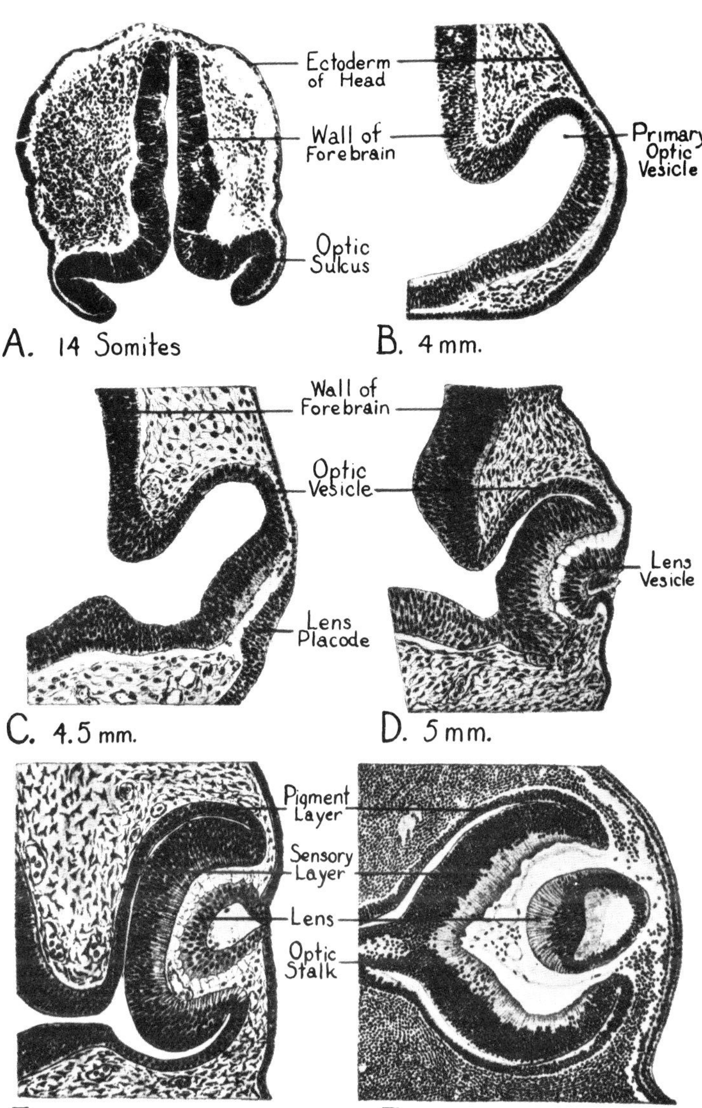

Figure 1.2. Early development of the optic cup. The optic pit (sulcus) is in the lateral wall of the neural groove. This deepens to form vesicles which differentiate and invaginate to form the optic cup.

Table 1.1.
Embryologic Derivation of Ocular Structures

CORNEA:	Epithelium—surface ectoderm
	Stroma— ⎫
	Endothelium— ⎭ cranial neural crest
CONJUNCTIVA:	Epithelium and surface ectoderm
SCLERA:	Cranial neural crest
LENS:	Surface ectoderm
IRIS:	Pigment epithelium—neural ectoderm
	Sphincter and dilator muscles—neural ectoderm
	Stroma—cranial neural crest
CILIARY BODY:	Epithelium—neural ectoderm
	Muscles—cranial neural crest
RETINA:	Neural ectoderm
CHOROID:	Cranial neural crest
OPTIC NERVE:	Neural ectoderm
SHEATH OF OPTIC NERVE:	Cranial neural crest

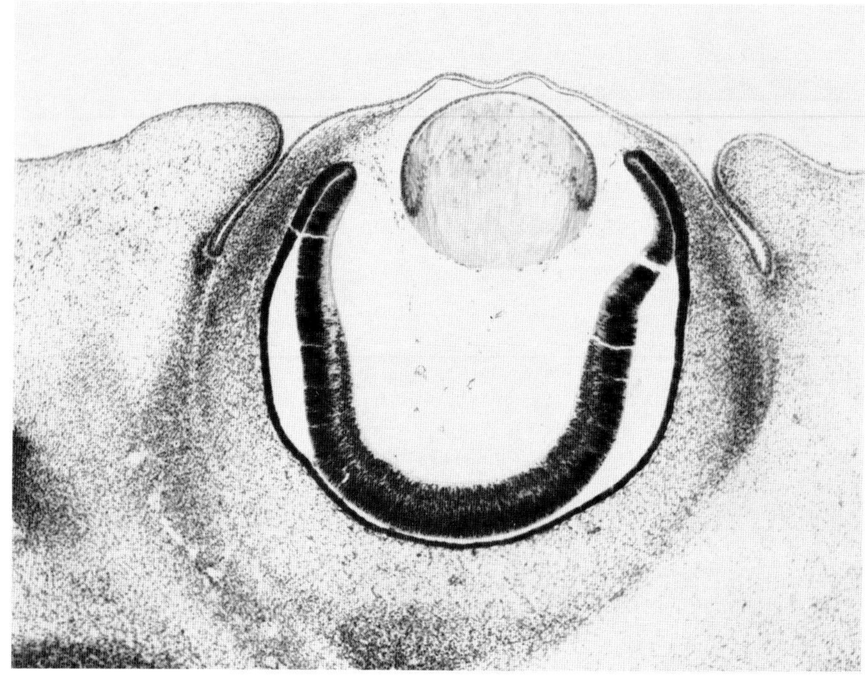

Figure 1.3. The eyelid folds form a circular palpebral aperture at the seventh week of gestation. (Reproduced with permission from W. R. Green.)

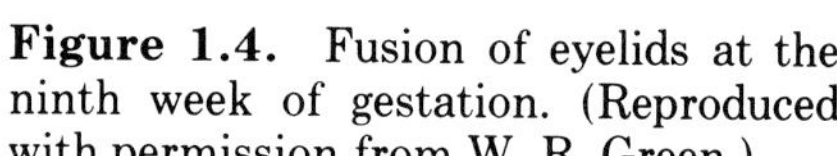

Figure 1.4. Fusion of eyelids at the ninth week of gestation. (Reproduced with permission from W. R. Green.)

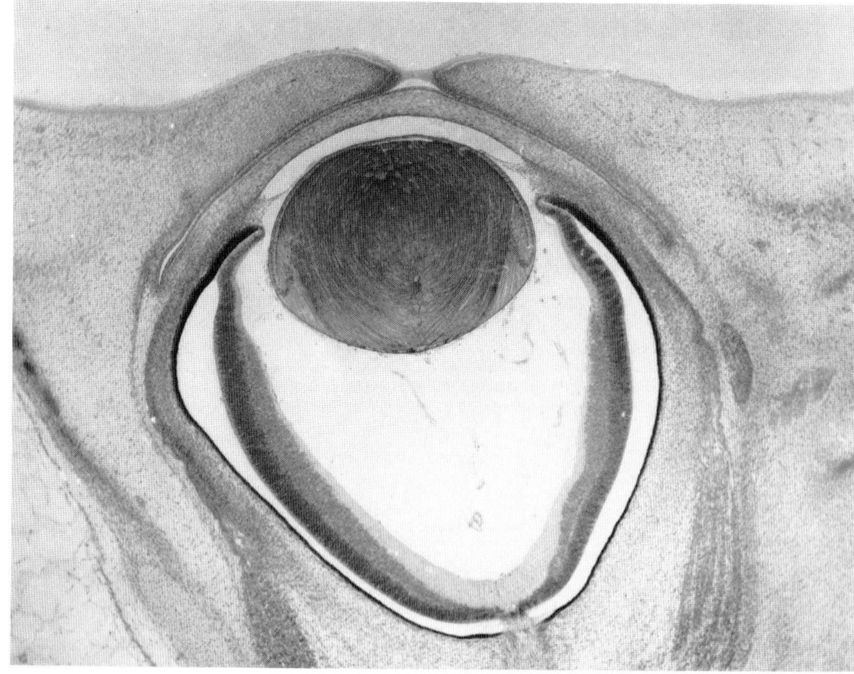

At this stage the eyelids consist primarily of undifferentiated mesenchymal tissue, although early development of the orbicularis muscle is already present. The surface ectoderm becomes the skin, while the inner layer of ectoderm becomes the conjunctiva. The space between the eyelids and anterior surface of the globe produces the conjunctival sac.

Fusion of the eyelid margins at the ninth week of gestation permits differentiation of the eyelid (Fig. 1.5). Failure of eyelid fusion inhibits the development of the lid margin, which in turn produces a coloboma or localized defect of the lid. The first structures to develop are the specialized sebaceous glands (Meibomian and Zeis) and the tarsus. The tarsal glands originate as a solid ectodermal ingrowth from the eyelid margin. The ectodermal ingrowth eventually develops a central lumen and secretes sebum by the end of the fifth month of gestation. The secretion of the Meibomian glands may facilitate the separation of the fused eyelids during the sixth month of gestation. The tarsus consists of consolidated mesenchyme enveloping the Meibomian glands. The development of the tarsus is actually coordinated with the invagination of ectodermal cords which serve as precursors of the Meibomian glands.

Eyelid cilia develop shortly after the tarsus and tarsal glands. The cilia arise from an epithelial bud and develop

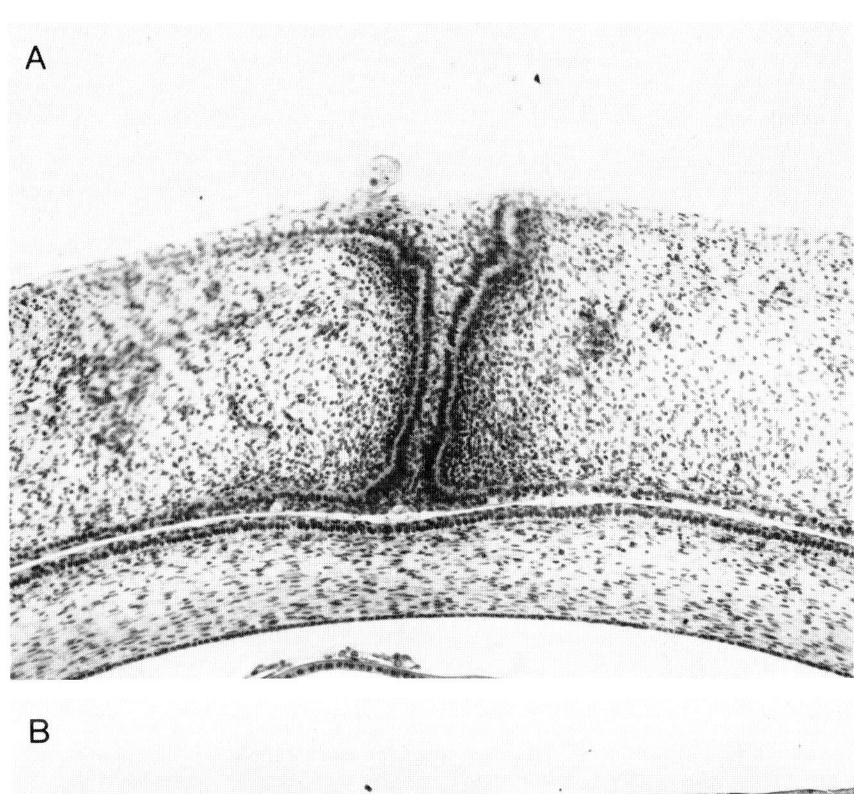

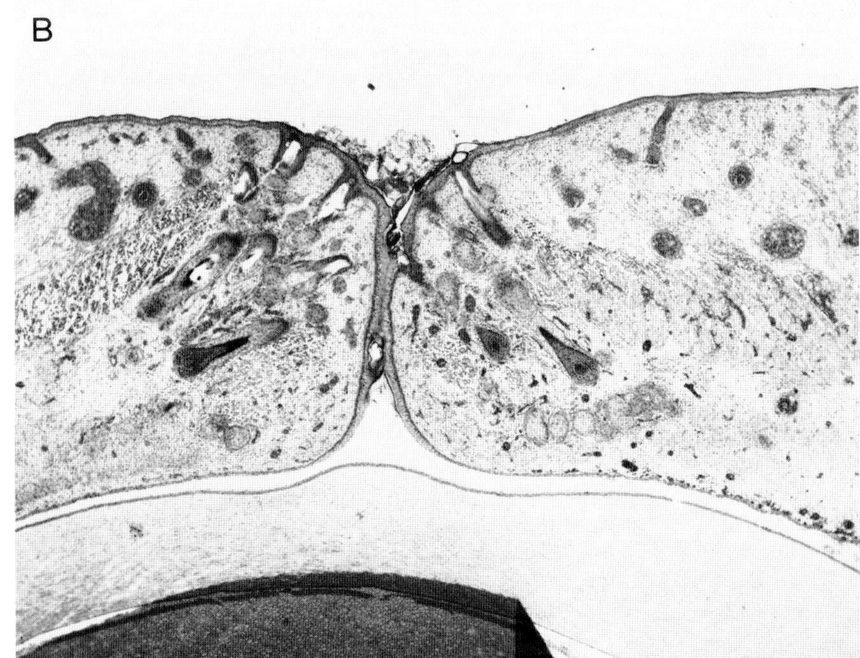

Figure 1.5. Differentiation of eyelid margin. (*A*) About 10 weeks gestation, shortly after eyelid fusion, the tissues are still undifferentiated. (*B*) At 4½ months gestation the differentiation of the eyelid margin is apparent. (Reproduced with permission from W. R. Green.)

shortly after the eyelids have been fused. Two or three rows of cilia are formed, and are more numerous on the upper than the lower eyelid. As the lash follicles develop, a marginal strip of orbicularis muscle is isolated, thus forming the muscle of Riolan. The cilia are associated with small sebaceous glands (Zeis). In addition, the cilia are contiguous with modified apocrine sweat glands (Moll) which open into the lumen of the lash follicle.

The lacrimal gland is first evident at the eighth week of gestation. This gland develops from epithelial buds which arise on the superolateral side of the conjunctival cul-de-sac. These epithelial buds, generally eight in number, proliferate and migrate to the eventual fossa for the lacrimal gland in the superior temporal quadrant of the orbit. As the levator muscle and its aponeurosis develop, they constrict and divide the lacrimal gland to form an orbital and a palpebral lobe. The migratory path of the lacrimal gland, from the conjunctiva to its final position in the orbit, is evident by the excretory ducts of the gland. The excretory ducts enter the conjunctiva approximately 4–5 mm superior to the lateral aspect of the upper lid tarsus. Occasionally an excretory duct will enter the conjunctiva in the lateral canthal region. Despite the fact that the lacrimal gland has reached considerable size by the time of birth, the gland is not capable of secreting tears. The lacrimal gland continues maturation and begins tear secretion in the postnatal period.

The accessory lacrimal glands of Krause and Wolfring develop from epithelial invagination from the tarsal conjunctiva. The glands of Krause are located in the conjunctival fornix while the glands of Wolfring are located at the border of the tarsus. The glands of Krause are more densely distributed in the upper eyelid (8–20), than in the lower eyelid (2–8). Those glands present in the lower eyelid are concentrated in its lateral aspect. These glands are larger, but fewer in number, than the glands of Krause. The glands of Krause and Wolfring contribute to the aqueous portion of the tear film.

Classically, the caruncle was thought to develop from seclusion and isolation of the nasal portion of the lower eyelid, by the development of the lacrimal excretory system. The sequestration of the medial aspect of the lower eyelid explained the presence of sebaceous glands and cilia in the caruncle. A second theory of embryogenesis postulated an independent development of the caruncle from surface ectoderm. The independent origin theory is supported by the presence of the caruncle in patients with congenital absence of the canaliculi.

The plica semilunaris develops as a semilunar epithelial fold originating at the inner aspect of the globe. Mesenchymal elements are contained within the fold, which is well-defined by the fourth month of gestation. The plica semilunaris represents a vestigial element analogous to the nictating membrane or third eyelid in animals.

By the sixth month of gestation, the eyelids are fully developed and begin to separate. The mechanism governing separation of the eyelids is speculative, but three plausible theories are offered: action of the lid retractors (levator muscle and lower lid retractors); tarsal gland secretions; or keratinization of the lid margin epithelium. A combination of these three features is likely responsible for eyelid opening.

CLINICAL NOTE

Congenital anomalies of the eyelids result from aberrations in the various developmental stages. As noted, the embryologic derivation of the eyelids can be envisioned to occur in three basic stages. The first stage is the development of the primordial eyelid folds. The second stage occurs following the fusion of the lids at the ninth week of gestation. The third stage of eyelid development is the differentiation of tissues of the eyelid (Table 1.2).

The most severe eyelid anomalies are those that result from faulty eyelid fold development at the seventh week of gestation. Developmental failure of the lid folds may be localized or diffuse in nature. The diffuse failure of lid folds to develop produces cryptophthalmos. In the absence of the eyelids, the surface epithelium, which would normally develop into the cornea, becomes keratinized and forms the anterior surface of the eye. The corneal stroma shows defective development. Additional facial abnormalities may be associated with cryptophthalmos. Extensive anterior segment anomalies occur with cryptophthalmos; however, the retina and posterior pole may be normal. The degree of disorganization of the anterior segment has resulted in a second theory of pathogenesis of cryptophthalmos—that the eyelid folds are formed but obliterated by an inflammatory process.

Microblepharon is a shortening of the vertical dimension of the eyelid which prevents the lids from totally closing. Generally, microblepharon occurs in a milder form as nocturnal-lagophthalmos. Failure of full development of the eyelid fold constitutes the pathogenesis of microblepharon. More advanced forms of microblepharon may result from a failure of lid margin fusion, or a diffuse coloboma.

Colobomas, which may be unilateral or bilateral, present as full-thickness defects of the eyelid margin resulting from failure of eyelid margin fusion. Defects in the upper eyelid are generally nasal, while those in the lower eyelid are temporal (Fig. 1.6). Somewhat rarely, the lid defect is filled by scar tissue which forms adhesions between the eyelid and the globe. If the eyelid defect is extensive, secondary corneal exposure and ulceration may occur (Kiskadden and McGregor, 1947). Some cases of coloboma are accompanied by multiple facial defects, which are attributed to amniotic bands which constrict and prevent the development of the lids.

The second basic group of embryologic anomalies results from the failure of the eyelid margin to differentiate. Palpebral aperture abnormalities are the most

Table 1.2.
Embryologic Abnormalities of the Eyelids

STAGE I:	DEVELOPMENT OF LID FOLDS cryptophthalmos, microblepharon, coloboma
STAGE II:	DIFFERENTIATION OF LID MARGIN ankyloblepharon, ectropion, entropion, euryblepharon, epicanthal folds, epiblepharon
STAGE III:	DIFFERENTIATION OF LID TISSUES nevus, hemangioma, dermoids

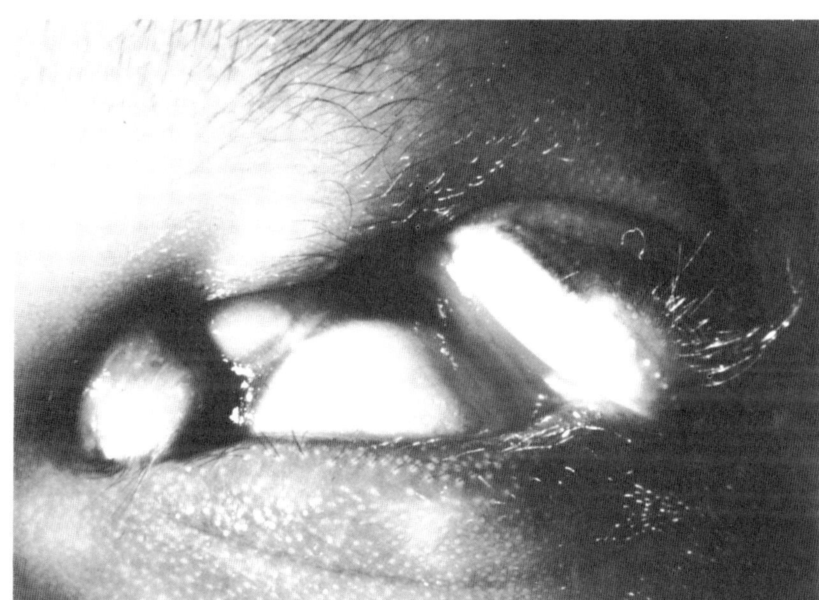

Figure 1.6.　Coloboma of the eyelid.

common in this group. Again, a wide range of disorders are present. However, most are less severe than the disorders resulting from failure of lid folds to develop.

Failure of the lid margin to separate will result in ankyloblepharon. Normally the lids separate at the sixth month of gestation. However, the lids may remain fused over some portion of the fissure, narrowing the palpebral aperture. Ankyloblepharon may occur in the nasal or temporal border of the eyelids and may be genetically determined. The lid margins may actually be fused together or may be tenuously attached by fine bands of extensile tissue.

Euryblepharon is defined as an enlargement of the horizontal palpebral fissure produced by the inferior temporal displacement of the lateral canthus (Fig. 1.7). These patients will have a widened palpebral fissure, especially laterally, and have intermittent problems with exposure keratopathy secondary to lagophthalmos.

Abnormalities in the eyelid position may result in congenital entropion or ectropion. These conditions are sometimes secondary to tarsal abnormalities. Congenital entropion may be produced by a defect in the retractor aponeurosis (Tse and Anderson, 1983). Rare instances of ectropion are associated with microblepharon which appears to mechanically evert the lashes.

Additional abnormalities in the palpebral apertures may be manifested as congenital folds in the eyelids. Eyelid folds may occur at the lid margin, (epiblepharon) or in the inner canthal region (epicanthus). Epiblepharon is characterized by a horizontal fold of skin at the lid margin, which rolls the lashes vertically, or posteriorly toward the eye. Epiblepharon is frequently encountered in Oriental eyelids. Epicanthus is a semilunar fold of skin at the side of the nose with its concavity directed at the inner canthus. Epicanthus is normally found in the infant, however development of the nasal bridge generally smooths the epicanthal folds (see "Eyelids" in Chapter 4.)

The final group of eyelid abnormalities is due to faulty differentiation of the lid tissues. The faulty development of eyelid tissues can produce a variety of disorders. Congenital ichthyosis with hyperkeratosis of the horny layers of the lid is one manifestation of an epithelial abnormality. This excessive scaling of the skin may be present at birth, or develop shortly thereafter.

In addition, there may be abnormalities in the development of the sweat glands of the eyelids. Congenital hyperhidrosis presents as an excessive number of sweat glands present in the ocular adnexal region. Sweating in these areas on minor exertion or emotional stimulation is readily apparent.

Ectodermal dysplasia is characterized by widespread abnormalities in ectodermal structures. Generally, a typical triad is encountered clinically: hypohidrosis, hypotrichosis, and hypodontia. Abnormalities in sebaceous glands and abnormal lacrimal secretions will result in the production of an abnormal precorneal tear film. However, despite the abnormalities in the precorneal tear film, corneal involvement is rare (Beckerman, 1973; Wilson, et al, 1973).

Congenital pigmentary disturbance may present in a variety of ways. Congenital hyperpigmentation of the episclera of one eye, often involving the uveal tract, is recognized as melanosis oculi. The darker pigmentation of the affected eye is secondary to an increase in the number of melanocytes in the affected tissues. Acquired melanosis oculi is a premalignant condition of the skin and conjunctiva and should be differentiated from the congenital condition (Reese, 1966). Oculodermal melanocytosis (nevus of Ota) involves the skin around the orbit plus congenital melanosis oculi. Hyperpigmentation in oculodermal melanocytosis results from typical branching melanocytes deep in the dermis of the skin, and in melanosis oculi the nevus cells are in the episclera. Curiously, nevocellular lesions have a tendency to occur in the eyelid margin (Doxanas, et al, 1977). These may occur

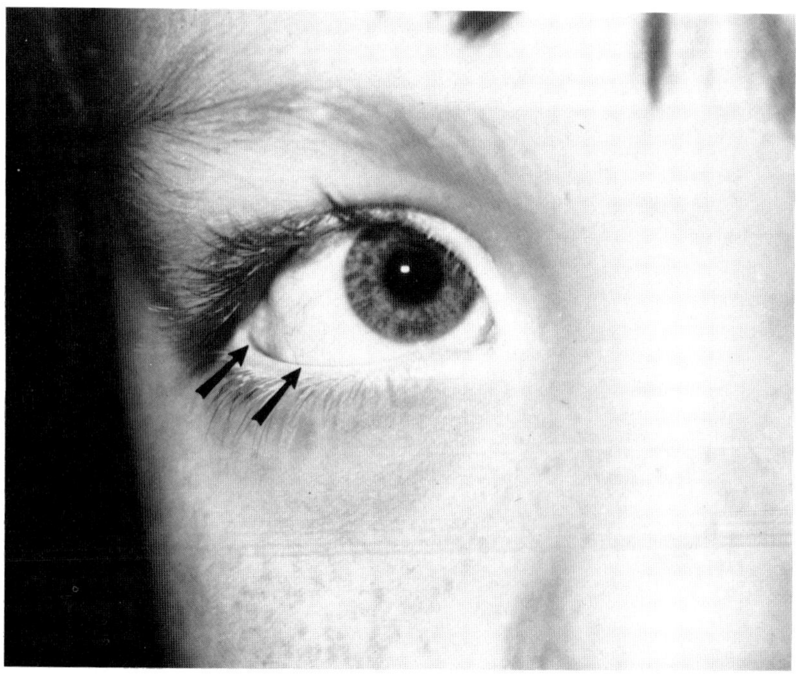

Figure 1.7. Euryblepharon: Produced by the inferior-temporal displacement of the lateral canthus. The lateral portion of the eyelid does not approximate the globe (*arrows*).

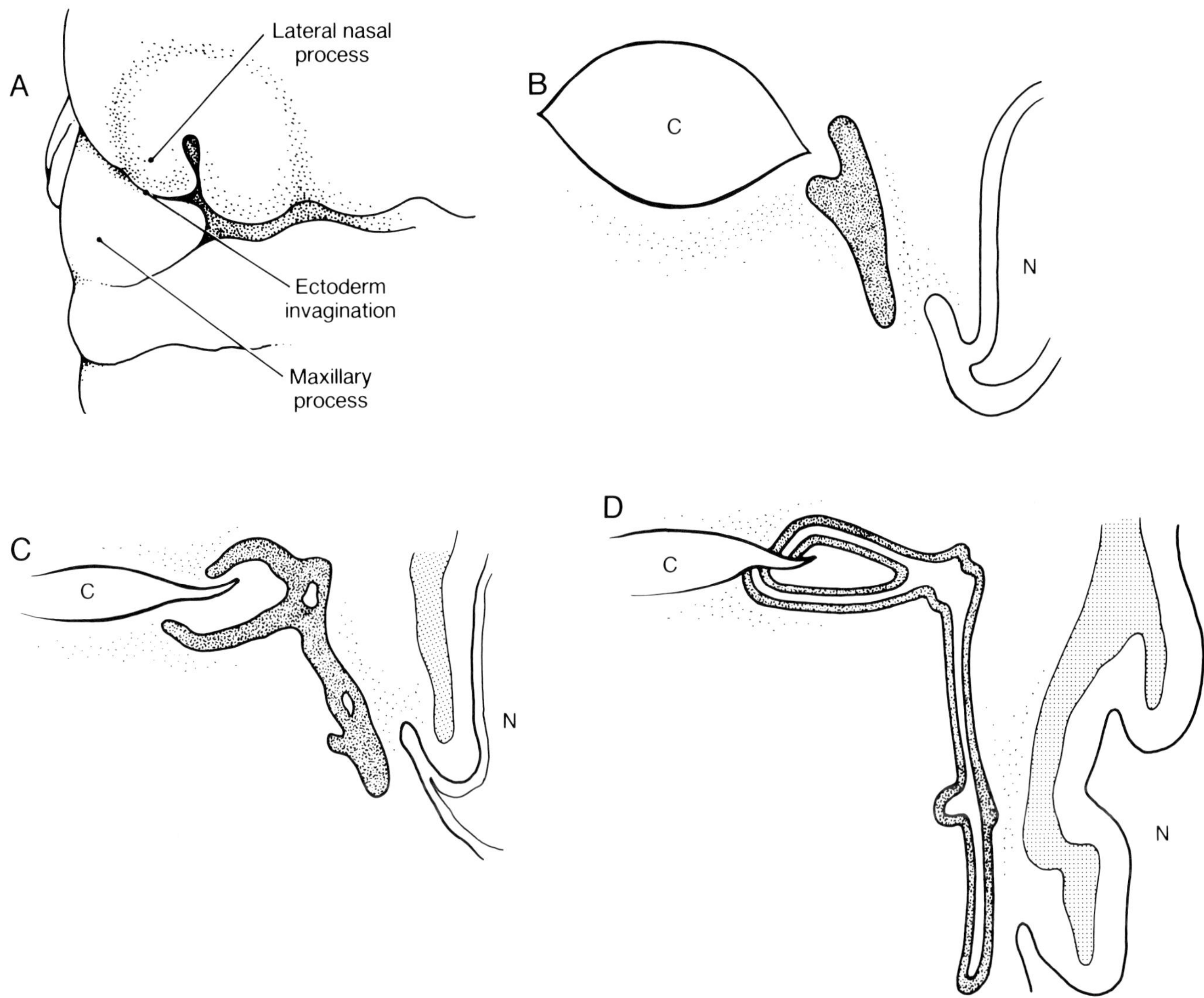

Figure 1.8. Development of the lacrimal excretory system. (*A*) Ectodermal invagination between the lateral nasal process and the maxillary process. (*B*), (*C*), (*D*), Development of the lacrimal excretory system, transforming from a solid ectodermal bud (*B*) to a canalized system (*D*). *A*, 5½ weeks; *B*, 6 weeks; *C*, 12 weeks; *D*, 3½ months gestation. c, conjunctiva surface; N, nasal cavity. (*B*, *C*, *D* reproduced from Duke-Elder S, Cook C: Embryology, In Duke-Elder S (ed): *System of Ophthalmology*, Vol. III, Part 1. © 1963, St. Louis, C.V. Mosby.)

on corresponding portions of the upper and lower eyelids producing kissing nevi, which are a result of the fused eyelids during embryologic development. In addition to pigmentary tumors, hemangiomas, lipomas, neurofibromas of the eyelids may occur.

Abnormalities in eyelid motility may result from faulty development of the levator muscle. Berke and Wadsworth (1945, 1955) examined the levator in congenitally ptotic patients and correlated the severity of levator muscle abnormalities with the degree of ptosis. Those patients with the most severe degree of ptosis had a levator muscle infiltrated with fatty or fibrous tissue. More recently aponeurotic defects have been demonstrated to account for some cases of congenital ptosis with good function (Anderson and Gordy, 1979; Doxanas and Dryden, 1981).

NASOLACRIMAL SYSTEM

Precursors of the lacrimal excretory system first appear in the sixth week of gestation along the cleft between the lateral nasal and maxillary processes. The excretory system appears as a cord of epithelial cells. The solid cord extends downward into the mesenchyme to form the naso-optic fissure and detaches from the surface (Fig. 1.8). The sequestered epithelial cord lies between the precursors of the medial canthus and the nose. During the 6–12 week gestation period, the lacrimal ectodermal bud proliferates to extend toward the inner canthus and nasal cavity. As the epithelial cord approaches the medial canthus, it splits, conforming to the medial canthal angle. The precursors of the canalicular system are thereby established. At approximately the eighth or ninth week of gestation, the epithelial cord reaches the eyelid margin. During this time the epithelial cells also proliferate forming the rudiment of the lacrimal sac, and extend inferiorly to approach the nasal activity.

During the third month of gestation the solid epithelial cords, which comprise the lacrimal excretory system at that point, undergo central disintegration or necrosis (Fig. 1.9). Isolated cavities appear within the system which, with time, become confluent. The canalization process is completed by the sixth month of gestation. However, the puncta do not

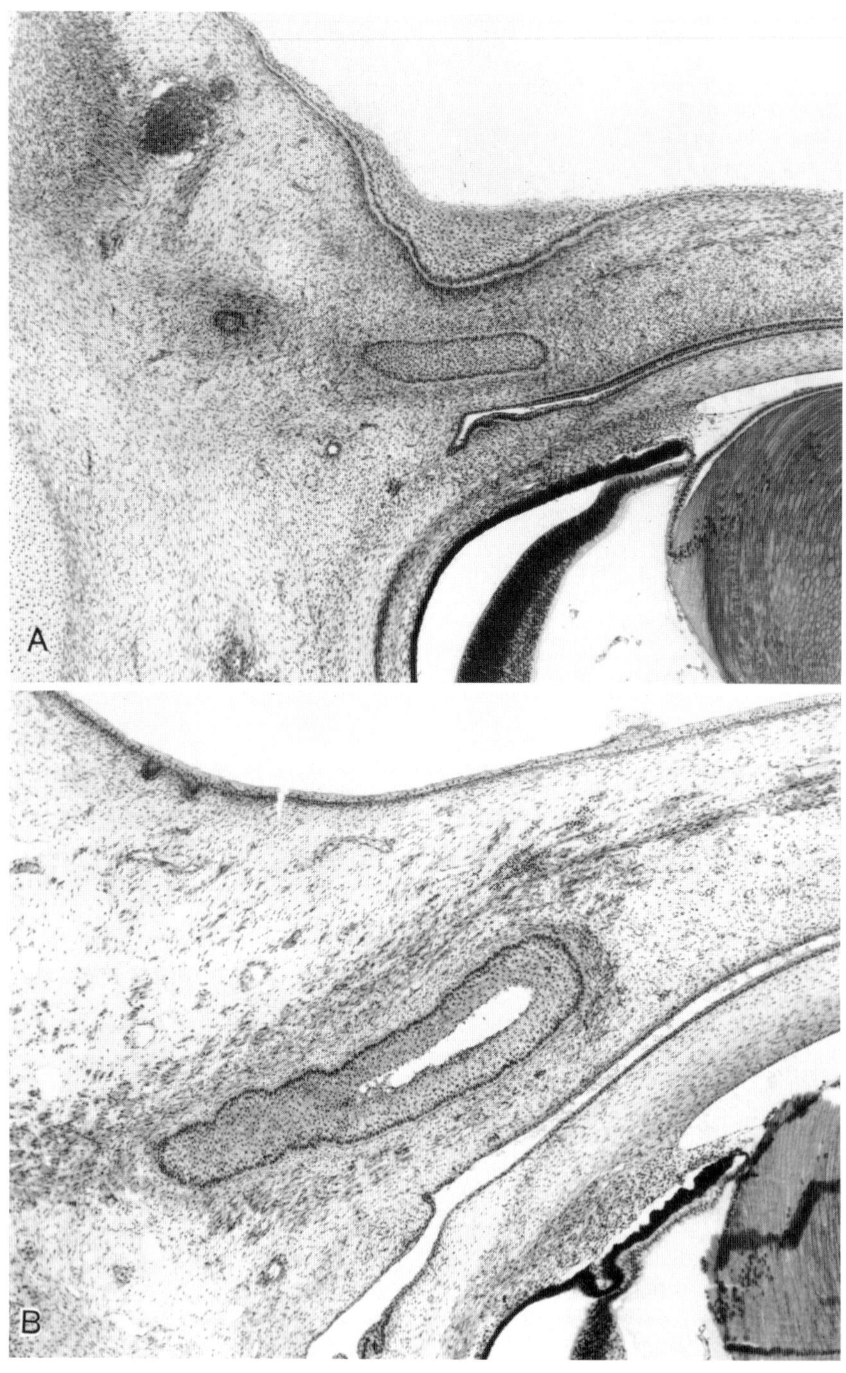

Figure 1.9. Histopathology of the lacrimal excretory system. (*A*) At 9 weeks gestation the excretory system is a solid epithelial cord. (*B*) At approximately 12–15 weeks the solid epithelial cords undergo central disintegration to form a lumen. (Reproduced with permission from W. R. Green.)

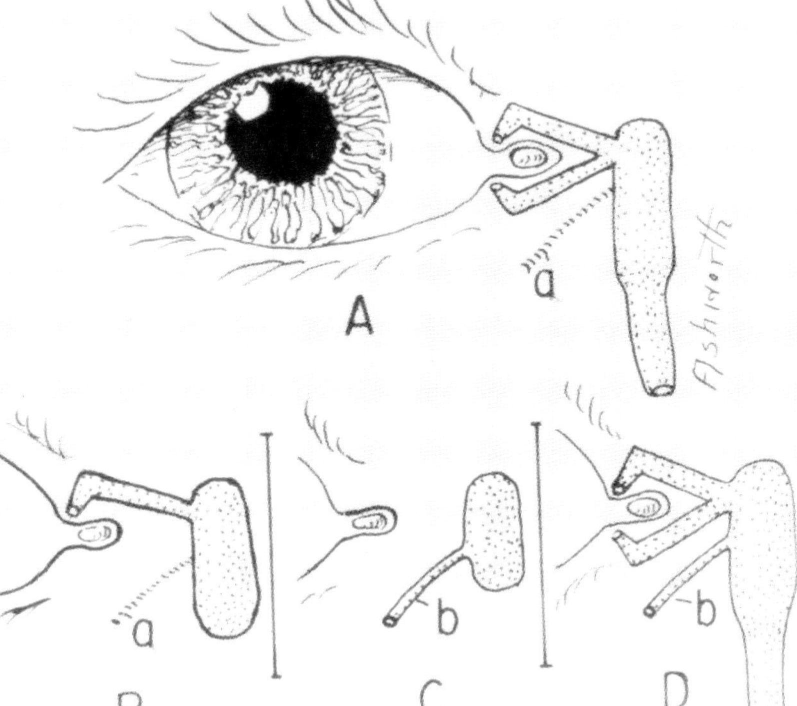

Figure 1.10. Schematic of the development of the lacrimal excretory system. (*A*), Normal development of the lacrimal excretory ducts with involution of lacrimal anlage duct (*a*). (*B*), Development of tear sac and upper canaliculus with involution of the anlage duct (*a*). (*C*), Persistent anlage of the lacrimal duct (*b*) and a partial tear sac. (*D*), Normal lacrimal excretory system, with persistence of the anlage duct (*b*). (Reproduced from Jones LT, Wobig JL: *Surgery of the Eyelids and Lacrimal System.* © 1976, Birmingham, Aesculapius Publishing Co.)

become patent until the seventh month, and the lower end of the duct opening into the inferior meatus is frequently occluded by a thin membrane at birth.

CLINICAL NOTE

The incidence of congenital obstruction of the lacrimal system has been reported to vary between 6–73% (Guerry and Kending, 1948; Cassidy, 1948; Ffooks, 1962). Congenital abnormalities of the lacrimal system may be related to its various stages of development. The most severe analomies would result from the failure of fusion between the lateral nasal and maxillary processes. In this event, ectodermal invagination would not occur, resulting in the absence of the lacrimal excretory system. Additional abnormalities may result from an abnormal course taken by the epithelial buds which would produce ectopic or duplicated puncta and canaliculi. The final group of abnormalities results from the failure of the ectodermal buds to canalize.

Abnormalities in the canalicular system and puncta may range from the failure of the system to develop to the formation of multiple systems. Agenesis or atresia of the canalicular system may be secondary to the failure of the anlage in the lacrimal passages to proliferate in the sixth to twelfth week of gestation. In addition, agenesis of the canalicular system may result from the failure of canalization of the solid epithelial cords which serve as precursors of the system. Atresia of the puncta is a condition produced by a failure in the opening of the conjunctival epithelium. The latter condition may be transmitted as an autosomal dominant feature (Sorsby, 1931; Town, 1943).

Supernumerary puncta and canalicular systems may also occur. These result in the abnormal budding or proliferation of the superior portion of the epithelial cord of the lacrimal system. Abnormalities in the canaliculus may be associated with other orbital and/or facial anomalies (Kirk, 1956).

Abnormalities of the lacrimal excretory system are much more common in its inferior portion, that is, from the lacrimal sac and the nasolacrimal duct. Lacrimal sac fistulas generally extend from the lacrimal sac or common canaliculus to the cheek. Fistulas may result from inflammation, but are more likely to occur secondary to a residual anlage of the lacrimal duct (Fig. 1.10). This duct results from the original ectodermal invagination between the lateral nasal and maxillary processes at the sixth week of gestation. With the fusion of the facial processes and sequestration of the ectodermal precursors of the lacrimal excretory system, the anlage of the lacrimal duct involutes. Some patients, generally those with a hereditary tendency, will have a persistence of the anlage of the duct, which produces a lacrimal sac fistula (Jones and Wobig, 1976; Masi, 1969).

As previously noted, the opening of the nasolacrimal duct into the nose is frequently blocked by a membrane at birth. Jones and Wobig (1976) recognized eight variations of obstruction of the lower end of the nasolacrimal duct (Fig. 1.11). The most common variation is the nasolacrimal duct's failure to penetrate the nasal mucosa. The persistent valve of Hasner at the opening of the nasolacrimal duct into the nose may be quite thin (Busse, et al, 1980) (Fig. 1.12). The nasolacrimal duct can also terminate on the floor of the nose lateral to the nasal mucosa. In addition, there

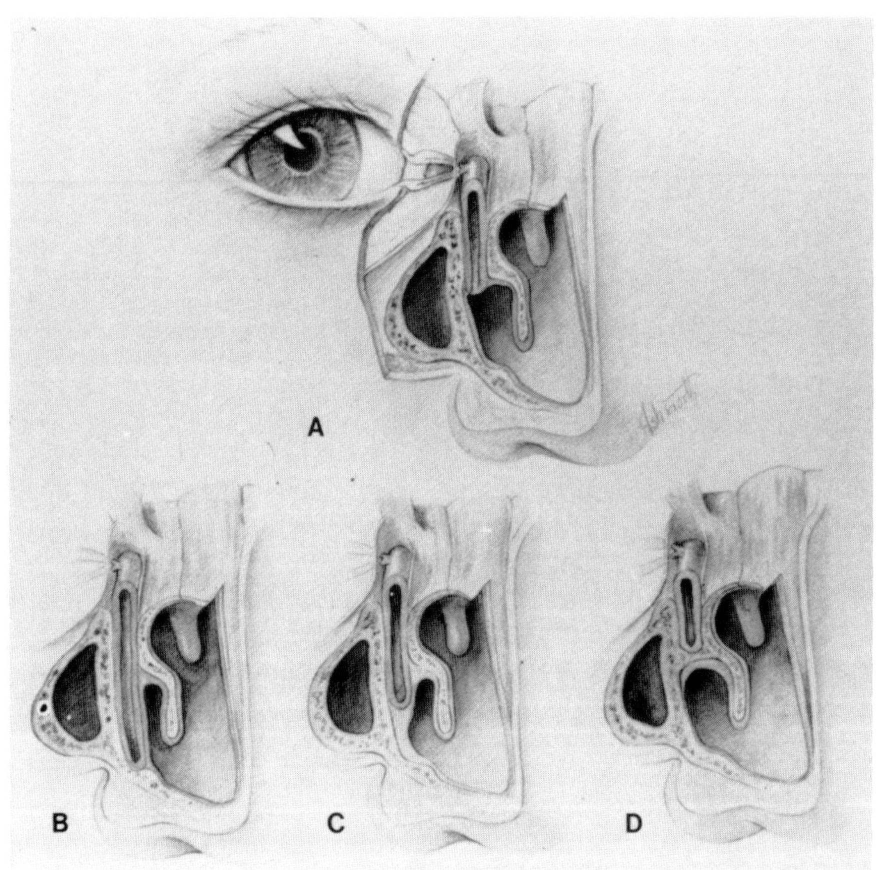

Figure 1.11. Congenital obstruction of the nasolacrimal duct. (*A*), Nasolacrimal duct normally entering the nose beneath the inferior turbinate. (*B*), The nasolacrimal duct extending to the floor of the nose. (*C*), The nasolacrimal duct lateral to the nasal mucosa. (*D*), No nasolacrimal duct due to failure of development of the bony nasolacrimal canal. (Reproduced from Jones LT, Wobig JL: *Surgery of the Eyelids and Lacrimal System.* © 1976, Birmingham, Aesculapius Publishing Co.)

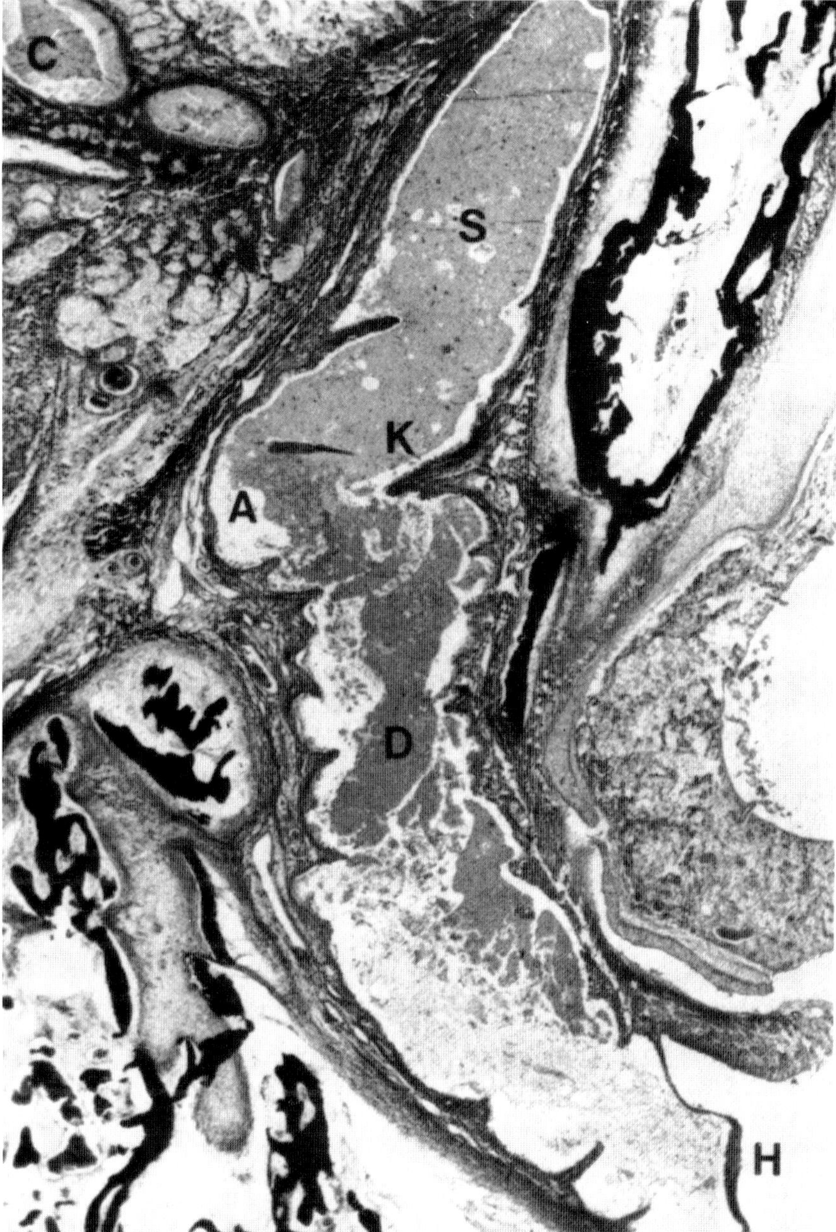

Figure 1.12. Congenital nasolacrimal duct obstruction. The most common abnormality is the persistence of the thin valve of Hasner. *C*, Canaliculus; *S*, lacrimal sac; *K*, valve of Krause; *A*, sinus of Arlt; *D*, nasolacrimal duct; *H*, persistent valve of Hasner. (Von Geison stain × 6) (Reproduced from Busse H, et al: *Archives of Ophthalmology*, 98:528, 1980. American Medical Association.)

may be a failure of development of the bony nasolacrimal canal, which produces an osseous obstruction of the system. The bony nasolacrimal canal may extend to the floor of the nose without an opening into the nose. The nasolacrimal duct may be compromised by approximation of the inferior turbinate to the lateral wall of the nose, thereby obliterating the inferior meatus, or it may terminate blindly in the inferior turbinate. The final abnormality producing nasolacrimal duct obstruction is a misdirection leading to its terminating in the maxillary sinus.

The nasolacrimal duct obstruction may be manifested at birth as an amniotocele, or mucocele of the lacrimal sac. The amniotic fluid is pumped into the sac producing grayish, soft fluctuant enlargement of the sac (Fig. 1.13). Divine and co-workers (1983) reported the intranasal extension of mucoceles of the lacrimal sac beneath the inferior turbinate, and encouraged careful nasal evaluation in these patients. Amniotoceles will frequently become secondarily infected (mucopyoceles) and require early probing, unlike other congenital obstructions of the nasolacrimal duct (Weinstein, et al, 1982; Scott, et al, 1979). Obstruction of the lacrimal system may also manifest as epiphora or chronic dacryocystitis. Generally conservative measures such as massage of the lacrimal sac or antibiotics will eliminate the need for probing in a majority of infants (Kushner, 1982). If conservative

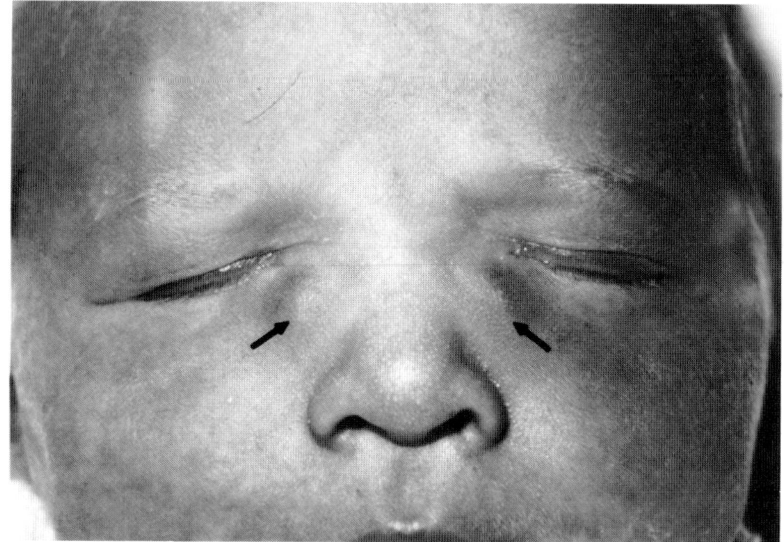

Figure 1.13. Congenital dacryocystocele (amniotocele). An active lacrimal pump in utero fills and distends the lacrimal sac.

management of congenital lacrimal excretory system obstruction is not effective, probing or even a dacryocystorhinostomy may be indicated.

ORBIT

The orbit originates from mesenchymal folds which envelop the developing globe (Fig. 1.14). The optic pit first appears early in the embryonic period, at 3½ weeks gestation. The optic pits develop on the lateral portions of the head, creating a 180° orbital axis. Above the developing stomatodeum (primitive mouth), the mesenchyme with its overlying ectoderm proliferates and condensates forming the olfactory pits. Continued mesenchymal proliferation, which is particularly prominent around the olfactory pits, produces medial and lateral nasal folds. The two medial nasal folds coalesce to form the frontonasal process, while the lateral nasal process is formed between the olfactory pit and the optic vesicle.

The maxillary process originates as a triangular extension of the dorsal end of the mandibular arch. The maxillary process proliferates and extends medially till it meets with the lateral nasal fold. This occurs at approximately 5–6 weeks gestation. As noted in the embryology section on the lacrimal excretory system, ectoderm buried at the junction of the maxillary and lateral nasal process forms the anlage of the nasolacrimal duct.

Following fusion with the lateral nasal process, the maxillary processes rapidly proliferate in a forward and upward direction to fuse in the midline. The development of the maxillary processes forces the orbital axis to a more anterior aspect.

Prior to the sixth week of gestation, the optic vesicles are in the same coronal plane, directed 180° apart. Development of the maxillary process forces the orbital axis anteriorly (Fig. 1.15). At 6–7 weeks gestation the angle formed by the orbital axis is 160°. By the seventh week of gestation the angle of orbital axis is reduced to 120°. By ten weeks gestation the angle of orbital axis is further reduced to a 72°

angle. The angle of the orbital axis approaches 45° at birth, and remains stable thereafter.

Orbital Walls

The bony orbital walls are developed from the mesenchymal processes which surround the globe. Classically, the orbital bones were thought to be derived from mesoderm. However, recent experimental data has shown that osseous structures of the orbit are primarily derived by cranial neural crest (Ozanics and Jakobiec, 1982). The lateral wall and orbital floor are formed by the maxillary process. The zygomatic and maxillary bones (except the frontal process of the maxilla) are derived from the maxillary process. The medial orbital wall develops from the lateral nasal process. The lateral nasal process forms the frontal process of the maxilla, the nasal bone, lacrimal bone and the lateral portion of the ethmoids. The roof of the orbit develops from the paraxial mesenchyme, which forms the capsule of the forebrain. Fusion of the orbital bones is incomplete at the junction of the maxillary process and the lateral nasal contribution of the frontal process of the maxilla. The patent channel thus produced is the nasolacrimal canal.

Posteriorly the sphenoid bone develops from the base of the skull. At approximately the fifth week of gestation the sphenoidal mesenchyme fans out to produce the precursors of its greater and lesser wings. A gap in the body of the sphenoid bone produces the optic foramen, which later develops into a canal.

The orbital walls are apparent by the fourth week of gestation. By the sixth week of gestation, centers of ossification are apparent. The development of the globe and the orbital contents proceed as independent functions. The orbital contents, such as muscles, nerves, and vessels, evolve even in the absence of the globe (anophthalmia). At 6–7 months gestation, the bones comprising the orbit, although not fused, are united by periosteum.

At birth, the orbital margin is circular, with equal hori-

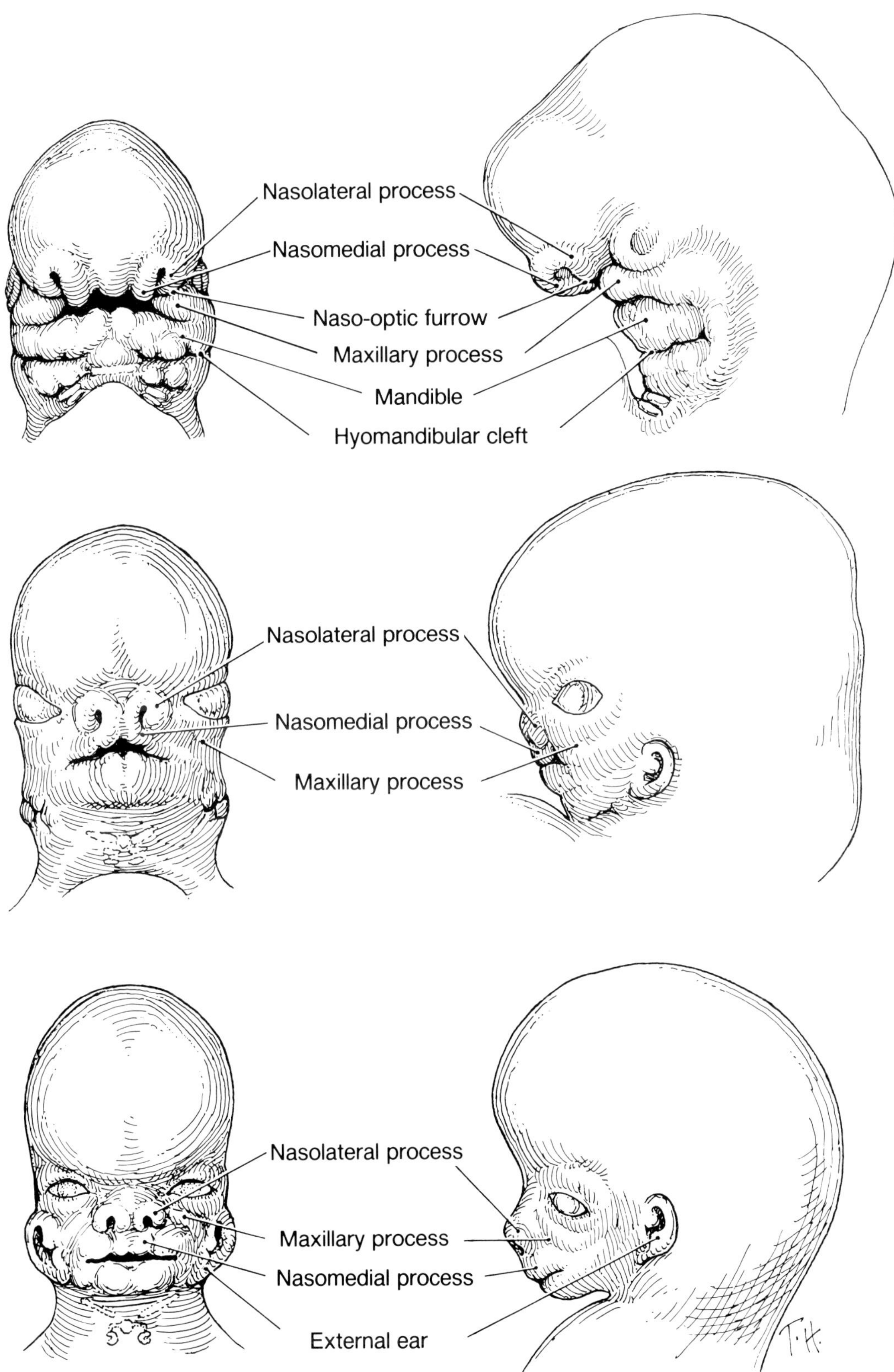

Figure 1.14. Frontal and lateral views of the developing face. *Top*, 5½ weeks; *middle*, 7 weeks; *bottom*, 8 weeks gestation. (Reproduced from Patten B.M.: *Human Embryology* (ed 3). © 1968, New York, McGraw Hill Book Company.)

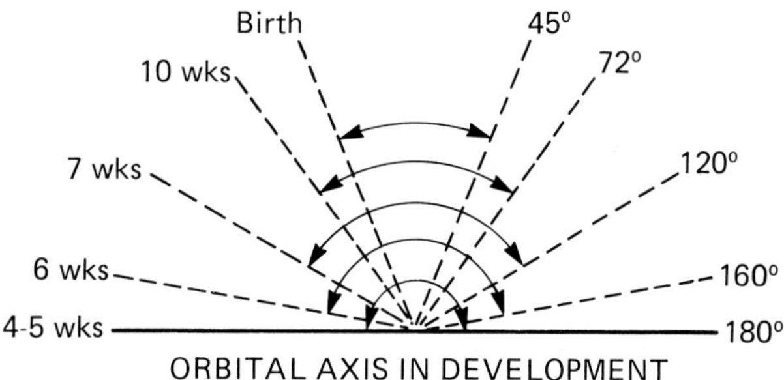

Figure 1.15. Orbital axis in development. At approximately 4–5 weeks of gestation the orbital axis is directed 180°. At birth the axis bisecting the central orbits forms a 45° angle. (Reproduced from Duke-Elder S, ed: *System of Ophthalmology*, Vol. III, Part 1, Embryology. 1963, St. Louis, C.V. Mosby, p 215.)

zontal and vertical dimensions. The orbital opening comprises almost 50% of the height of the face. With advancing age, the maxilla and mandible develop, changing the superficial orbital configuration. Growth of the facial bones produces a disparate horizontal enlargement of the orbital margin which transforms it into a rectangle. Maxillary and mandibular development will also reduce the proportion of face occupied by the orbital opening. In the adult the orbit occupies approximately 30% of the height of the face.

CLINICAL NOTE

Congenital defects in the orbital bones are not uncommon. These may result from the failure of ossification of the orbital bones or from the protrusion of cranial structures into the orbit. The bones developed by the ossification of the primitive nasal capsule, particularly the lamina papyracia of the ethmoid and the orbital surface of the maxilla, will frequently produce defects of the orbital walls. In addition, defects of the orbital walls may occur at the poorly ossified lacrimal bone, or the ethmoidomaxillary suture. The latter defect will produce continuity of the ethmoidal and maxillary sinus. Orbital defects may be associated with orbital meningoceles and encephaloceles, dermoids, congenital angiomas, and neurofibromas.

Orbital cephalocele occurs when a portion of the cerebral contents protrude into the orbit through a defect in its bony wall. With time, atrophy of the cerebral contents results in the appearance of a meningocele (cystic space lined by meninges). Orbital cephaloceles are more likely to be located anteriorly, manifested as fullness in the eyelids. Upon palpation, one may encounter a fluctuant cyst which can be reduced with pressure. Posterior cephaloceles extend from the optic foramen or sphenoid fissure, and are more extensive than those occurring anteriorly. Large orbital defects transmit increased intracranial pressure, so that a Valsalva maneuver, for example, will produce an increase in proptosis.

Dermoid cysts are frequently encountered in the orbit. Dermoid cysts are produced by the sequestration of ectodermal elements along the lines of closure of a fetal fissure. As such, these occur most frequently at the superior temporal quadrant of the orbit, corre-

sponding to the zygomaticofrontal suture. These tumors may also occur in the superior nasal orbit at the frontoethmoid suture. Progressively enlarging, orbital dermoids may assume a prominent size, extending well into the orbit posteriorly or anteriorly.

EXTRAOCULAR MUSCLES

The extraocular muscles are derived from mesodermal condensations situated on each side of the developing head. There have been numerous theories concerning the development of the extraocular muscles. Lewis (1910) felt that extraocular muscles were derived from a single mass of cells at the 24th day of gestation. These cells were initially posterior, but migrated anteriorly around the globe. The individual muscles could be identified by the 5th or 6th week of gestation.

Gilbert (1947, 1957) examined multiple specimens and felt that the extraocular muscles were developed from three mesodermal condensations. The mesodermal condensations, which serve as precursors for the extraocular muscles, are derived from the premandibular and maxillomandibular condensations. The single premandibular condensation is the precursor for the superior rectus (24th day), medial rectus, inferior rectus, and inferior oblique (28–30 day). Lateral to the premandibular condensation, two maxillomandibular condensations form. These buds form the lateral rectus and the superior oblique. The levator muscle is the last "extraocular muscle" to form. It buds from the medial aspect of the superior rectus muscle. This theory was accepted since it readily explained the innervation of the extraocular muscles. The oculomotor nerve supplies the superior rectus, medial rectus, inferior rectus, inferior oblique, and the levator palpebrae superioris. The superior oblique and the lateral rectus are innervated by the trochlear (C. N.-IV), abducen (C. N.-VI) nerves, respectively.

At the 6th week of gestation, mesoderm condenses between the extraocular muscles in the anterior orbit to form the muscle cone. The intermuscular septum is formed anteriorly, blending with scleral condensations. Posteriorly, the intermuscular fibrous septum does not form, and is not recognized even in the fully developed orbit. The formation of Tenon's capsule appears to be coordinated with the development of the insertion of extraocular muscles.

Sevel (1981) reappraised the development of the extraocular muscles by analyzing 54 specimens ranging from 8 mm to term. Prior to Sevel's study, the migratory theory of extraocular muscle development was accepted (Mann, 1964; Fink, 1953, 1956). The migratory theory implied that muscles originated posteriorly, and migrated anteriorly in the orbit (Fig. 1.16). Sevel felt extraocular muscles each developed independently from a superior and inferior mesoderm condensation (Fig. 1.17). The superior mesoderm complex forms the superior rectus, levator palpebrae superioris, and the superior oblique muscles. The inferior complex forms the inferior rectus and oblique muscles, while the medial and lateral recti form from both complexes.

In addition, Sevel documented the simultaneous development of the origin, belly, and insertion of the extraocular muscle. Six stages of muscle development were recognized: (1) mesenchymal cell; (2) early myoblast cells; (3) myoblast

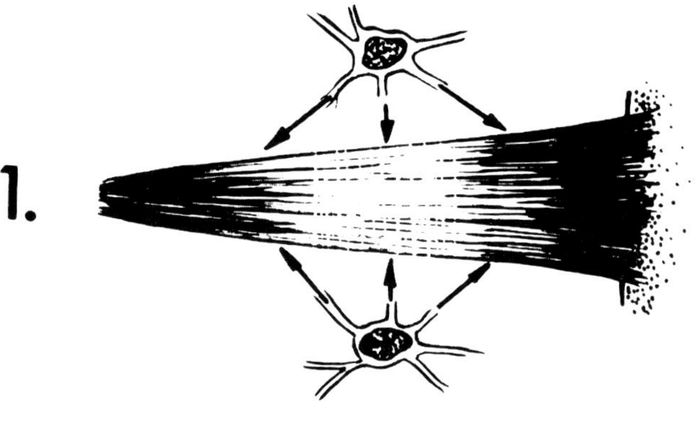

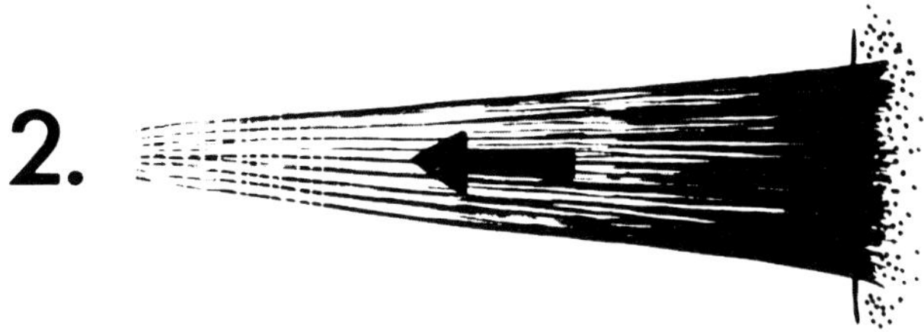

Figure 1.16. Development of the extraocular muscles: (1) Muscle developing from orbital tissue (Sevel's theory). (2) Previous theory of extraocular muscle development. The muscle originates posteriorly at the orbital apex then migrates to the globe. (Reproduced from Sevel D: *Ophthalmology*, 88:1330, 1981.)

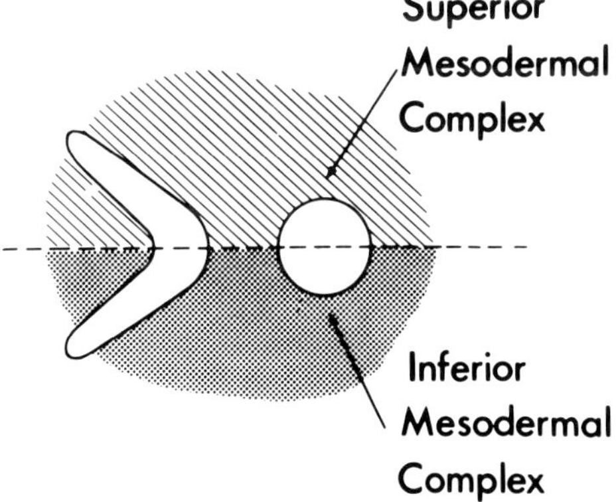

Figure 1.17. Mesodermal complexes serving as origins of extraocular muscles. (Reproduced from Sevel D: *Ophthalmology*, 88:1330, 1981.)

cell; (4) fusion of the myoblast cell; (5) myotube cell and (6) mature muscle cell.

CLINICAL NOTE

The segmental development and fusion of the extraocular muscles has distinct clinical applications. The aponeurotic approach to ptosis surgery has revealed levator aponeurotic defects in both acquired and congenital ptosis. The aponeurotic defect noted at the time of surgery is usually at the insertion of the levator aponeurosis on the anterior surface of the tarsus, inferiorly. This rather consistent location of the aponeurosis defect implicates an inherent weakness at this point. Sevel (1981) reported that the levator palpebrae superioris and its aponeurosis may develop simultaneously from multiple foci. Failure of fusion of the multiple foci of the levator muscle and its aponeurosis then might produce a defect or inherent weakness. An inherent weakness in the levator aponeurosis would account for the occurrence of aponeurotic defects in congenital ptosis. Another causal theory of aponeurotic defects in congenital ptosis is a failure of complete inferior migration and fusion of the levator aponeurosis on the anterior surface of the tarsus.

References

Anderson RL, Gordy DD: Aponeurotic defects in congenital ptosis. *Ophthalmology* 86:1493, 1979.

Beckerman BL: Lacrimal anomalies in anhidrotic ectodermal dysplasia. *Am J Ophthalmol* 75:728, 1973.

Berke RN: Blepharoptosis. *Arch Ophthalmol* 34:434, 1945.

Berke RN, Wadsworth JAC: Histology of levator muscle in congenital and acquired ptosis. *Arch Ophthalmol* 53:413, 1955.

Busse H, Müller KM, Kroll P: Radiological and histological findings of the lacrimal passages of the newborn. *Arch Ophthalmol* 98:528, 1980.

Cassidy JV: Dacryocystitis in infancy. *Am J Ophthalmol* 31:773, 1948.

Divine RD, Anderson RL, Bumsted RM: Bilateral congenital lacrimal sac mucoceles with nasal extension and drainage. *Arch Ophthalmol* 101:246, 1983.

Doxanas MT, Dryden RD: Levator aponeurotic defects in congenital ptosis. Presented at the Twelfth Annual Symposium of the American Society of Ophthalmic Plastic and Reconstructive Surgery, Atlanta, Nov. 6, 1981.

Doxanas MT, Green WR, Arentsen JJ, Elsas FJ: Lid lesions of childhood: A histopathologic survey at the Wilmer Institute (1923–1974). *J Ped Ophthalmol* 13:1, 1976.

Duke-Elder S: *System of Ophthalmology*, Vol III. Normal and Abnormal Development. Part 1. Embryology. St. Louis, CV Mosby, 1963.

Fink WH: Development of the extrinsic muscles of the eye. *Am J Ophthalmol* 36:10, 1953.

Fink WH: Development of the orbital fascia. *Am J Ophthalmol* 42:269, 1956.

Ffooks OO: Dacrocystitis in infancy. *Br J Ophthalmol* 46:422, 1962.

Gilbert PW: The origin and development of the extrinsic ocular muscles in the domestic cat. *J Morphol* 81:151, 1947.

Gilbert PW: The origin and the development of the human extrinsic ocular muscles. *Contrib Embryol* 36(No. 246):59, 1957.

Guerry D, Kendig EL: Congenital impatency of the lacrimal duct. *Arch Ophthalmol* 39:193, 1948.

Jakobiec FA, Tannenbaum M: Embryological perspectives on the fine structure of orbital tumors. *Int Ophthalmol Clin* 15:85, 1975.

Johnston MC, Noden DM, Hazelton RD: Origins of avian ocular and periocular tissues. *Exp Eye Res* 29:27, 1979.

Jones LT, Wobig JL: *Surgery of the Eyelids and Lacrimal System*. Birmingham, Aesculapius, 1976, p 169.

Kirk RC: Developmental anomalies of the lacrimal passages. *Am J Ophthalmol* 42:227, 1956.

Kiskadden WS, McGregor MW: Coloboma of the eyelids. *Plast Reconstr Surg* 2:60, 1947.

Kushner BJ: Congenital nasolacrimal system obstruction. *Arch Ophthalmol* 100:597, 1982.

Le Lievre C, Le Douarin N: Mesenchymal derivatives in the neural crest: Analysis of chimaeric quail and chick embryos. *J Embryol Exp Morphol* 34:125, 1975.

Lewis WH: The development of the muscular system. In Kiebel F, Mall FP (eds): *Manual of Human Embryology*, Vol 1. Philadelphia and London, JB Lippincott, 1910, p 519.

Mann IC: *Developmental Abnormalities of the Eye*. Philadelphia, JB Lippincott, 1957.

Mann IC (ed): *The Development of the Human Eye*, ed 3. New York, Grune & Stratton, 1964.

Masi AV: Congenital fistula of the lacrimal sac. *Arch Ophthalmol* 81:701, 1969.

Noden DM: Periocular mesenchyme: Neural crest and mesodermal interactions. In Duane TD, Jaeger EA (eds): *Biomedical Foundations of Ophthalmology*, Vol 1. Philadelphia, Harper & Row, 1982.

O'Rahilly R: The optic, vestibulocochlear and terminal vomeronasal neural crest in staged human embryos. In Rohen (ed): *Eye Structure II Symposium*. Stuttgart, Schattauer Verlag, 1965.

Ozanics V, Jakobiec FA: Prenatal development of the eye and its adnexa. In Duane TD, Jaeger EA (eds): *Biomedical Foundations of Ophthalmology*, Vol 1. Philadelphia, Harper & Row, 1982.

Reese AB: Precancerous and cancerous melanosis. *Am J Ophthalmol* 61:1272, 1966.

Scott WE, Fabre JA, Ossoinig KC: Congenital mucocele of the lacrimal sac. *Arch Ophthalmol* 97:1656, 1979.

Sevel D: A reappraisal of the origin of human extraocular muscle. *Ophthalmology* 88:1330, 1981.

Sorsby A: Congenital absence of all four puncta. *Proc R Soc Med* 25:692, 1931.

Torczynski E, Jakobiec FA, Johnston MC: Synophthalmia and cyclopia: A histopathologic, radiographic and organogenetic analysis. *Doc Ophthalmol* 44:311, 1977.

Town AE: Congenital absence of lacrimal puncta in three members of a family. *Arch Ophthalmol* 29:767, 1943.

Tse DT, Anderson RA: Aponeurosis disinsertion in congenital entropion. *Arch Opthalmol* 101:436, 1983.

Weinstein GS, Biglan AW, Patterson JH: Congenital lacrimal sac mucoceles. *Am J Ophthalmol* 94:106, 1982.

Wilson FM, Grayson M, Pieroni D: Corneal changes in ectodermal dysplasia. Case report, histopathology and differential diagnosis. *Am J Ophthalmol* 75:17, 1973.

CHAPTER 2

Osteology

It is recommended that one have a skull available for reference while reading the orbital osteology chapter. A disarticulated skull and sphenoid bone are also very useful.

The paired orbital cavities are located on each side of the sagittal plane of the skull. The orbits which are formed by the facial bones serve as sockets for the eyes. The orbits also contain the nerves and vessels supplying the eyes, as well as other areas of the face. The bony orbit and the structures contained within the orbit act to support, protect, and maximize the function of the eye.

The orbital cavity is formed by seven bones: frontal, sphenoid, zygomatic, maxilla, ethmoid, lacrimal, and palatine. Superiorly, the orbit is bordered by the anterior cranial fossa and the frontal sinus (Fig. 2.1). Nasally, the ethmoid sinus is separated from the medial orbital wall by the thin lamina papyracea of the ethmoid bone. Inferiorly, the maxillary sinus lies beneath the orbital floor. The lateral orbit is bordered anteriorly by the temporalis fossa, while posteriorly it borders the middle cranial fossa. Anteriorly, the orbit is bounded by the orbital septum.

The key to understanding the bony orbit is cognizance of the sphenoid bone. The greater and lesser wings of the sphenoid comprise the orbital apex and a significant portion of the roof and lateral walls of the orbit. In addition, the

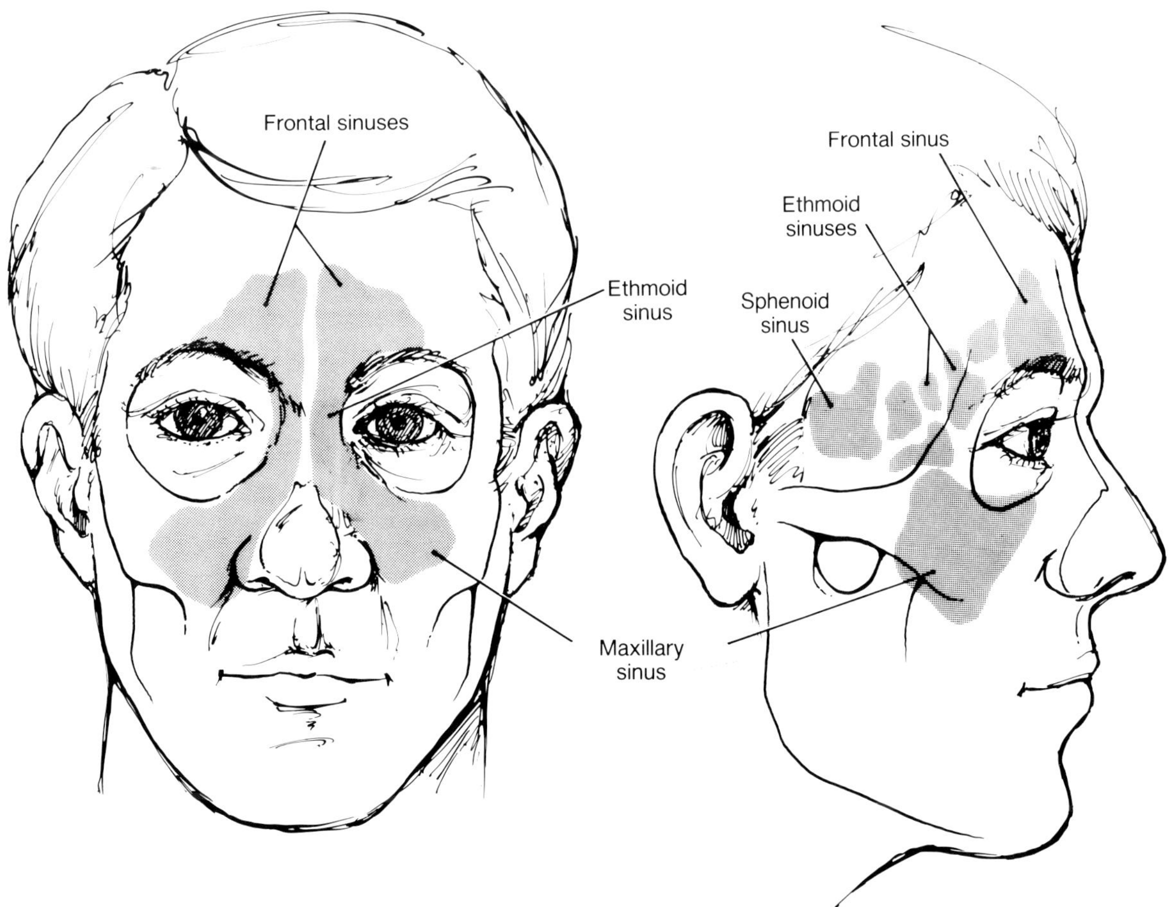

Figure 2.1. Relationship of the orbits to the paranasal sinuses.

cranial nerves and vascular structures are transmitted through this bone.

The orbital cavity approximates the configuration of a pear. The facial, or anterior, aspect of the orbit has an enlarged, prominent bony margin. Anteriorly, the orbit is roughly a quadrangular pyramidal shape. However, the greatest diameter of the orbit is at that portion immediately within the orbital margin. In addition, the floor of the orbit does not extend to the orbital apex. The medial walls of the orbit are almost parallel, whereas the lateral orbital walls are divergent, approximating a 90° angle with each other (Fig. 2.2). The position of the lateral orbital walls accounts for the apparent lateral orientation of the orbit. The central axis of the orbits are directed 45° from one another. The extension of the central axis of the orbits, beyond the orbital apex, converge near the center of the skull.

The continuity of the orbital cavity is interrupted by the large optic foramen and smaller foramina, through which run nerves and vessels. In addition, the large superior and inferior orbital fissures allow the entrance of the cranial nerves, as well as vascular structures, into the orbital cavity. The nasolacrimal canal in the anterior portion of the orbital floor allows tears to drain into the nose.

There are great variations in the orbital dimensions. Those listed in Table 2.1 are averages, and do not represent absolute measurements. Many discrepancies arise as to the depth of the orbit. This is, in part, due to the failure of the orbital floor to extend to the orbital apex. Warwick (1976) records the orbital depth at 40 mm. However, this measure-

Table 2.1.
Average Orbital Dimensions

Height of orbital margin	40	mm
Width of orbital margin	35	mm
Depth	40–50	mm
Interorbital distance[a]	25	mm
Volume	30	cc

[a] Interorbital distance is the distance between the medial orbital walls.

ment reflects the posterior extension of the orbital floor (posterior wall of the maxillary sinus). The distance from the inferior orbital margin to the orbital apex (optic foramen) is at least 50 mm (Whitnall, 1932).

CLINICAL NOTE

The orbital dimensions are of obvious importance when performing orbital surgery. The orbital dimensions are highly variable, so extreme caution should be exercised while performing surgery. For example, the exploration of an orbital floor fracture beyond 40 mm from the inferior orbital margin should be performed very cautiously. Entrapment of orbital tissues more posterior than 40 mm is rare and must occur in the inferior portion of the ethmoids.

Orbital volume is important when one considers enucleation and the evaluation of an anophthalmic socket. The orbital volume approaches 30 cc, while the volume of a 25 mm globe is 6.5–7.0 cc. Thus, the ratio of volume of the orbit to the globe is approximately 4.5:1. An enucleated 25 mm eye is usually replaced by a 16–18 mm orbital implant (volume 2.4 cc). The loss in orbital volume is, therefore, approximately 4.0 cc. In addition, the removal of the globe produces atrophy of orbital fat which accentuates the reduction in orbital volume. The orbital volume is partially but not totally replaced by the ocular prosthesis. The loss of orbital volume will be evident by a sunken superior sulcus, which frequently accompanies the anophthalmic socket.

To simplify the evaluation of the orbit, it will be reviewed in a step-wise fashion. First, the orbital margin will be studied, followed by the walls of the orbit. Relationships of the bony structures will be stressed. Due to the complex interrelationships of the sphenoid bone, this bone will be specifically evaluated. Finally, fissures and foramina of the orbit will be reviewed, noting their clinical significance.

ORBITAL MARGIN

The orbital margin is nearly rectangular in shape, with rounded corners. The orbital margin does not form a complete rectangle since it is discontinuous at the lacrimal fossa. As noted previously, the horizontal and vertical openings of the orbit are 40 mm and 35 mm respectively.

The superior orbital margin is formed by the frontal bone (Fig. 2.3). The superior orbital rim is interrupted in the nasal one-third by the supraorbital notch, which is totally enclosed to form a foramen in 25% of orbits. The supraorbital notch, or foramen, marks the point of transition of the contour of the superior orbital margin, from a sharp border

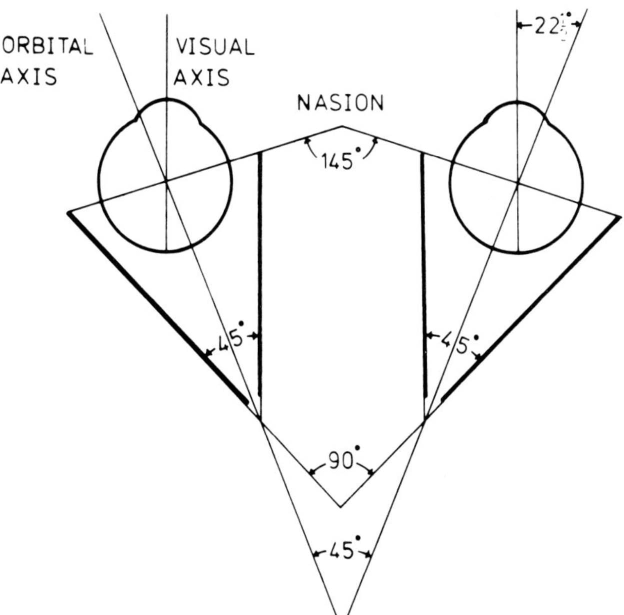

Figure 2.2. Geometry of the orbits: The medial orbital walls are parallel, while the lateral orbital walls form a 90° angle with one another. The central orbital axis is directed 45° from one another. (Reproduced from Eggers HM: Functional anatomy of the extraocular muscles. In Duane TD, Jaeger EA: *Biomedical Foundations of Ophthalmology.* © 1982, Philadelphia, Harper and Row Publishers.)

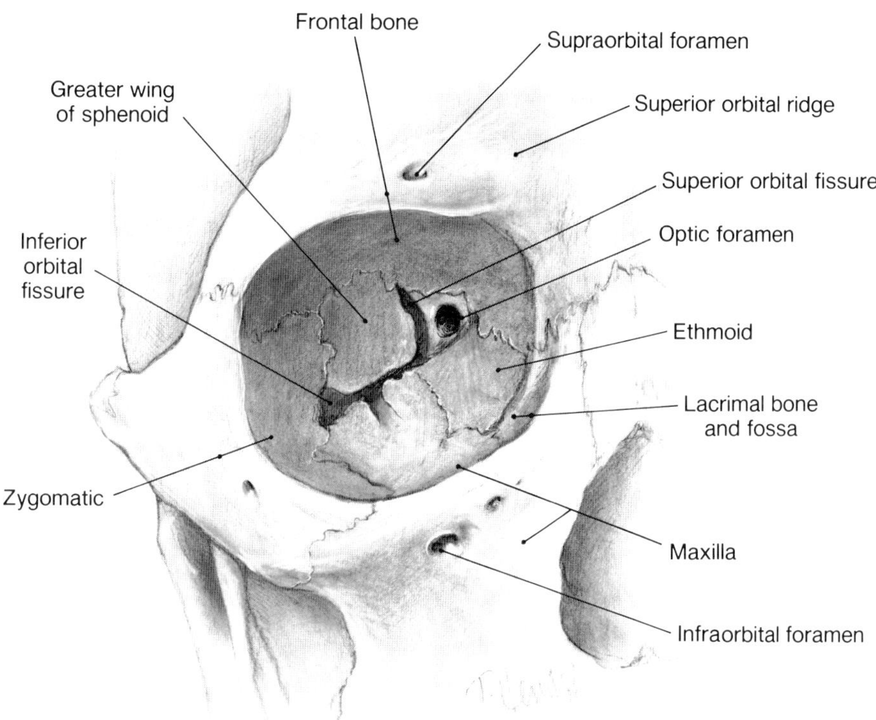

Figure 2.3.   Frontal view of the orbit.

laterally, to a rounded one medially. The rounded nasal orbital margin is crossed by the nerves and vessels transversing this area. Medial to the supraorbital notch, a smaller notch is created by the supratrochlear nerve and vessels.

Variability exists in the pattern of the supraorbital nerves and vessels. As noted, the vessels may produce a notch in the orbital margin, or produce a foramen in 25% of instances. The supraorbital nerves may also groove the orbital roof and frontal bone comprising the forehead. The supraorbital ridge is an elevation of the nasal supraorbital area. This is produced primarily by the frontal sinuses. Lempke and Stasior (1982) also attribute the supraorbital ridge to the attachment of the brow fat pad, which they feel with time, produces reactive bony hyperplasia. The supraorbital ridge is more developed in males than in females.

The medial orbital rim consists of the maxillary process of the frontal bone (pars nasalis), lacrimal bone, and the frontal process of the maxilla (Fig. 2.3). The orbital rim is discontinuous at the inferior nasal border (Fig. 2.4). If the medial margin is traced from its superior aspect, it is continuous with the posterior lacrimal crest. From the inferior orbital margin, the medial rim is continuous with the anterior lacrimal crest. The discontinuity of the medial orbital margin produces the fossa for the lacrimal sac. The inferior portion of the lacrimal fossa is continuous with the nasolacrimal canal.

A vascular groove, parallel to the anterior lacrimal crest, is located in the frontal process of the maxilla. The groove is called the sutra nothra, or sutra longitudinal imperfecta of Weber, and lodges a branch of the infraorbital artery.

The frontal process of the maxilla serves as the point of attachment of the anterior portion of the medial canthal tendon.

The inferior orbital margin is formed by the maxilla medially, and the zygoma (malar) bone, laterally (Fig. 2.3). Since each bone comprises approximately one-half of the inferior orbital margin, the zygomaticomaxillary suture lies nearly in the middle of the orbital margin. At the zygomaticomaxillary suture, a fine elevation is frequently present forming the infraorbital tubercle.

Approximately 10 mm below the zygomaticomaxillary suture is the infraorbital foramen. The foramen can be followed posteriorly to the infraorbital sulcus on the orbital floor. Through the infraorbital foramen pass the infraorbital nerve (main branch of the maxillary division of the trigeminal nerve) and the infraorbital artery (branch of the internal maxillary artery). The maxilla below the infraorbital foramen forms the anterior wall of the maxillary sinus.

The lateral orbital margin is formed by two bones: the zygomatic process of the frontal bone and the frontal process of the zygomatic (malar) bone (Fig. 2.3). These bones meet in the superior portion of the rim at the zygomaticofrontal suture. The lateral orbital rim is recessed slightly posterior and is concave, as to enhance the lateral field of gaze.

On the orbital surface of the lateral orbital rim, about 10 mm below the zygomaticofrontal suture, is the lateral orbital tubercle (Whitnall's tubercle) of the zygomatic bone. This tubercle is easily palpable and marks the point of attachment of the lateral canthal tendon, the check ligament of the lateral rectus muscle, Lockwood's suspensory ligament, and

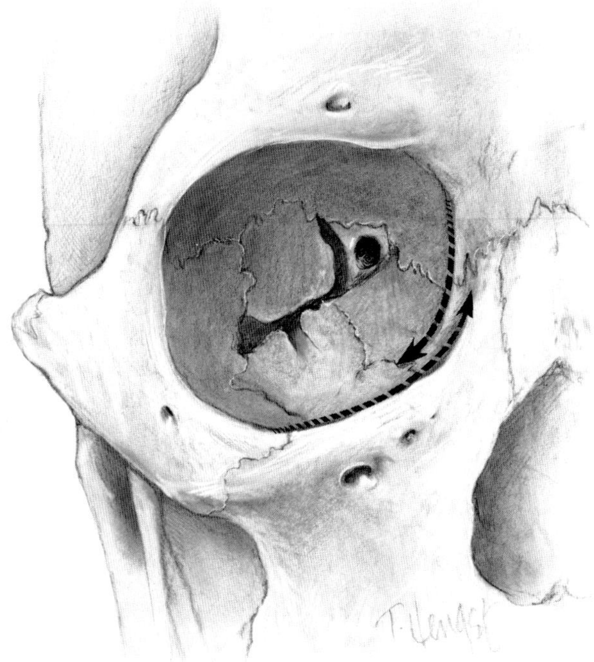

Figure 2.4. Discontinuity of the inferior nasal orbital margin. The nasal orbital margin traced from its superior aspect is continuous with the posterior lacrimal crest. The nasal margin traced from the inferior orbital margin is continuous with the anterior lacrimal crest. The discontinuity of the orbital margin creates the fossa for the lacrimal sac.

the lateral horn of the levator aponeurosis. These confluent attachments produce the lateral retinaculum of Hesser.

As with the supraorbital ridge, there appears to be a similar sexual distinction in the lateral orbital rim. The lateral orbital rim is characteristically strong and prominent in the male, while it is less conspicuous in the female. This feature, coupled with the less prominent supraorbital ridge in the female, produces a "softer" orbital contour in the female.

CLINICAL NOTE

The orbital margin produces a protective boundary for the globe. The margin is bold and stronger than the orbital walls. If the orbit is struck with a round object which diffuses its impact, the orbital rim will withstand considerable force. However, compression of the orbital contents will produce a "blow-out" fracture of the inferior or medial orbital walls (Fig. 2.5A) (Smith and Regan, 1957; Dodick, et al, 1971; Rumelt and Ernest, 1972; Mirsky and Saunder, 1979). Direct blows will most frequently be localized to the zygoma, which as noted previously, is the strongest bone of the orbital margin. The zygomatic bone will fracture at sites of potential weakness, namely the zygomaticofrontal and zygomaticomaxillary sutures (Fig. 2.5B). Clinically, this fracture is evident by depression or flattening of the orbital rim, inferior deviation of the lateral canthus, localized step or discontinuity in the inferior orbital rim at the zygomaticomaxillary suture, tenderness over the zygomatic arch, and pain with mastication and ecchymosis of the buccal mucosa. Various amounts of floor fracture and displacement may be associated with tripod fractures. Appreciation of the anatomic relationships of the orbit and orbital rim is essential for the diagnosis and repair of such fractures. Failure to reduce tripod (trimalar) fractures may produce significant functional and cosmetic deformities.

ORBITAL WALLS

The walls of the orbit are indefinite due to their rounded borders. The superior, inferior, and lateral orbital walls are triangular in shape, while the medial orbital wall is rectangular. The orbital floor extends approximately two-thirds of the depth of the orbit, while the other walls extend to the orbital apex. The lateral orbital wall is most clearly defined due to its posterior definition by the superior and inferior orbital fissures. The walls of the orbit will each be discussed separately, emphasizing anatomic concepts. In addition, the sphenoid bone will be discussed separately due to its complexity and structural importance and the transmission of orbital nerves and vessels.

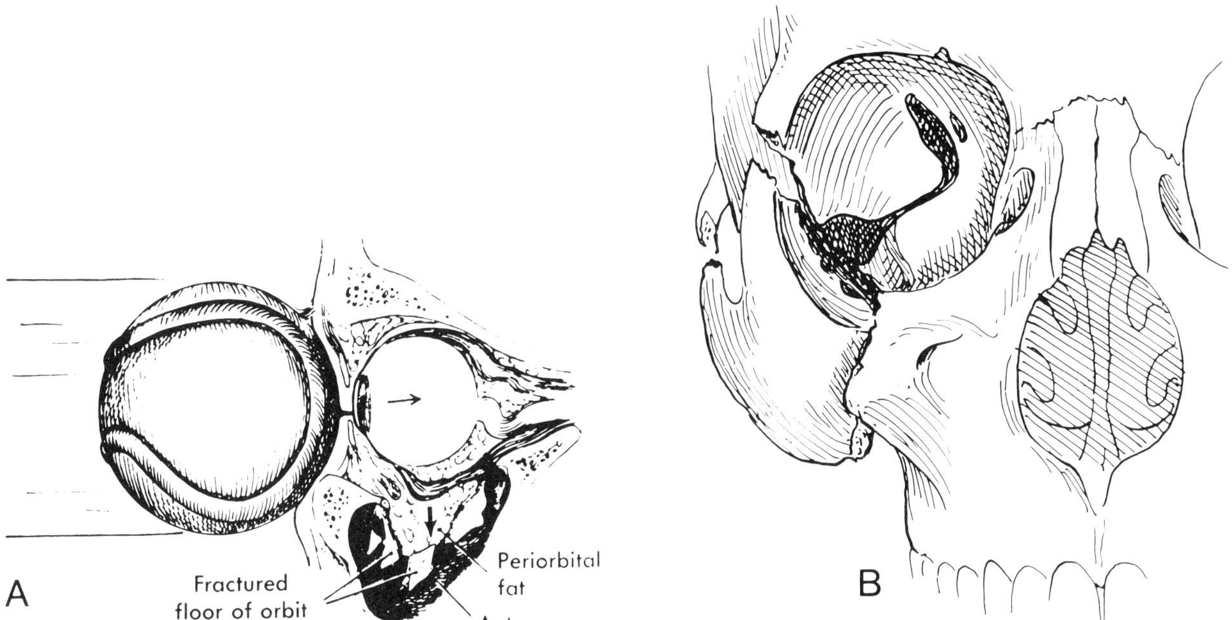

Figure 2.5. Pathophysiology of orbital fractures: (*A*), Blow-out fracture created by a blunt object compressing the orbital contents. (*B*), Tripod fracture produced by direct injury at the lateral orbital rim. The fracture is produced at the zygomaticofrontal and zygomaticomaxillary sutures. Sites of potential weakness. (Reproduced from Zizmor J, Noyek AM: Radiographic diagnosis of orbital fractures. In Aston SJ, et al (eds): *Third International Symposium of Plastic and Reconstructive Surgery of the Eye and Adnexa.* © 1982, Baltimore, William & Wilkins.)

Superior Orbital Wall

The superior wall of the orbit is triangular in shape and is formed by two bones (Fig. 2.6). The bulk of the orbital roof consists of the orbital plate of the frontal bone with the lesser wing of the sphenoid making up the posterior portion. The anterior one-half of the orbital floor is thick and forms the floor of the frontal sinus. Posteriorly, the roof is thinner, separating the orbit from the anterior cranial fossa.

The roof of the orbit is joined to the medial wall by the frontoethmoidal suture. Within the frontoethmoidal suture are the anterior and posterior ethmoidal canals. The roof is separated from the lateral wall by the zygomaticofrontal suture anteriorly and the superior orbital fissure posteriorly. The frontosphenoid suture crosses the orbital roof obliquely extending from the superior orbital fissure toward the posterior ethmoidal foramen on the medial wall. This suture may be obliterated in the adult.

The trochlea and the fossa for the lacrimal gland constitute two important features of the orbital roof. The fovea trochlearis is a small circular dimple in the superior nasal angle of the orbit, approximately 4–5 mm from the orbital margin. The fovea trochlearis marks the point of attachment of the trochlea, a cartilaginous pulley of the superior oblique tendon.

The fossa for the lacrimal gland is a depression in the frontal bone in the anterior superior lateral angle of the orbit. Most often, the fossa for the lacrimal gland is merely a deeper concavity of the orbital roof. The normal inferior limit is marked by the ridge at the zygomaticofrontal suture at the lateral orbital rim.

The roof of the orbit is usually smooth. However, numerous small holes, or depressions, may be present, producing a spiculated appearance. These are especially notable at the medial side of the anterior portion of the fossa for the lacrimal gland and are known as the cribra orbitalia. The porous condition may be produced by faulty development of the orbital plate of the frontal bone. Venous drainage extends to veins of the periorbita, which also prevents the bone from developing as a compact, smooth layer.

Medial Orbital Wall

The medial orbital wall is oblong, and is composed of four bones: frontal process of the maxilla, lacrimal bone, lamina papyracea of the ethmoid, and the sphenoid (Fig. 2.7). The bulk of the medial wall consists of the extremely thin and delicate lamina papyracea. This bone articulates with the frontal bone above, the maxilla below, and the lacrimal bone anteriorly.

Anteriorly, the medial wall of the orbit consists of the maxillary process of the frontal bone and the fossa for the lacrimal sac. The fossa for the lacrimal sac is formed by the frontal process of the maxilla and the lacrimal bone. The lacrimal bone is divided into two portions by a vertical ridge, the posterior lacrimal crest. Posterior to the crest, the lacrimal bone is flat and articulates with the lamina papyracea and the ethmoidal air cells. Anterior to the posterior lacrimal crest, the lacrimal bone becomes thinner and forms a portion of the fossa for the lacrimal sac. The lower end of the

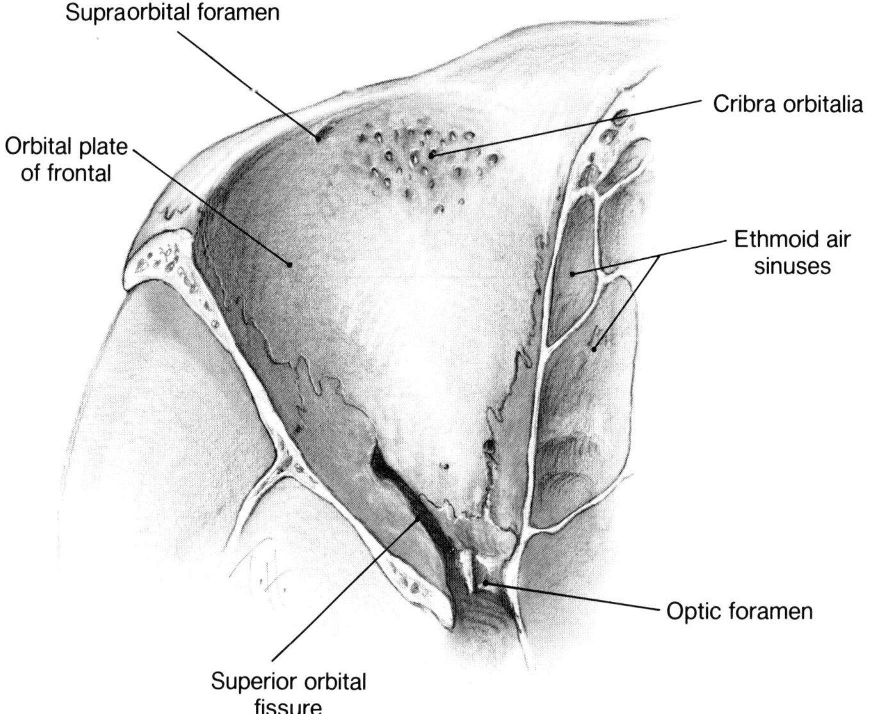

Figure 2.6.   Superior orbital wall (roof).

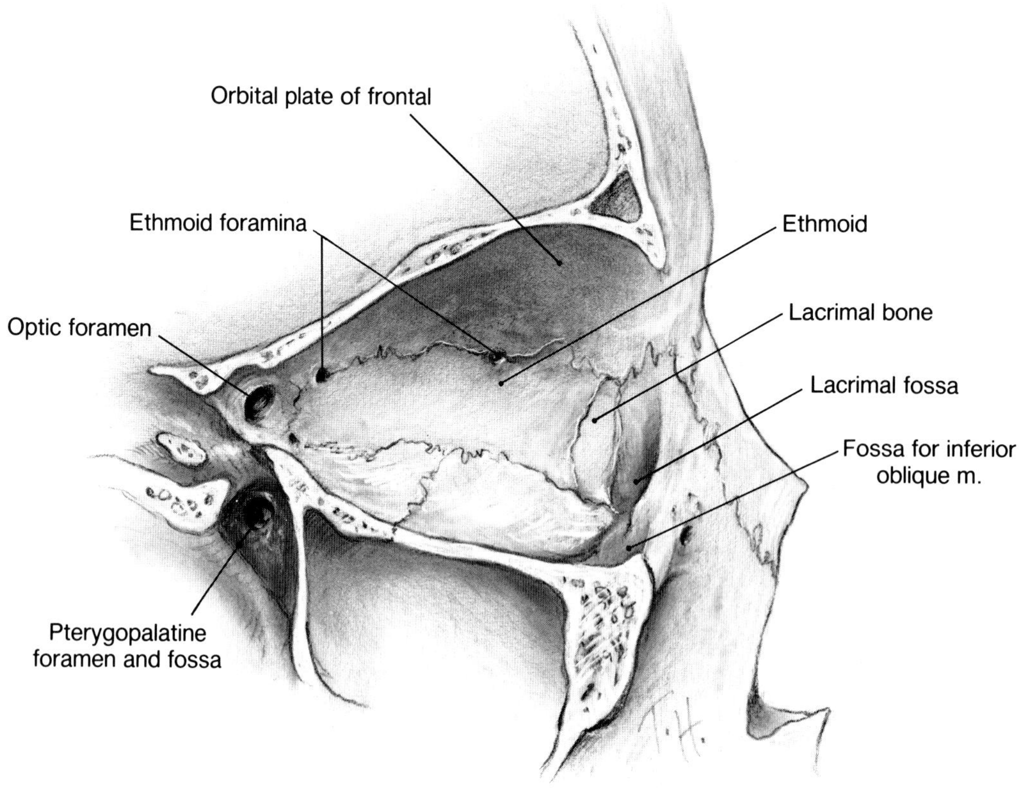

Figure 2.7.   Medial orbital wall.

posterior lacrimal crest terminates in a hook-like projection, the hamular process. The hamular process articulates with the lacrimal notch of the maxilla to define the opening of the nasolacrimal canal. The lacrimal bone is fragile orbital since it is incompletely ossified. The fragility of the lacrimal bone is clearly evidenced by the ease with which it is perforated during a dacryocystorhinostomy.

The anterior portion of the fossa for the lacrimal sac is formed by the frontal process of the maxilla, which forms the anterior lacrimal crest. As noted previously, the anterior lacrimal crest becomes continuous with the inferior orbital margin. The posterior lacrimal crest is continuous with the medial orbital margin.

The posterior aspect of the medial orbital wall is comprised of a portion of the body of the sphenoid bone. The orbital apex is a confluence of the body and greater and lesser wings of the sphenoid.

The lamina papyracea of the ethmoid is, as the name suggests, extremely thin. As such, it provides a poor anatomic barrier and will allow the intrusion of infection and tumors from the sinus into the orbit. The anterior and posterior ethmoidal vessels transverse this bone through their respective foramen.

CLINICAL NOTE

The thin medial wall produces a poor anatomic barrier to confine infections of the paranasal sinuses. Serious consequences, even death, may result from diseases of the paranasal sinuses, such as orbital cellulitis, orbital abscess, optic neuritis, cavernous sinus thromboses (Jarrett and Gutman, 1969; Krohel, et al, 1982; Watters, et al, 1976; Schramm, et al, 1982).

Infections of the paranasal sinuses are the most frequent cause of orbital cellulitis (Fig. 2.8). Watters and co-workers (1976) demonstrated radiographic evidence of sinusitis in 77 of 91 patients evaluated with orbital cellulitis.

Mucoceles of the frontal and ethmoidal sinuses develop when the normal communication from the sinuses to the nose is obstructed. Pressure, produced by the inspissated mucus, produces bony atrophy and erosion along the path of least resistance. Mucoceles may, therefore, extend into an adjoining sinus, the orbit, or the skin through the frontal bone. Mucoceles present most frequently in the superior nasal portion of the orbit (Iliff, 1973).

Inferior Orbital Wall

The inferior orbital wall is comprised of three bones: maxilla, zygomatic, and palatine (Fig. 2.9). The majority of the orbital floor is formed by the orbital plate of the maxilla. Posteriorly, the palatine bone is hard to identify since the suture uniting it with the maxilla is frequently obliterated in the adult. The shape of the inferior orbital wall approximates an equilateral triangle. As noted previously, the orbital floor extends only about two-thirds of the depth of the orbit, and therefore, does not extend to the orbital apex.

The infraorbital sulcus extends from the lateral portion of the orbital apex toward the central portion of the inferior orbital margin. Approximately 15 mm prior to reaching the orbital margin, the infraorbital sulcus is converted to a canal. The canal bends inferiorly beneath the orbital margin to exit on the anterior surface of the maxilla at the infraorbital foramen. The infraorbital canal contains the infraorbital artery and nerve. The sulcus and canal are lined by periorbita so the infraorbital nerve and vessels do not actually extend into the orbital tissues. The orbital floor is thinnest along and medial to the infraorbital sulcus.

A great deal of variability exists in the course of the infraorbital sulcus, canal, and foramen. This is a reflection of the embryonic development of the structure. Initially, the infraorbital nerve lies on the orbital floor and extends over the orbital margin. Toward the end of the second month of gestation, the nerve produces a groove in the orbital floor. With time, the anterior portion of the groove is covered with bone, which forms the infraorbital canal.

The fossa for the origin of the inferior oblique muscle is a shallow depression which is situated in the extreme anteromedial angle of the floor just behind the orbital margin, and lateral to the nasolacrimal canal. The fossa for the inferior oblique muscle may be difficult to visualize in the skull, but may be palpable as a subtle depression on the orbital floor.

CLINICAL NOTE

The inferior orbital rim and the adjacent anterior portion of the orbital floor are strong in comparison to the thinner portion of the orbital floor. The orbital floor is thinnest near the infraorbital sulcus, and may measure only ½–1 mm thick at that point. It is not surprising that blunt trauma to the orbit which compresses the orbital contents can produce fracture of the orbital floor (Fig. 2.5A).

Early signs and symptoms of a "blow-out" fracture of the orbital floor are restriction of extraocular motility (particularly upward gaze), infraorbital anesthesia (including anesthesia in the canine teeth), and orbital swelling and ecchymosis. Conversely enophthalmos may be noted early if the fracture is very large (Fig. 2.10 A and B). Restriction of extraocular motility may be secondary to a hematoma of the inferior rectus muscle, or incarceration of the periorbita or orbital septa in the inferior orbit (Fig. 2.10C). Rare instances of incarceration of the inferior rectus muscle have been documented (Fig. 2.10D). Forced duction testing and time will differentiate between inferior rectus hematoma and entrapment of orbital tissues. Enophthalmos secondary to a blow-out fracture is due to a significant herniation of orbital contents into the maxillary sinus. In addition, late enophthalmos may be secondary to atrophy of orbital fat. As noted, the thinnest portion of the orbital floor surrounds the infraorbital groove, so involvement of the nerve and resultant infraorbital anesthesia is not infrequent. Involvement of the anterior-superior alveolar nerve produces anesthesia of the canine teeth. Surgical indications for blow-out fractures are diplopia in functional positions with positive forced duction testing and enophthalmos. Most blow-out fractures do not require surgical repair (Putterman, et al, 1974). In most cases repair is best carried out approximately 7–10 days after injury when swelling has decreased and one has followed the condition for approximately one week to see that improvement has not occurred.

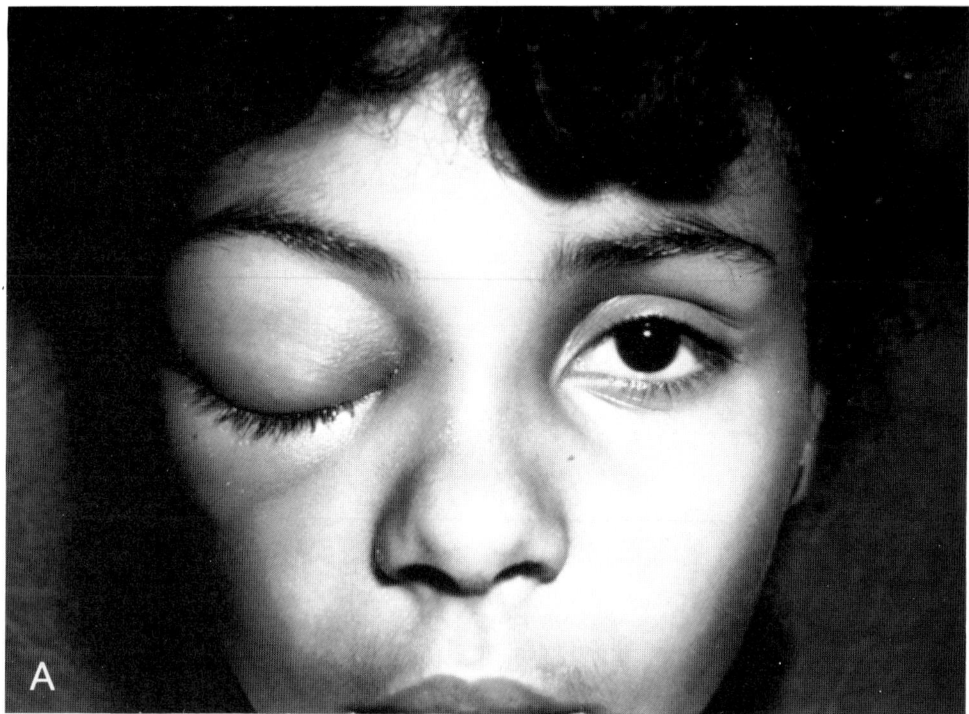

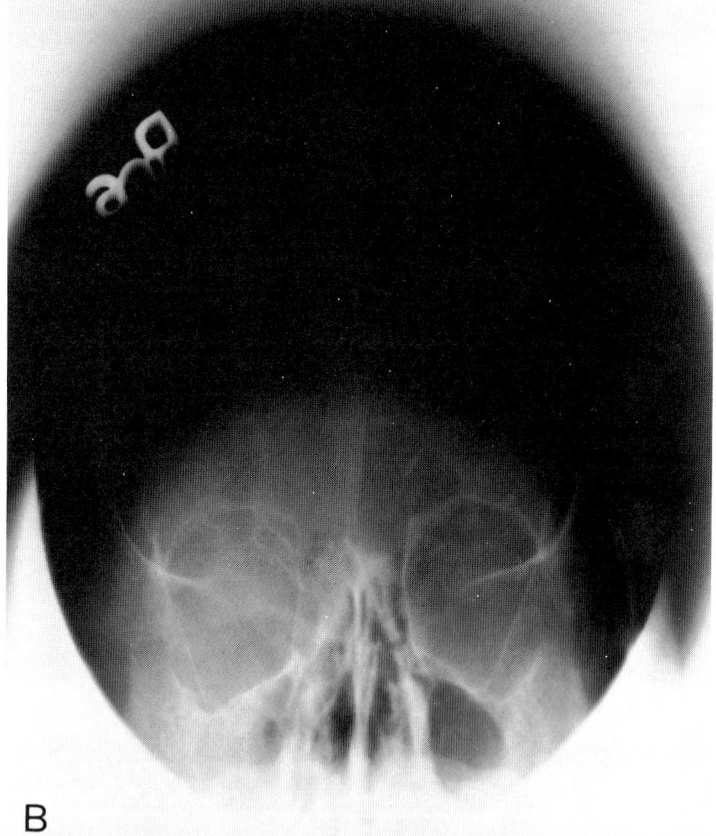

Figure 2.8.   Orbital cellulitis. (*A*), Clinical appearance of a patient with orbital cellulitis and a subperiorbital abscess. (*B*), Caldwell projection of the above patient demonstrating acute frontal, ethmoidal, and maxillary sinusitis on the involved right side.

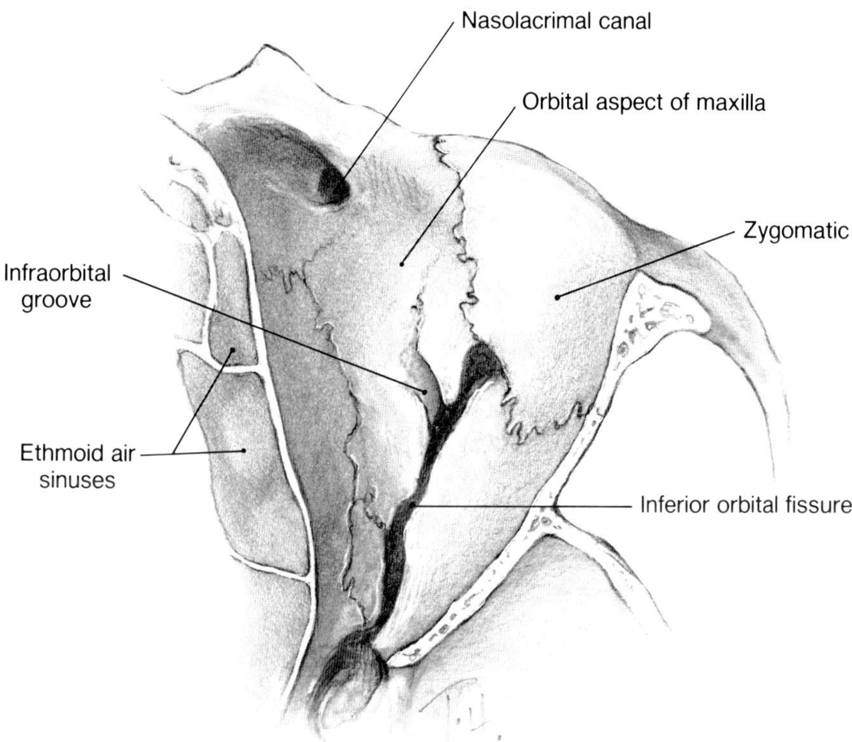

Figure 2.9.   Inferior orbital wall (floor).

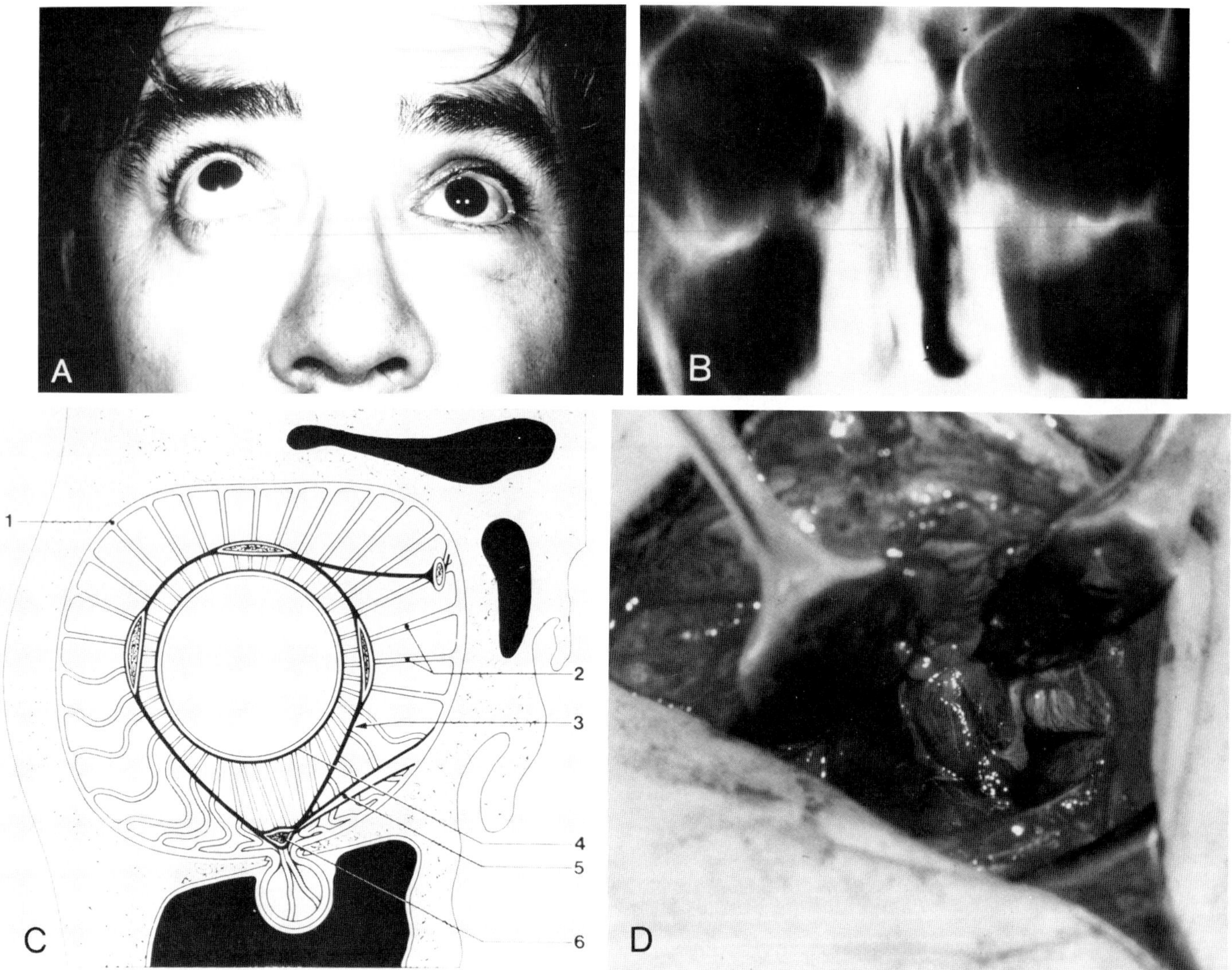

Figure 2.10.   Orbital blow-out fracture: (*A*), Clinical appearance of patient with restricted upgaze and enophthalmos. (*B*), Tomographic radiologic evaluation demonstrating orbital floor fracture and herniation of orbital contents into the maxilla. (*C*), Schematic drawing illustrating entrapped orbital connective tissue within the fracture producing restricted ocular motility. (*D*), Operative view of entrapped inferior rectus muscle in a blow-out fracture. (*C*), reproduced from Koornneef L: Orbital connective tissue. In Duane TD, Jaeger EA (eds): *Biomedical Foundations of Ophthalmology*, Vol. 1, Chapter 32. © 1982, Philadelphia, Harper and Row Publishers.)

Lateral Orbital Wall

The lateral orbital wall is triangular in shape (Fig. 2.11). The lateral wall consists of the zygomatic (malar) bone, and the greater wing of the sphenoid. Its margins are well defined. In the posterior orbit, the boundaries of the lateral orbital wall are defined by the superior and inferior orbital fissures. The lateral orbital wall is flat and is directed at a 45° angle from the medial orbital wall.

The lateral orbital wall is the strongest of the orbital walls, due to the prominent zygomatic bone. However, beyond the orbital rim, the lateral wall becomes quite thin prior to articulating with the greater wing of the sphenoid bone. The temporalis fossa is separated from the orbit by the thin lateral plate of the zygomatic bone. The lateral orbital wall thickens following the junction of the zygoma with the greater wing of the sphenoid. However, posteriorly the greater wing of the sphenoid again thins and only a small amount of bone separates the orbit from the middle cranial fossa. The lateral wall of the orbit is, therefore, a series of alternating strong and weak areas. The strong buttresses consisting of the orbital margin and the anterior portion of the greater wing of the sphenoid are separated by thinner areas of the zygomatic bone in the temporalis fossa, and the posterior portion of the greater wing of the sphenoid.

The lateral orbital tubercle (Whitnall's tubercle) is located just inside the orbital margin. The spina recti lateralis is a small spur at the orbital apex which serves as the site of

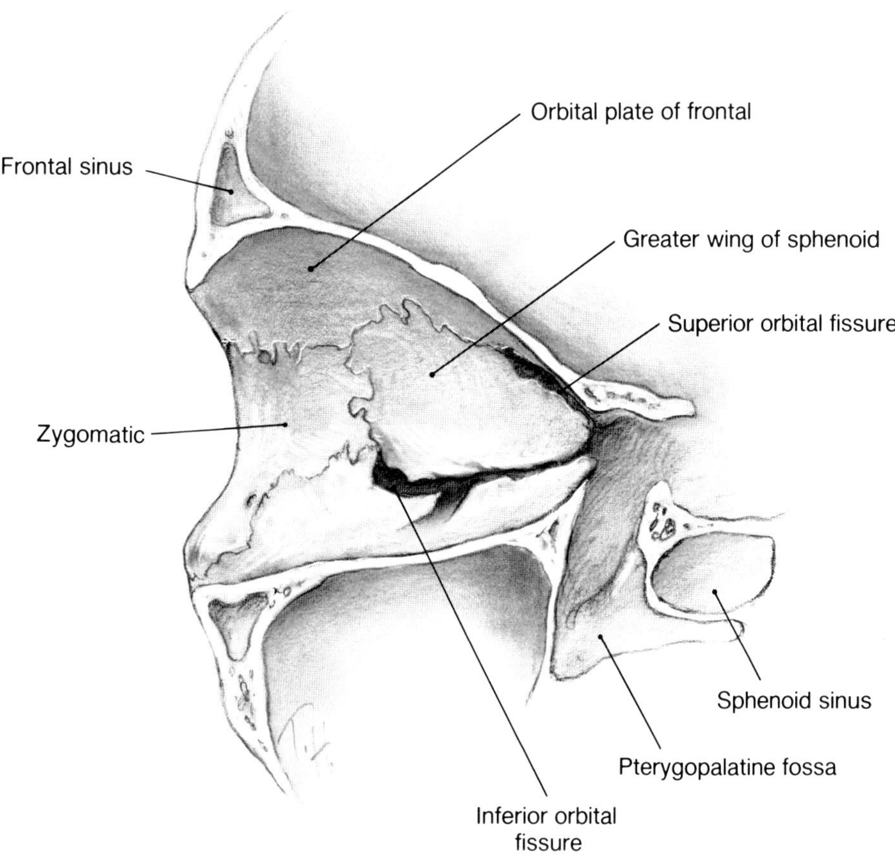

Figure 2.11.   The lateral orbital wall.

origin of the lateral rectus muscle. This is located at the widest portion of the superior orbital fissure opposite the optic foramen.

SPHENOID BONE

Because of its central role in orbital and cranial osteology, the sphenoid bone will be considered in detail. The sphenoid bone has four distinct components: the body, greater wings, lesser wings, and the pterygoid processes. The greater and lesser wings and the pterygoid processes are paired structures. When viewed from the anterior aspect, the sphenoid bone resembles a bat with its wings extended (Fig. 2.12 *A* and *B*).

The sphenoid body is cuboidal in shape and is hollow because of the sphenoidal air cells. The superior surface of the body forms the central portion of the floor of the intracranial surface. Anteriorly, the sphenoid body articulates with the lamina cribosa of the ethmoid bone. The olfactory lobes of the brain are cradled in concavities of the anterior aspect of the sphenoid. Posteriorly, there is a ridge which forms the chiasmatic (optic) groove, which extends anteriorly to the optic foramen and canal. The chiasmatic groove extends posteriorly directly superior to the tuberculum sellae and still more posteriorly to the sella turcica. The sella turcica houses the pituitary gland. Its anterior boundary is formed by the tuberculum sellae and the anterior

clinoid process. The posterior boundary is formed by a plate of bone, the dorsum sellae, terminating at its superior margin, as the posterior clinoid process. On either side of the dorsum sellae is a notch for the passage of the abducens nerve. Inferior to this notch is the petrosal process, which articulates with the petrous portion of the temporal bone. A sharp angle is created by the petrous ridge of the temporal bone and is responsible for abducens palsy associated with elevated intracranial pressure (trauma, subdural hematoma, or pseudotumor cerebri).

The lateral surfaces of the body of the sphenoid are united with its greater wing and pterygoid plates. Adjacent to the attachment of the greater wing is a broad groove, shaped like an italic letter "*F*", which forms the carotid sulcus.

The anterior surface of the body of the sphenoid forms the posterior wall of the nasal cavity. A midline sphenoid crest articulates with the perpendicular plate of the ethmoid and forms part of the nasal septum. The posterior portion of the sphenoid articulates with the basilar portion of the occiput.

The greater wings of the sphenoid are strong bones which contribute to the lateral orbital wall. As noted previously, the gap between the greater and lesser wings forms the superior orbital fissure. Inferiorly, the inferior orbital fissure prevents articulation of the greater wing of the sphenoid with the maxillary and the palatine bones. The superior portion of the greater wing forms part of the middle cranial

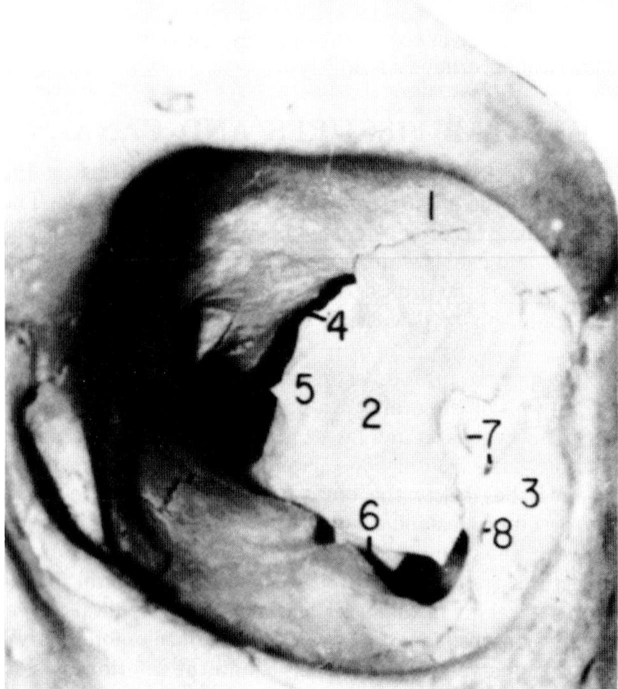

Figure 2.13. Lateral orbital wall with orbital fissures. (*1*) Frontal bone; (*2*) greater wing of the sphenoid; (*3*) zygomatic bone; (*4*) supraorbital fissure; (*5*) spina recti lateralis; (*6*) infraorbital fissure; (*7*) foramen for anastomotic branch of the zygomatico-orbital nerve to the lacrimal nerve; (*8*) zygomatico-orbital foramen. (Reproduced from Jones LT, Wobig JL: *Surgery of the Eyelids and Lacrimal System.* © 1976, Birmingham, Aesculapius Publishing Co.)

sympathetics, is formed by the union of two roots of the lesser wings of the sphenoid bone. The optic canals extend posteriorly and medially to middle cranial fossa, continuing the direction of the lateral orbital wall. If the optic canals were projected posteriorly, they would converge in the center of the dorsum sellae (Fig. 2.2).

Above, the optic canal is separated from the frontal lobe and olfactory tract by a plate of bone. Medially, the sphenoid sinus and the ethmoidal air cells border the optic canal. The medial wall of the canal actually indents the superior lateral wall of the sphenoid sinus. The optic strut forms the inferolateral border of the optic canal. The optic strut is a bony ridge joining the lesser wing to the body of the sphenoid. Cares and Bakay (1971) have demonstrated the optic strut to be a major diagnostic landmark in the radiographic evaluation of the optic canal. Erosion and enlargement of the optic foramen is reflected by abnormalities of the optic strut.

The total length of the optic canal ranges from 5.0–11.0 mm. Variability in length of the optic canal is in part due to the oblique openings of the canal. The proximal (intracranial) opening of the canal is elliptical in shape, with its greatest diameter directed horizontally. The distal (orbital) opening of the canal is also elliptical, but its widest diameter is vertically oriented. The distance between the two orbital openings averages 28 mm and the distance between the cranial openings is 14.7 mm (Miller, 1982). Measurements

of the optic canal have been tabulated by Maniscalco and Habal (1978), following the review of 83 specimens. The figures are summarized in Table 2.2.

The optic nerve is firmly fixed within the optic canal. The dura is adherent to the bone of the optic canal and the optic nerve itself (Fig. 2.15). The bony confinement of the intracanalicular optic nerve predisposes it to compression and damage. This may occur with small, radiographically imperceptible lesions.

CLINICAL NOTE

Recognition of compressive lesions of the retrobulbar optic nerve are of extreme importance. Early diagnosis and decompression of the optic nerve will prevent permanent damage to the nerve, permitting a return of visual function. Unfortunately, since intracanalicular and intracranial compression lesions rarely produce optic nerve head swelling, diagnosis is based upon the characteristic clinical history and physical signs. These include a painless reduction of visual acuity, dyschromatopsia, and the presence of afferent pupillary defect. Virtually any type of visual field anomaly may exist in patients with compressive optic neuropathy (Miller, 1982).

Retrobulbar optic nerve compression may be secondary to a variety of intraorbital and intracranial conditions. These pathologic conditions confined to the optic canal are: optic nerve glioma, meningioma, inflammation (particularly sarcoidosis), arachnoid hyperplasia, and the anterior extension of intracranial tumors. These processes may be evidenced by radiographically detected optic canal enlargement. Radiographic evidence of optic canal enlargement is either a discrepancy in size or contour of the optic canal size and configuration, or a canal measuring greater than 7 mm.

Optic nerve gliomas of childhood will involve its anterior intracanalicular portion in approximately 70% of cases. The enlarged optic canal is round and without signs of bony erosion (Evans, et al, 1963). The large optic canals are likely to represent congenital changes since only one documented case exists where a previously normal optic canal subsequently enlarged secondary to an optic nerve glioma (Spencer, 1972).

Compressive processes compromising the optic nerve within its canal are usually best treated by surgical decompression. The two most common causes of compressive optic neuropathy are dysthyroid ophthalmology and acute orbital trauma. The dramatic increase in orbital volume in dysthyroid patients compromises the nerve at the orbital apex. Decompression of the orbit, by removal of its walls (Naffziger, 1931; Dollinger, 1911; Sewell, 1936; Hirsch and Urbanek, 1930; Walsh and Ogura, 1957) is indicated in patients with optic neuropathy nonresponsive to steroid therapy, and severe exposure keratopathy. Recent reports (Anderson and Linberg, 1981; Kennerdell and Maroon, 1982) emphasize decompression of the orbital apex in patients with dysthyroid optic neuropathy.

Traumatic intracanalicular fractures producing optic nerve compression are an unusual problem. Immediate visual loss following trauma is a result of irreversible nerve damage. However, progressive visual loss following an injury is generally due to optic nerve

Figure 2.14. The orbital apex. The optic canal formed by the union of two roots of the greater wing of the sphenoid. On the lateral wall, the spina recti lateralis serves as an origin of the lateral rectus muscles. The medial orbital wall has the anterior and posterior ethmoidal canals.

Table 2.2.
Optic Canal Measurements[a]

	Average	Range
	mm	*mm*
Optic canal length	9.22	5.5–11.5
Horizontal canal width		
Proximal	7.18	5.0–9.5
Distal	4.87	4.0–6.0
Roof thickness	2.09	1.0–3.0
Medial canal wall thickness	0.21	0.1–0.31
Optic ring thickness	0.57	0.4–0.74
Anterior to posterior ethmoidal foramen	12.25	8.0–19.0
Posterior ethmoidal foramen to optic ring	6.78	5.0–11.0

[a] From Maniscalco JE, Habal MB: Microanatomy of the Optic Canal. *Journal of Neurosurgery* 48:402, 1973.

compression. Kennerdell and co-workers (1976) reported transantralethmoidal decompression of the optic canal for selected fractures with gratifying results. Anderson and co-workers (1982) recently reported large dose steroids as an effective modality to medically decompressing the optic canal in cases of acute trauma with progressive visual loss.

The anterior ethmoidal canal is formed at the junction of the ethmoids and the orbital plate of the frontal bone about 25 mm posterior to the medial orbital rim. The canal opens into the superior ethmoids inferior to the cribriform plate. The canal transmits the anterior ethmoidal nerve and artery which, if inadvertently severed at medial orbitotomy, may cause an orbital hemorrhage.

The posterior ethmoidal canal is about 12 mm posterior to the anterior ethmoidal canal. The canal is situated in the frontoethmoidal suture and opens into the posterior ethmoids. Through it are transmitted the posterior ethmoidal

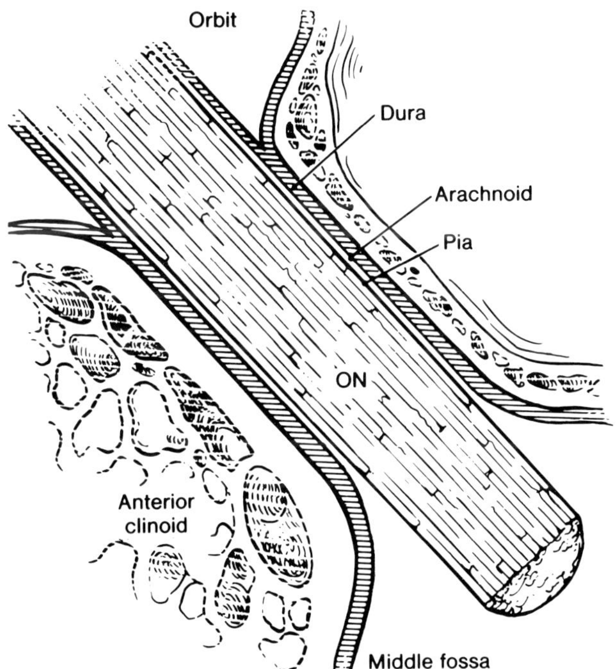

Figure 2.15. Schematic drawing of the optic nerve sheaths. The dura is tightly adherent to bone within the optic canal. Within the orbit the dura splits, forming the outer sheath of the optic nerve and the periorbita. (Reproduced from Miller NR: *Walsh and Hoyt's Clinical Neuro-Ophthalmology* (ed 4). © 1982, Baltimore, Williams & Wilkins.)

artery and, occasionally, a small nerve (sphenoethmoidal nerve of Luschka).

References

Anderson RL, Linberg JV: Transorbital approach to decompression in Grave's disease. *Arch Ophthalmol* 99:120, 1981.

Anderson RL, Panje WR, Gross CE: Optic nerve blindness following blunt forehead trauma. *Ophthalmology* 89:445, 1982.

Cares HL, Bakay L: The clinical significance of the optic strut. *J Neurosurg* 34:355, 1974.

Dodick JM, Galin MA, Littleton JT, Sod LM: Concomitant medial wall fracture and blowout fracture of the orbit. *Arch Ophthalmol* 85:273, 1971.

Dollinger J: Die Druchenllastung der Augenhöhle durch Entfernung der ausseren Orbitalwand bei hochgradigen Exophthalmus und konsekotiver Hornhauterkantung. *Dtsch Med Wochenschr* 37:1888, 1911.

Evans RA, Schwartz JF, Chutorion AM: Radiographic diagnosis in pediatric ophthalmology. *Radiol Clin North Am* 1:459, 1963.

Hirsh VO, Urbanek J: Behandlung eines exzessives Exophthalmus (Basedow) durch Entfernung von Orbitalfett von der Kieferhöhle aus. *Monatsschr Ohrenheilkd Laryngorhinol* 64:212, 1930.

Jarrett WH, Gutman FA: Ocular complications of infection in the paranasal sinuses. *Arch Ophthalmol* 81:683, 1969.

Iliff CE: Mucoceles in the orbit. *Arch Ophthalmol* 89:392, 1973.

Kennerdell JS, Amsbaugh GA, Myers EN: Transantral-ethmoidal decompression of optic canal fracture. *Arch Ophthalmol* 94:1040, 1976.

Kennerdell JS, Maroon JC: An orbital decompression for severe dysthyroid exophthalmos. *Ophthalmology* 89:467, 1982.

Krohel GB, Krauss HR, Winnick J: Orbital abscess: Presentation, diagnosis, therapy and sequelae. *Ophthalmology* 89:492, 1982.

Lempke BN, Stasior OG: The anatomy of eyebrow ptosis. *Arch Ophthalmol* 100:981, 1982.

Maniscalco JE, Habal MB: Microanatomy of the optic canal. *J Neurosurg* 48:402, 1978.

Miller NR: *Walsh and Hoyt's Clinical Neuro-Ophthalmology*, ed 4. Baltimore, Williams & Wilkins, 1982, p 48.

Mirsky RG, Saunder RA: A case of isolated medial wall fracture with medial rectus entrapment following seemingly trivial trauma, *J Pediatr Ophthalmol Strab* 16:287, 1979.

Naffziger HC: Progressive exophthalmos following thyroidectomy: Its pathology and treatment. *Ann Surg* 94:582, 1931.

Putterman AM, Stevens T, Urist MJ: Nonsurgical management of blow-out fractures of the orbital floor. *Am J Ophthalmol* 77:232, 1974.

Renn WH, Rhoton AL: Microsurgical anatomy of the sellar region. *J Neurosurg* 45:288, 1975.

Rumelt MB, Ernest JT: Isolated blow-out fracture of the medial orbital wall with medial rectus muscle entrapment. *Am J Ophthalmol* 73:451, 1972.

Schramm VL, Curtin HD, Kennerdell JS: Evaluation of orbital cellulitis and the results of treatment. *Laryngoscope* 92:732, 1982.

Sewell EC: Operative control of progressive exophthalmos. *Arch Otolaryngol* 24:621, 1936.

Smith B, Regan WF, Jr: Blow-out fracture of the orbit: Mechanism and correction of internal orbital fracture. *Am J Ophthalmol* 44:733, 1957.

Spencer WH: Primary neoplasma of the optic nerve and its sheaths: Clinical features and current concepts of pathogenetic mechanisms. *Trans Am Ophthalmol Soc* 70:490, 1972.

Walsh TE, Ogura JH: Transantral orbital decompression for malignant exophthalmos. *Laryngoscope* 67:544, 1957.

Warwick R: *Eugene Wolff's Anatomy of the Eye and Orbit*, ed 7. Philadelphia, WB Saunders, 1976, p 19.

Watters EC, Wallar PH, Hiles DA, Michaels RH: Acute orbital cellulitis. *Arch Ophthalmol* 94:785, 1976.

Whitnall SE: *Anatomy of the Human Orbit and the Accessory Organs of Vision*, ed 2. London, Oxford University Press, 1932, p 95.

CHAPTER 3

Radiographic Evaluation of the Orbit

JONATHAN J. DUTTON, M.D., PH.D.

Radiographic analysis of the orbit is essential to the evaluation of any patient with suspected orbital disease. Such studies not only contribute to the diagnosis of specific pathologic processes, but also guide the ophthalmologist in planning appropriate medical and surgical management.

Plain x-rays of the orbit may be difficult for the nonradiologist to interpret because of the superimposition of multiple shadows cast by overlapping bony structures. Because of the depth of the skull and orbit even minor alterations in head position lead to significant changes in the relationships of these shadows. Such difficulties may be largely overcome with hypocycloidal tomography, a complex movement of x-ray tube and film that provides images in a single plane as little as 1 mm in thickness, with blurring of adjacent levels.

Thin-section, high-resolution computerized tomographic scanning (CT) allows even more precise analysis by imaging soft-tissue relationships to surrounding orbital bones. Special techniques of window alteration allow density contrasts suitable for examination of specific tissues, while contrast enhancement with intravenous iodinated compounds may provide additional information.

It is the purpose of this chapter to review the plain x-ray and CT findings in the normal orbital anatomy as a basis for the recognition of pathologic states.

PLAIN ROENTGENOGRAMS OF THE ORBIT

Complete radiographic evaluation of the orbit should include as a minimum five views as follows:
1. Caldwell's posteroanterior projection
2. Waters' posteroanterior projection
3. Lateral projection
4. Submentovertex projection
5. Oblique apical projections of the optic canals

Depending upon specific findings in the above routine series, additional views and special techniques may be ordered as indicated.

For plain films of the skull the x-rays are directed with reference to Reid's orbitomeatal baseline. The latter is an imaginary line drawn between the lateral canthal angle and the external auditory meatus. Positive (+) angulation to this baseline refers to an upward direction of the x-ray beam; negative (−) angulation indicates a downwardly directed beam.

Caldwell's Projection

First described by Caldwell in 1918, this view was intended for evaluation of the frontal and ethmoid sinuses. This projection represents the skull in the frontal or coronal plane and yields the single most informative film in the orbital series. The x-ray beam is directed in the posterior-anterior incidence and is usually angled downward at −15 to 20° to Reid's orbitomeatal baseline. It thus projects the petrous pyramids to just below the inferior orbital rim. Variation in technique may yield extended Caldwell's views at up to −28°. This projection gives excellent visualization of the frontal and ethmoid sinuses, the anterior and posterior aspects of the medial orbital wall, the entire rim, superior and inferior orbital fissures, and oblique orbital line (Yanagisawa, et al, 1968; Newton and Potts, 1971; Potter, 1982). (Fig. 3.1). Also seen well are the sphenoid ridges, anterior clinoid processes, lateral expansions of the sphenoid sinuses, and the zygomaticofrontal suture. Hyrtl's canal, which transmits a branch of the middle meningeal artery into the orbit, is seen in the superolateral orbital shadow, and the foramen ovale projects onto the upper medial half of the maxillary sinus.

Waters' Projection

In 1915 Waters and Waldron described this projection, designed to view the maxillary antrum and ethmoid sinuses without obstruction by the petrous bones. The x-ray beam is directed from posterior to anterior and is angled downward at −35 to 45° to Reid's baseline (Merrell and Yanagisawa, 1968). An extended Waters' projection at −45 to 50° is sometimes employed for better visualization of the maxillary antrum (Fig. 3.2).

The Waters' view provides good imaging of the maxillary sinus, orbital rims, lesser wings of the sphenoid, anterior rim of the medial orbital wall, orbital floor, oblique orbital line, zygomatic arches, the infraorbital foramen, foramen ovale, foramen rotundum, and zygomaticofacial foramen. The petrous pyramids are projected well below the level of the antrum. With the mouth open the sphenoid sinus may

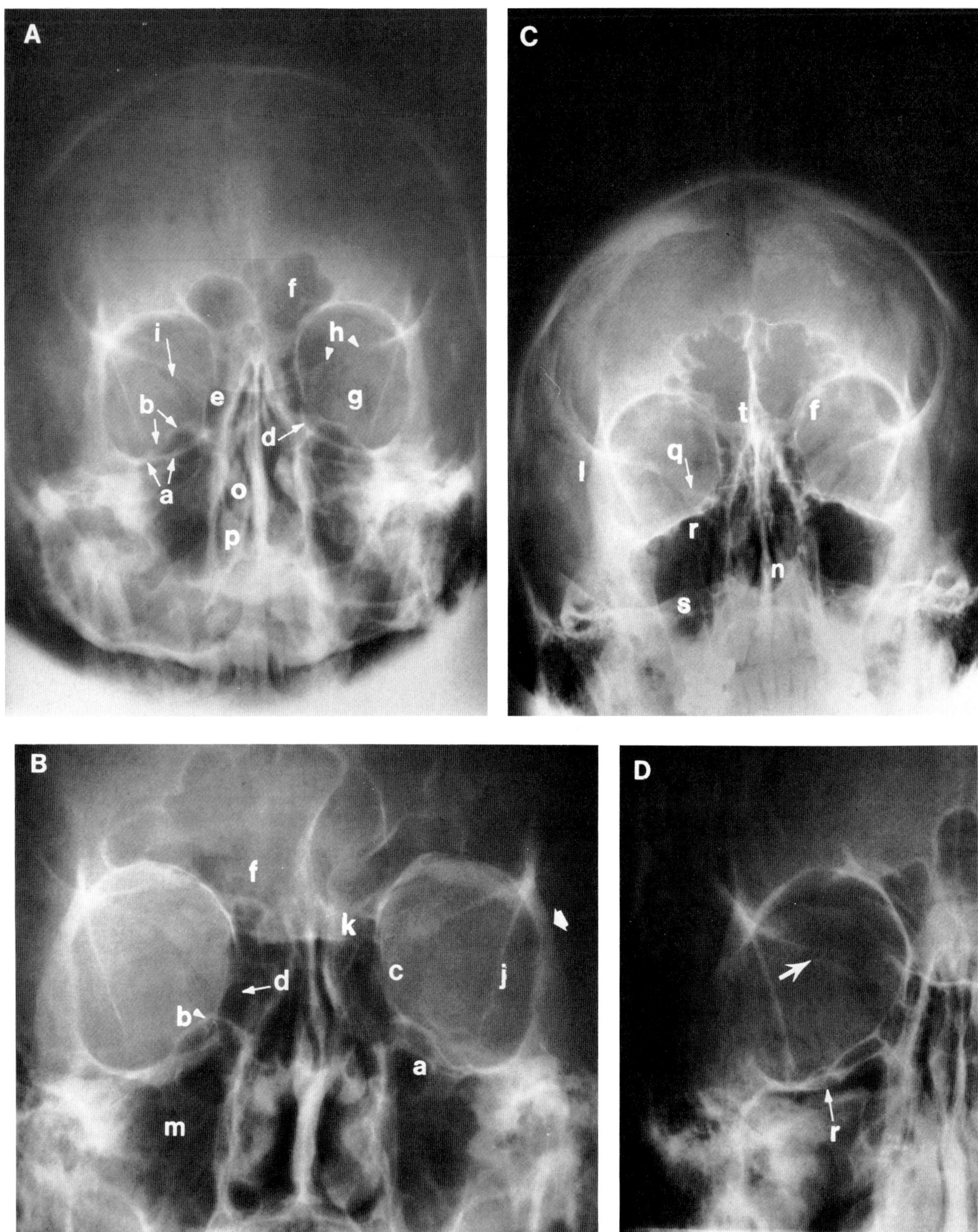

Figure 3.1. Radiographic examination of the orbit by Caldwell's projection. In (*A*) the x-ray tube is angled downward at approximately −20° to Reid's canthal-meatal baseline, and the petrous pyramids lie at the level of the orbital floor. In (*C*) the x-ray tube is angled at about −28° to Reid's baseline with the petrous pyramids lying low in the maxillary sinus. (Abbreviations: *a*, inferior orbital rim, anterior orbital floor; *b*, posterior orbital floor; *c*, lamina papyracea; *d*, anterior portion of the medial orbital wall; *e*, ethmoid sinus air cells; *f*, frontal sinus; *g*, greater wing of the sphenoid bone; *h*, lesser wing of the sphenoid bone; *i*, superior orbital fissure; *j*, oblique orbital line; *k*, planum sphenoidale; *l*, lateral orbital rim; *m*, maxillary sinus; *n*, nasal septum; *o*, middle turbinate; *p*, inferior turbinate; *q*, anterior clinoid process; *r*, infraorbital foramen; *s*, petrous pyramid of temporal bone; *t*, crista galli; *small arrow head*, frontozygomatic suture; *large arrow*, Hyrtl's foramen.)

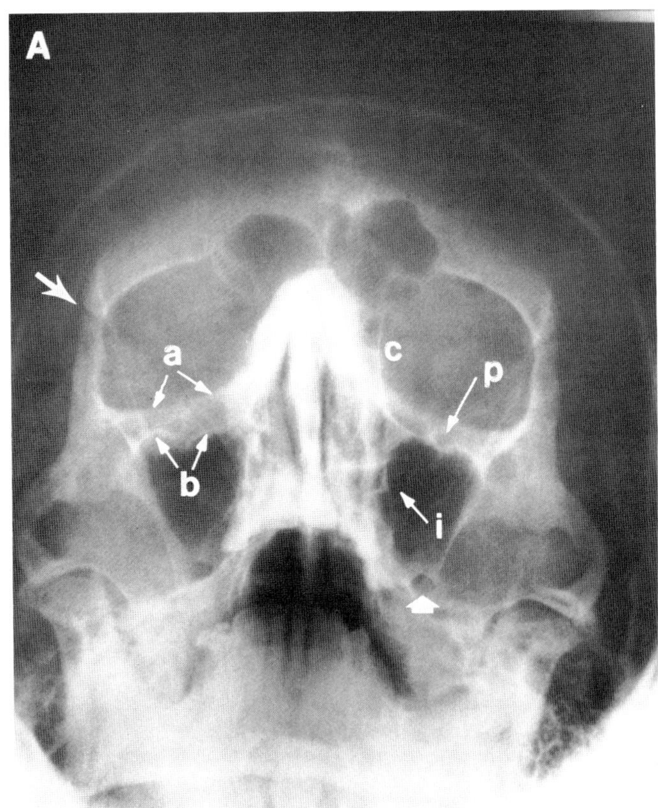

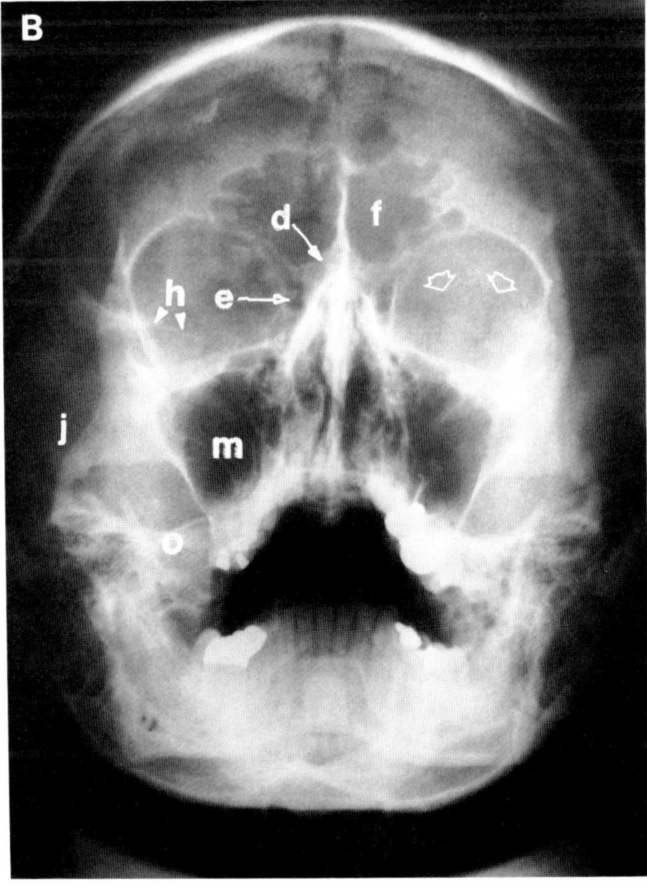

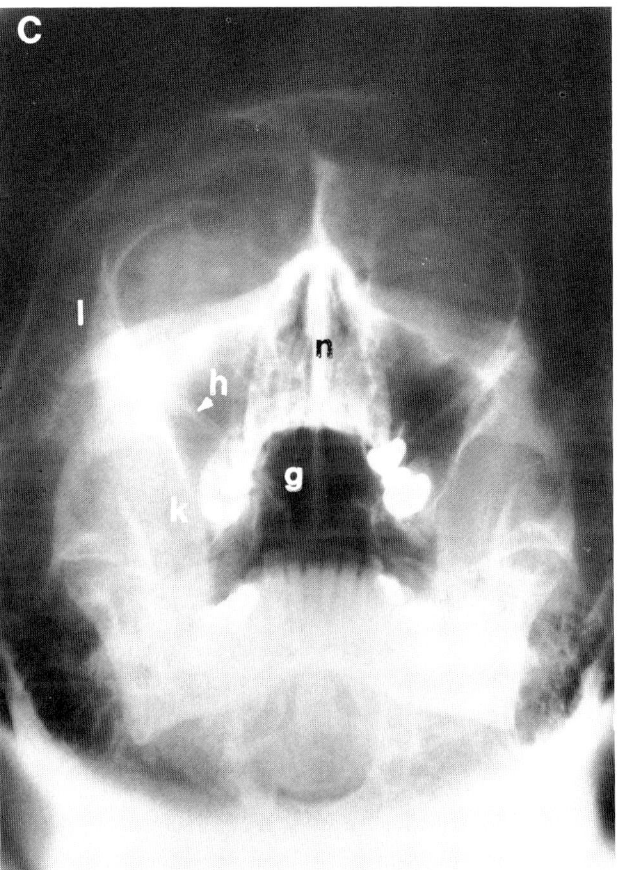

Figure 3.2. Radiographic examination of the orbit by Waters' projection. The x-ray beam is angled downward at approximately −35° to Reid's baseline in (*A*) so that the petrous pyramids project to just below the maxillary sinus. In (*C*) the beam is directed at approximately −45° to Reid's baseline giving an extended Waters' projection. (Abbreviations: *a*, inferior orbital rim, anterior orbital floor; *b*, posterior orbital floor; *c*, anterior portion of the medial orbital wall; *d*, nasal bone; *e*, ethmoid sinus; *f*, frontal sinus; *g*, sphenoid sinus; *h*, lesser wing of the sphenoid bone; *i*, superior orbital fissure; *j*, zygomatic arch; *k*, infratemporal fossa; *l*, lateral orbital rim; *m*, maxillary sinus; *n*, nasal septum; *o*, petrous pyramid of temporal bone; *p*, infraorbital foramen; *closed arrow head*, foramen ovale; *open arrow head*, palpebral fissure.)

be seen (Fig. 3.2*C*). The Waters' view is the best projection for visualization of the orbital floor in its relationship to the underlying maxillary sinus in blow-out fractures, although a better, more perpendicular shadow of the floor itself is obtained on the Caldwell's projection. Neither the posterior portion of the medial wall nor the optic canals can be seen on this view.

Lateral Projection

In this projection the x-ray beam is directed horizontally across the head and is centered just in front of the external auditory meatus on Reid's baseline. It represents the skull in the sagittal plane (Fig. 3.3). It provides the best visualization of the orbital roof. Good views are also obtained of the sphenoid, frontal and ethmoid sinuses, the anterior clinoid processes, nasopharynx, cribriform plate, infraorbital canal, and the lateral portion of the orbital floor (Yanagisawa, et al, 1968). However, this view gives a poor image of most of the medial floor because of the latter's more oblique orientation to the x-ray beam. Both left and right lateral projections are required for adequate evaluation of the orbits and for comparison of symmetry. Good views of the orbital floor may be obtained with hypocycloidal tomography in the lateral incidence.

Submentovertex Projection

Also referred to as the base view, in this projection the x-ray beam is directed from the skull vertex downward toward the submental triangle at an angle of −90° to Reid's baseline, thus representing the skull in the axial plane. This projection gives the best view of the lateral orbital wall and is also utilized for examination of the sphenoid and ethmoid sinuses, nasopharynx, lamina papyracea, lateral and posterior walls of the maxillary sinus, the greater wings of the sphenoid, and the middle cranial fossa (Zizmor and Lombardi, 1973) (Fig. 3.4).

Axial hypocycloidal tomography in the submentovertex projection is a valuable supplemental study that provides excellent visualization of the paranasal sinuses and their relation to the orbit, particularly the ethmoid air cells and the medial orbital wall. It is also good for evaluation of the body of the sphenoid and for fractures through the optic canal.

Oblique Apical Projection

Also known as optic canal views, in this projection the optic foramen is projected into the inferolateral quadrant of the orbital shadow, and the optic canal is oriented perpendicular to the film plane (Hartman and Gilles, 1959; Potter and Trokel, 1971) (Fig. 3.5). The x-ray beam is directed from posterior to anterior at 38° to the midsagittal plane, and downward at −30 to 35° to Reid's baseline in adults. In children, because of the shallower inclination of the optic canals, the beam is angled at −20°.

This projection provides good views of the optic foramen, optic strut, anterior clinoid processes, planum sphenoidale, superior ethmoid air cells, and posterior frontal sinus.

RADIOGRAPHIC ANATOMY OF THE ORBIT
Orbital Roof

Composed of the orbital plate of the frontal bone anteriorly and the lesser wing of the sphenoid posteriorly, this is best seen in lateral projection. The left and right roof shadows are usually seen separately, although with variation in technique they may be completely superimposed. They appear as curved, convex upward, wavy lines extending from the superior orbital rim to the anterior clinoids (Fig. 3.3). Irregularities in the superior surface of these lines results from variable thickness caused by convolutions of the overlying frontal lobes. The flat inferior border represents the orbital surface. Below the roof shadows is the smooth upper wall of the ethmoid sinuses and the cribiform plate, which are easily distinguished from the orbital roof.

Tomograms in the frontal and sagittal planes give better visualization of the orbital roof and are necessary for complete evaluation.

Lateral Orbital Wall

Posteriorly this is formed by the greater wing of the sphenoid, and anteriorly by the zygomatic process of the frontal bone and orbital process of the zygomatic bone. Since the greater wing lies at an oblique angle to the sagittal plane, it is not seen well in either frontal or lateral projections, and in oblique views is largely obscured by the lateral orbital rim.

The oblique orbital or innominate line is seen as a dense linear shadow in frontal projections of Caldwell and Waters extending from the superolateral orbital rim to the lateral portion of the orbital floor behind the maxillary antrum (Figs. 3.1 and 3.2). This is produced by a vertical depression along the lateral plate of the greater wing where it joins the orbital plate at the anterior extent of the temporalis fossa. Inferiorly the oblique line may end in a medial right angle turn representing the roof of the infratemporal fossa and the outline of the pterygoid plates. When this latter structure projects just beneath the orbital floor it can be mistaken for a depressed fracture (Merrell and Yanagisawa, 1968; Potter, 1971).

The best evaluation of the lateral wall is in the submentovertex projection where the orbital plate of the greater wing of the sphenoid appears as a straight line (Fig. 3.3). A small swelling along this line anteriorly represents the sphenozygomatic suture where the greater wing is thickened for insertion of the anterior origin of temporalis muscle fibers. Superimposed upon this lateral orbital wall is an S-shaped curvilinear line that begins laterally at the zygomatic arch and curves medially at its posterior aspect to join the lateral orbital wall. This curved line is the lateral and posterior wall of the maxillary antrum. In a true −90° submentovertex projection this antral line lies anterior to the straight lateral orbital wall, but on suboptimal projections may be superimposed upon, or lie behind, that structure.

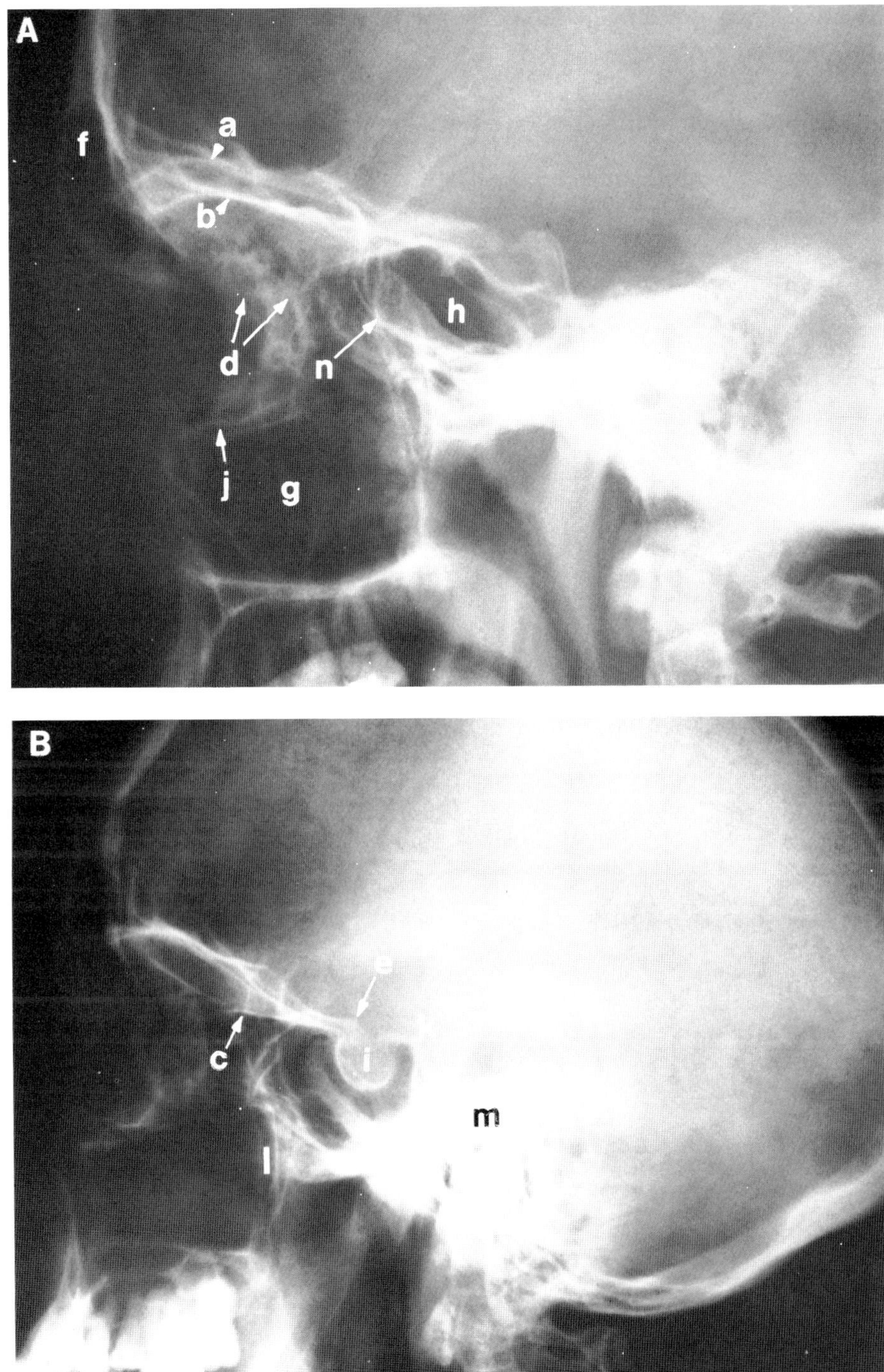

Figure 3.3. Radiographic examination of the orbit by lateral projection. (Abbreviations: *a*, cranial surface of orbital roof; *b*, orbital surface of orbital roof; *c*, roof of ethmoid sinus; *d*, cribiform plate; *e*, anterior clinoid process; *f*, frontal sinus; *g*, maxillary sinus; *h*, sphenoid sinus; *i*, sella turcica; *j*, orbital floor; *k*, dorsum sellae; *l*, posterior wall of maxillary sinus; *m*, petrous pyramid of temporal bone; *n*, greater wing of sphenoid bone forming the posterior wall of the orbit.)

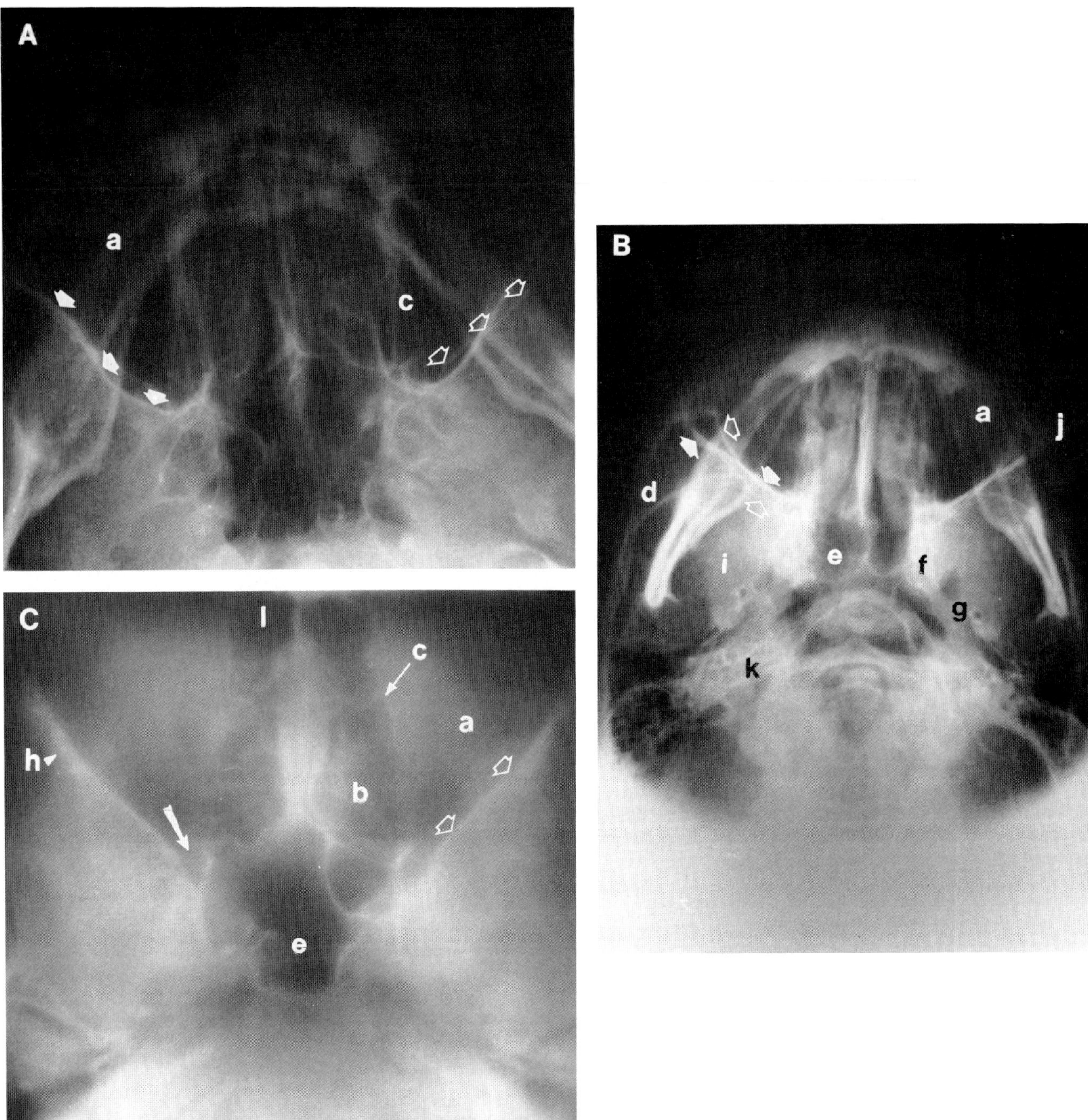

Figure 3.4. Radiographic examination of the orbit by submentovertex projection. (*A*) and (*B*) are plain films with the x-ray beam angled at approximately −90° to Reid's baseline. (*C*) is an hypocycloidal tomogram in the submentovertex projection, or axial plane. (Abbreviations: *a*, orbit; *b*, ethmoid sinus; *c*, lamina papyracea; *d*, greater wing of the sphenoid bone forming the anterior wall of the middle cranial fossa; *e*, sphenoid sinus; *f*, pterygoid plate; *g*, foramen ovale; *h*, zygomaticosphenoid suture; *i*, middle cranial fossa; *j*, zygomatic arch; *k*, petrous pyramid of temporal bone; *l*, nasal cavity; *open arrow heads*, lateral wall of the orbit; *solid arrow heads*, posterior wall of the maxillary sinus; *arrow*, orbital end of the optic canal.)

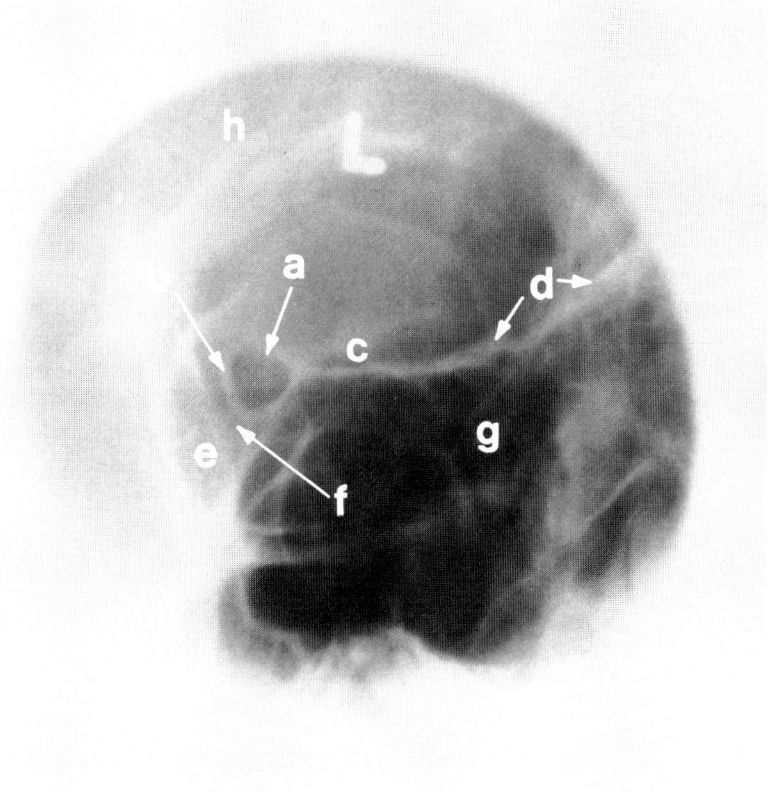

Figure 3.5. Oblique apical projection for examination of the optic canal. The x-ray beam is angled at 38° to the midsagittal plane and −35° to Reid's baseline. The optic canal in this projection is positioned into the inferolateral portion of the orbital shadow. (Abbreviations: *a*, left optic canal represented by its orbital rim; *b*, optic strut; *c*, planum sphenoidale; *d*, lesser wing of the sphenoid of the right side; *e*, greater wing of the sphenoid; *f*, superior orbital fissure; *g*, ethmoid sinus air cells; *h*, orbital roof.)

Medial Orbital Wall

This lies parallel to the sagittal plane and is formed by the lamina papyracea of the ethmoid bone posteriorly, and the lacrimal bone and frontal process of the maxillary bone anteriorly. It is well seen on Caldwell's projection (Fig. 3.1), but axial tomograms are required for best demonstration. The posterior portion of the medial wall cannot be seen well on the Waters view.

In Caldwell's frontal projection the medial wall appears as a double shadow. That portion overlying the posterior ethmoid air cells projects as an oblique line extending laterally and downward, and is continuous with the posterior orbital floor. The anterior portion of the medial wall is a less distinct vertical line joining the inferior orbital rim and represents the posterior lacrimal crest. On Waters' view this anterior line is more distinct (Fig. 3.2).

Immediately medial to the lamina papyracea are the ethmoidal air cells. The nasolacrimal canal cannot be seen clearly on standard projections, but may be visualized on frontal and lateral tomograms. The lacrimal bone is difficult to see on any plain film.

Orbital Floor

This portion of the orbit is formed by the orbital plate of the maxillary bone and is bounded laterally by the inferior orbital fissure, and posteriorly by the superior orbital fissure.

The floor is not well seen on lateral films because of its oblique orientation. On Caldwell's projection it usually produces two distinct shadows (Fig. 3.1). The upper shadow represents the posterior floor just in front of the inferior orbital fissure; the lower, denser shadow is produced by the inferior orbital rim. In some techniques of Caldwell's projection these two lines may be superimposed. In others there may be three distinct shadows—the denser orbital rim centrally, the posterior floor projecting above this, and the anterior floor (that portion just behind the rim) below it (Yanagisawa, et al, 1968).

On the Waters' projection these relationships are reversed: the thicker upper shadow represents the orbital rim, and the thinner lower line is the posterior floor. In the extended Waters' view the floor cannot be seen due to its marked obliquity to the x-ray beam.

On standard frontal projections the orbital floor slopes upward from lateral to medial. Laterally the posterior floor turns downward into the inferior orbital fissure which may appear as a notch that can be mistaken for a floor fracture.

Medially the floor shadows are continuous with the medial wall lines. The posterior floor line passes into the posterior medial wall shadow at the lamina papyracea; the more anterior floor line joins the anterior medial wall at the posterior lacrimal crest.

Evaluation of the orbital floor is significantly enhanced with frontal and lateral tomograms which are essential for adequate delineation of blow-out fractures.

Orbital Rim

The rim of the orbit is clearly seen on both Caldwell's and Waters' projections as a dense, well corticated shadow (Figs. 3.1 and 3.2). On the steeper projections the inferolateral portion of the rim forms an abrupt right-angle turn caused by the convex shape of the lower rim. Inferiorly the infraorbital canal can be seen on Waters' view projecting just below the rim. Along the superior rim the supraorbital notch is seen. The zygomaticofrontal suture is frequenlty visible on the lateral rim, and the zygomaticomaxillary suture may be distinguished inferolaterally. It is important not to confuse these sutures with rim fractures.

Superior and Inferior Orbital Fissures

The superior orbital fissure is a club-shaped opening between the middle cranial fossa and the orbital apex, and transmits numerous neurovascular elements. It lies between the lesser wing of the sphenoid above and the greater wing below, and is separated from the optic canal by the optic strut. Subject to wide variation in size and shape, the fissure is seen best on Caldwell's projection (Fig. 3.1), and on coronal tomograms as a lucency with well corticated margins running obliquely in the medial orbit just below the lesser wing shadow. Below, it turns to a more vertical orientation before joining the inferior orbital fissure.

In the Waters' projection most of the superior fissure lies behind the maxillary antrum and is difficult to distinguish. However, where it crosses the inferior orbital rim its relative lucency can be confused with a rim fracture.

The inferior orbital fissure lies in the lower portion of the orbit in Caldwell's view. Laterally it appears notched where the sulcus passes into the orbital portion of the infraorbital canal.

Optic Canals

The optic foramen is not normally seen on standard frontal projections because of its medial position and oblique orientation. The canal is bounded above by the lesser wing of the sphenoid, inferiorly and laterally by the optic strut, and medially by the body of the sphenoid (Fig. 3.5). It is best evaluated on oblique apical projections which place the foramen into the inferolateral orbit. In this view the orbital end of the canal predominates the image since the bone here is thicker than on the cranial end. This orbital end of the canal is vertically oval in shape. In 4% of the normal population the ophthalmic artery will notch the canal floor forming a "keyhole" deformity (Kier, 1966).

Pneumatization of the anterior clinoids which project just above the optic foramina, or within the lesser wing, may be mistaken for the canals. The true optic canals can be located by following the planum sphenoidale and lesser wing laterally where they continue into the roof of the optic canal (Fig. 3.5).

For best evaluation of the optic canal in its mid or cranial portions oblique apical tomograms at 1 mm intervals are required. These are taken at 38° to the midsagittal plane and at −35° to Reid's orbitomeatal line, which will give cross-sections of the canal. In this projection the vertically oval outline of the orbital end of the canal passes into the circular midportion, and then back to the horizontally oval cranial end (Goalwin, 1922). A significant portion of the cranial end may be eroded or fractured and not show on routine foramen views, but will be evident on tomograms.

Simultaneous visualization of both canals may be achieved with axial tomograms taken in an hyperextended submentovertex projection.

Paranasal Sinuses

The frontal sinuses are best seen in the Caldwell and lateral projections (Figs. 3.1 and 3.3). They are typically asymmetric in size and shape, and may extend back into the orbital plate of the frontal bone producing a pathologic-appearing lucency above the orbital rim. The margins are well corticated and the sinuses are normally clear.

The ethmoid air cells lie between the orbits from which they are separated by the thin lamina papyracea (Figs. 3.2 and 3.4). The latter may be absent in some normal skulls (Yanagisawa, et al, 1968). Thin bony septa are often visible within the sinuses. Just above the ethmoid sinus and forming part of their roofs is a horizontal density seen on frontal projections joining the two sphenoid ridges, the planum sphenoidale (Fig. 3.1). Just below this, and difficult to see on routine frontal and lateral projections, is the thin cribriform plate forming the roof of the nasal cavity.

The sphenoid sinuses are largely obscured by the nasal cavity and ethmoid cells on frontal projections, but parts of them may be seen in the medial orbit where the sinus expands laterally into the greater wing. The lesser wing ridge traced medially leads to the roof of the sphenoid sinus. On the lateral view the sinus is clearly seen just anterior to the sella turcica (Fig. 3.3).

The maxillary sinus is best seen on the Waters' projection, and is well visualized also on the lateral and submentovertex views (Figs. 3.2, 3.3, and 3.4). On frontal films it appears as a lucent area just below the orbital floor. The oblique orbital line may project into it superiorly. Also extending into the sinus medially are the infraorbital foramen, and foramina rotundum and ovale. On base view the posterior sinus wall is clearly seen for its entire length (Fig. 3.4).

RADIOGRAPHIC EVALUATION OF ORBITAL PATHOLOGY

Plain x-rays of the orbits will reveal abnormalities in up to 70% of patients with orbital disease (Pfeiffer, 1943). A number of radiographic signs indicative of pathology have been discussed by Zizmor and elaborated upon by others (Lloyd, 1966; Newton and Potts, 1971; Potter, 1971; Zizmor and Raskind, 1971). The following signs should be evaluated for any patient suspected of having orbital disease.

1. Increased soft-tissue density. Abnormal soft-tissue densities may be observed within the orbital or sinus shadows as a result of hemorrhage or cellular infiltrates associated with inflammations or infections. Mucopolysaccharide infiltration in Graves' disease will increase soft-tissue density of extraocular muscles. Tumors within the orbit or sinuses may produce shadows, as will vascular anomalies such as varices, and cysts such as mucoceles and encephaloceles. Swelling of the eyelids and periorbital tissues, and thickening of sinus mucosa, also may be seen on routine radiographs.

2. Variations in size and shape. The two orbital outlines are normally symmetrical and do not differ by more than 2 mm in vertical height as measured from the infraorbital foramen (Potter, 1982). Increased size of the orbit with retention of cortication is indicative of long-standing increased intraorbital pressure (Fig. 3.6). Such enlargement may occur in children within months, but in adults may take several years. Masses within the muscle cone usually produce diffuse, symmetrical enlargement, while those outside the cone usually cause more asymmetric changes. Nondestructive enlargement may be seen with congenital disorders such as congenital glaucoma, megalophthalmos, serous cyst associated with microphthalmos, dermoid cysts, or with orbital dysplasia seen in neurofibromatosis. Benign mass lesions such as hemangiomas, mixed cell tumors of the lacrimal gland, lymphangiomas, neurofibromas, optic nerve gliomas, and meningiomas may also enlarge the orbit, as may varices or orbital inflammatory pseudotumor (Lombardi, 1967).

Small orbits are associated with anophthalmos and microphthalmos, enucleation during childhood with arrested growth and contraction (Kennedy, 1965), and with slowly enlarging adjacent sinuses with remolding of orbital walls as seen with some mucoceles.

The optic canals are normally symmetrical and a difference between the two sides of more than 1 mm, or a vertical height of greater than 6.5 mm is suggestive of pathology (Goalwin, 1922; Evans, et al, 1963; Potter, 1971). The canal attains adult size by the age of 3 years (Goalwin, 1922). Concentric enlargement without bony erosion, and a change to a circular outline may be seen with benign tumors of the optic nerve such as optic gliomas and meningiomas, but can also be seen with neurofibromas and optic nerve extension of retinoblastomas. Enlargement is also seen with aneurysms of the ophthalmic artery, arteriovenous malformations, or with chronic increased intracranial pressure, as well as with inflammatory granulomas of the optic nerve and with chiasmatic arachnoiditis.

Small optic canals are seen with developmental anomalies such as microphthalmos, the craniostoses, following enucleation in childhood, secondary to hyperostosis associated with meningioma or dysostosis seen in fibrous dysplasia or Paget's disease. It may also result from reactive osteoscle-

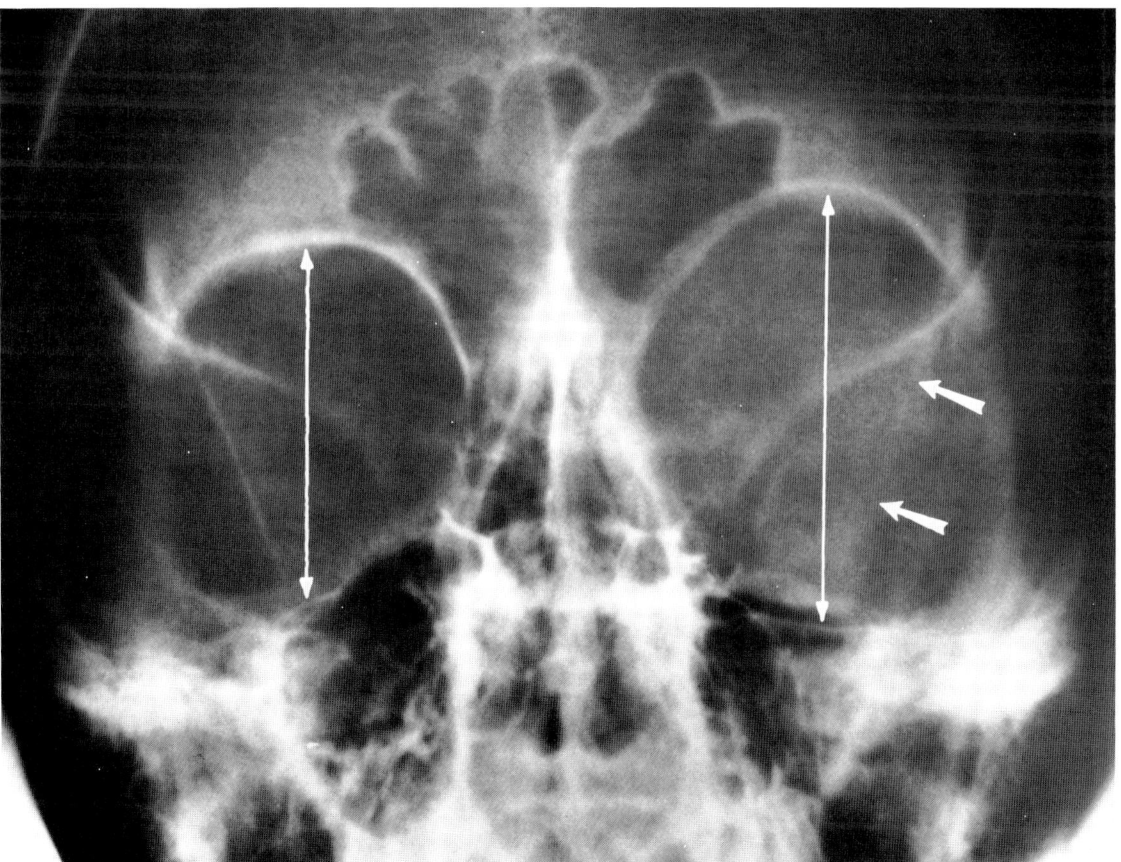

Figure 3.6. Caldwell's projection of a patient with a large optic nerve sheath meningioma of long duration. The left orbit is greatly enlarged with good preservation of the orbital rims. Note the erosion of the lateral orbital wall as indicated by absence of the left oblique orbital line (*heavy arrows*).

rosis following osteitis or inflammatory pseudotumor (Lloyd, 1976).

The superior orbital fissures are normally symmetrical. Smooth enlargement without bone destruction in the superior or inferomedial aspect of the fissure most commonly results from an intracavernous or internal carotid artery aneurysm (Rischbieth, 1958), and enlargement of the medial margin suggests a carotid-cavernous fistula (Shapiro and Robinson, 1967). Extension of a pituitary adenoma into the cavernous sinus with forward displacement of the internal carotid artery may also enlarge the fissure. Benign tumors such as hemangiomas, dermoids, and rarely meningiomas, as well as the histiocytosis-X syndromes may also enlarge this opening. Other causes include orbital varices, sphenoid sinus mucoceles, posterior encephalocele, neurofibromatosis, and chronic increased intracranial pressure (Kieffer, 1971).

Narrowing of the superior orbital fissure is associated with hyperostosis from meningiomas, with fibrous dysplasia and Paget's disease, and may be seen with osteoblastic metastatic tumors such as breast and prostate carcinomas. It can also occur with chronic hemolytic anemias of childhood.

3. *Tissue emphysema.* Air within the eyelids or orbit appears as a relative lucency on plain radiographs. Lid emphysema may result from dissection of air forward from the orbit, or from lacrimal bone fractures anterior to the orbital septum associated with lacerations of the lacrimal sac (Lloyd, 1966).

Orbital emphysema is seen in up to 50% of blow-out fractures, most commonly associated with medial wall disruption, but almost as common with floor fractures (Lloyd, 1966). It results from communication with paranasal sinuses and following increased pressure within the respiratory passages. Air in the orbit may be the only objective sign of orbital wall fracture. Emphysema may also be seen with fistulous tracts between the orbit and sinuses resulting from osteomyelitis or from erosion by tumors. Rarely, gas-forming bacteria in anaerobic orbital cellulitis will produce emphysema (Zizmor and Lombardi, 1973).

4. *Orbital calcification.* Abnormal calcification within the orbit is always of pathologic significance. It is associated with disease of vascular tissue, nerves, sclera and choroid, or is secondary to trauma, prolonged inflammation, or infection. Within the globe calcium deposition may be seen in pthisis bulbi as a result of metabolic changes, in 75% of retinoblastomas (Reese, 1963), in advanced retrolental fibroplasia, following severe endophthalmitis or long-standing retinal detachment, or in degenerated morgagnian cataracts. Orbital calcification is associated with extrascleral retinoblastomas, organized hematoma, 5–10% of hemangiomas (Lombardi, 1967), in some optic nerve gliomas, meningiomas, and dermoid cysts, and may be seen within the walls of some mucoceles, and in degenerated aneurysm walls (Alfano and White, 1955).

5. *Bony dehiscences.* Abnormal dehiscences in orbital bones are associated with benign mass lesions entering from adjacent intracranial cavities or paranasal sinuses. These include mucoceles of the frontal and ethmoid sinuses. They usually produce smooth oval to round areas of bone dehiscence along the medial and supermedial orbital walls, sometimes with areas of irritative osteoblastosis and increased density. Encephaloceles produce bony defects near the midline, or at the root of the nose. They may protrude through defects between the frontal, ethmoid, lacrimal, or maxillary bones. The cribiform plate may appear incomplete in lateral films.

Orbital dysplasia associated with neurofibromatosis results in absence of large portions of the greater wing of the sphenoid with elevation of the lesser wing and a lateral displacement of the oblique orbital line in frontal projection.

6. *Destruction of bone.* Irregular erosion of bone with loss of margins and cortication results from rapid destruction by expanding lesions or inflammatory processes (Fig. 3.6). Lacrimal gland tumors typically erode the superolateral orbital walls. Nasopharyngeal tumors invade the orbit by destruction of adjacent walls or inferior orbital fissure. Chordomas of the clivus may gain access through the superior orbital fissure with destruction of its margins.

Carcinomas of the eyelids can invade back into the orbits with erosion of the orbital rim or walls. Metastatic carcinomas rarely go to the orbit, but when present frequently involve the posterior orbital wall and oblique orbital line (Newton and Potts, 1971). Metastatic neuroblastoma in children produces a typical moth-eaten erosion of orbital and cranial bones. Rhabdomyosarcoma and lymphoma rarely produce bony destruction.

Inflammatory processes such as sinusitis commonly destroy bone and may simulate malignant disease. Adjacent sinuses are opacified. Orbital inflammation with osteomyelitis can also cause lytic lesions. The histiocytosis-X diseases produce irregular osteolysis more commonly in the lateral wall and orbital roof.

7. *Hyperostosis and osteoblastosis.* Increased density of bone may be seen in areas of healed fracture. Fibrous dysplasia, Paget's disease of bone, osteopetrosis, osteoblastic metastases, and hyperostosis associated with meningiomas, all produce thickened and hyperdense bony shadows on plain radiographs.

ORBITAL FRACTURES

Among 1,042 patients with facial injury, 60% were found to have facial/orbital fractures (Shultz, 1970), and of 610 patients with skull fractures 78% presented with ocular signs or symptoms (Blakeslee, 1929). In a series of 178 orbital fractures, 72% were multiple and 20% were bilateral (Pettinoti, 1955). Serious ocular injury is a common association with orbital fracture, seen in 15–40% of such patients (Milauskas and Fueger, 1966; Whyte, 1968; Emery, et al, 1971; Fradkin, 1971). Therefore, every patient suspected of having orbital fractures must have a complete ocular examination prior to radiographic or surgical manipulation.

Orbital fractures can be demonstrated on plain frontal projection radiographs in 55–75% of cases (Gould and Titus, 1966; ViGario, 1966; Emery, et al, 1971; Fradkin, 1971), and on tomography in 93–100% (Gould and Titus, 1966; Fradkin, 1971). The Waters' projection gives good visualization of the floor, together with the maxillary sinus, and is the standard initial view for evaluation of floor fractures. Caldwell's projection may give a better, more perpendicular view of the floor itself, and a slightly exaggerated Caldwell, taken at −25 to 27°, will place the petrous pyramids well below this level. Hypocycloidal tomography at 1–5 mm intervals in both the coronal and lateral projections should be obtained in any equivocal case, and for the definition of the size and extent of the fracture. Submentovertex tomograms may be required for medial wall fractures.

Rim Fractures

Fractures involving the orbital rims may occur alone or in association with wall breaks. They are best evaluated on frontal projections. Those of the superior rim are uncommon, but when present may involve damage to the lacrimal gland, levator aponeurosis, or trochlea. They may extend upward to involve the orbital roof and anterior cranial fossa (Zizmor and Raskind, 1971), and are frequently associated with optic canal fractures.

Disruption of the lateral rim is uncommon without associated floor fracture, and more usually involves complex tripod or LeFort type III fractures (Jurkiewicz and Nickell, 1971). Most commonly, such fractures involve separation of the zygomaticofrontal and zygomaticomaxillary sutures, with or without displacement of the zygomatic body.

Medial rim breaks are frequently found with nasofrontal, nasoethmoid, and nasomaxillary fractures, and tend to be of a more complex type. Fractures of the inferior rim are usually palpable as a distinct step-off, and frequently occur at the zygomaticomaxillary suture. They are commonly associated with more complex LeFort type II fractures.

Orbital Floor Fractures

Pure blow-out of the floor are the most common type of orbital fracture, representing 70–75% of all cases (Whyte, 1968; Zizmor and Raskind, 1971; Arger, 1977). This type was first described by MacKenzie in 1844, and a mechanism for its occurrence was proposed by Pfeiffer in 1943. This was later confirmed experimentally and clinically (Converse and Smith, 1957; Smith and Reagen, 1957; Converse and Smith, 1960).

The major radiographic signs of blow-out fracture include fragmentation and disruption of the orbital floor, depression of bony fragments into the maxillary sinus, and soft-tissue prolapse into the sinus (Figs. 3.7 and 3.8). Other corroborative signs are opacification of the sinus and orbital emphysema.

A blow-out fracture usually involves the posterior-medial portion of the orbital floor where the bone is thinnest. It appears on frontal projections as a depressed, regular, or irregular bone density in the upper portion of the maxillary sinus, usually associated with clouding of the sinus and with a globular herniation of soft orbital tissues. Enophthalmos, either early or late, is seen in up to 44% of such injuries (Pfeiffer, 1943). Stereoscopic Waters' views as well as frontal and lateral tomograms provide excellent visualization of such fractures.

Medial Wall Fractures

Although pure medial wall fractures can occur, they are most commonly associated with floor fractures and may be seen in 10–60% of such cases (Lombardi, 1967; Dodick, et al, 1969; Lerman, 1967). The ethmoid sinuses are frequently opacified and orbital emphysema is seen in about 20% of patients (Fueger and Milauskas, 1970). Medial wall fractures may be difficult to demonstrate radiographically without tomograms.

Optic Canal Fractures

Closed head trauma in the frontal region is a well-recognized cause of visual loss, the latter seen in 0.5–1.5% of such injuries (Pringle, 1922; Russell, 1940). Up to 5% of head injuries may be associated with fractures through the optic canals (Sugita, et al, 1970), and in cases of head trauma with visual loss canal fractures have been demonstrated in 6–92% of patients (Hughes, 1962; Fukado, 1975; Anderson, et al, 1982) (Fig. 3.8).

Radiographic demonstration of canal fractures may be difficult. In four patients studied by Potter and Trokel (1971), all had normal standard canal views, but fractures were evident in all on tomograms. Thus, any patient suspected of having a canal fracture but with normal plain x-rays of the optic canal, must have hypocycloidal tomograms in the oblique apical projection at 1 mm intervals. Both canals may be demonstrated in plain view in the submentovertex projection.

COMPUTERIZED TOMOGRAPHIC SCANNING OF THE ORBITS

Computerized tomography is a noninvasive technique that allows the simultaneous examination of bony structures and associated soft tissues. It has proven to be equal or superior to other methods of examination with a high diagnostic accuracy for orbital lesions (Gyldensted, et al, 1977; Niloskelainen, et al, 1977; Wende, et al, 1977). CT provides computer reconstructed visual images based on density averaging of tissues included within a certain volume of tissue space (the voxel), and assignment of density values proportional to their coefficient of absorption of x-rays. This tissue space is determined by the two-dimensional scanning array, which in the newer machines consists of a 500 by 700 matrix dividing the image into up to 350,000 units (the pixel). The voxel is determined by the pixel and the thickness of the scanned section, which may be as little as 1 mm. With fourth-generation CT units the volume of the voxel may be as little as 0.375 mm^3. Such high degrees of spatial resolution, plus changes in software that now allow multiplanar transformations of axial scans to coronal, frontal, and oblique axes, has dramatically improved the quality and diagnostic capabilities of this radiographic modality.

Orbital CT scanning should include both axial and coronal cuts. Special transformations in other planes can be reconstructed as indicated for more complete evaluation. Contrast enhancement is usually not needed for routine orbital scans since the lack of a blood-orbital barrier results in uniform

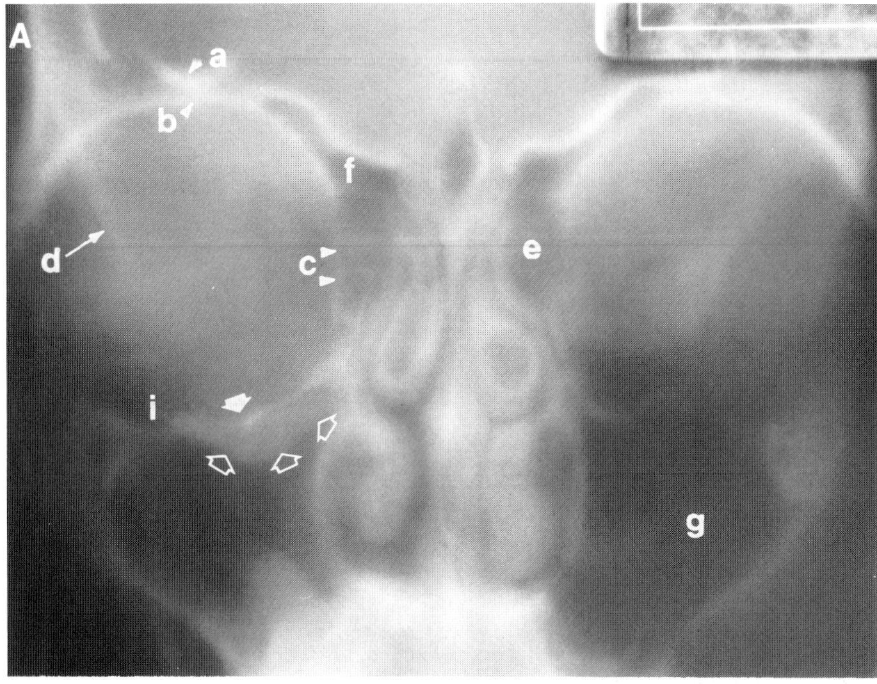

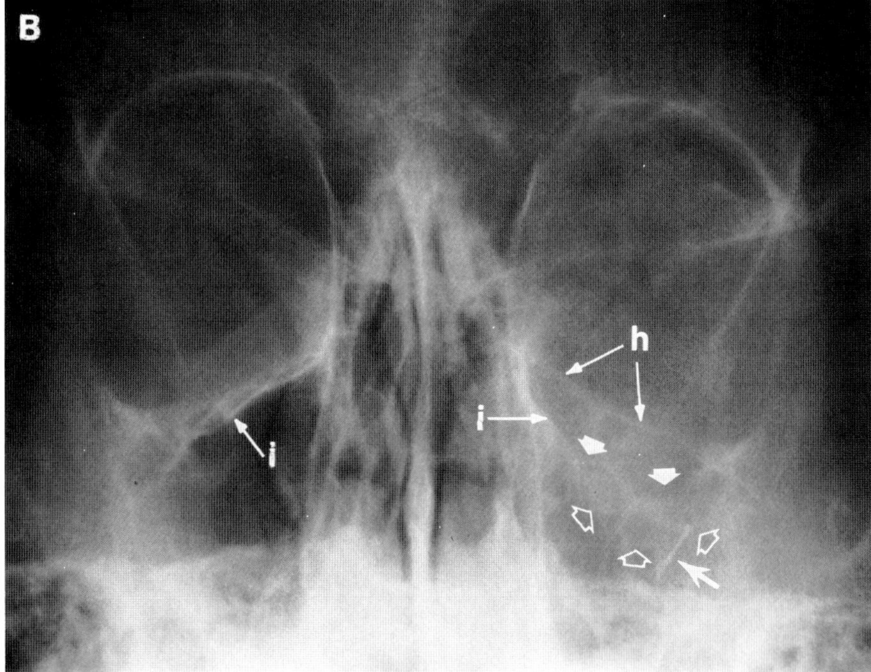

Figure 3.7. Blow-out fracture of the orbital floor. (*A*) is a frontal tomogram with the plane of section in the midorbit. (*B*) is Caldwell's projection. In (*C*) a fracture extends from the medial wall onto the orbital floor (*solid arrow heads*). (*D*) is a sagittal tomogram taken in the lateral projection demonstrating a fracture of the orbital floor with downwardly displaced bone (*solid arrow*). (Abbreviations: *a*, cerebral surface of the orbital roof; *b*, orbital surface of the orbital roof; *c*, lamina papyracea; *d*, oblique orbital line; *e*, ethmoid sinus air cells; *f*, frontal sinus; *g*, maxillary sinus; *h*, anterior orbital floor, inferior rim; *i*, posterior orbital floor; *open arrow heads*, orbital soft-tissues herniating into maxillary sinus; *solid arrow heads*, fractured and displaced orbital floor; *arrow*, spicule of bone in maxillary sinus.)

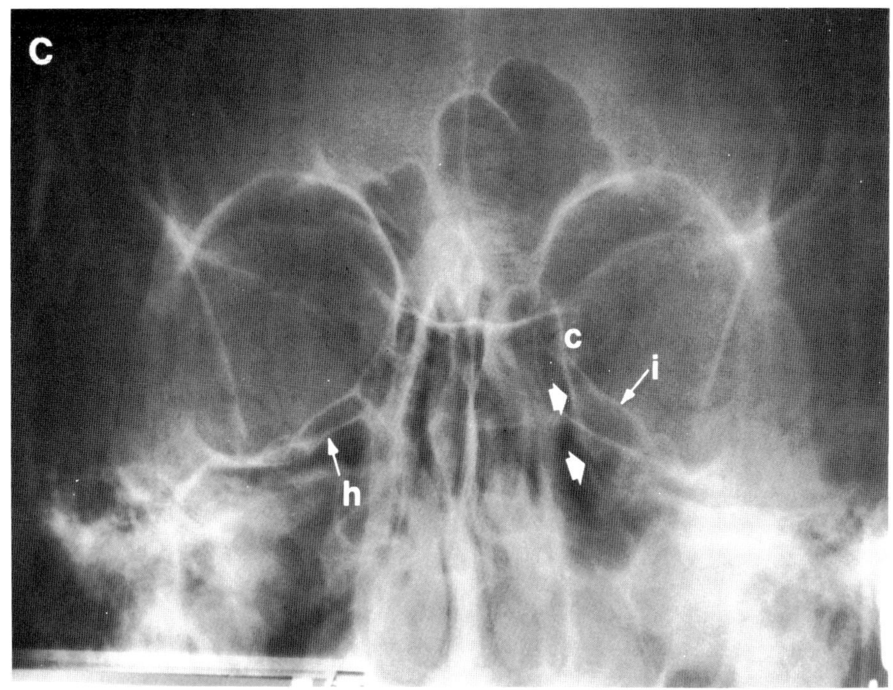

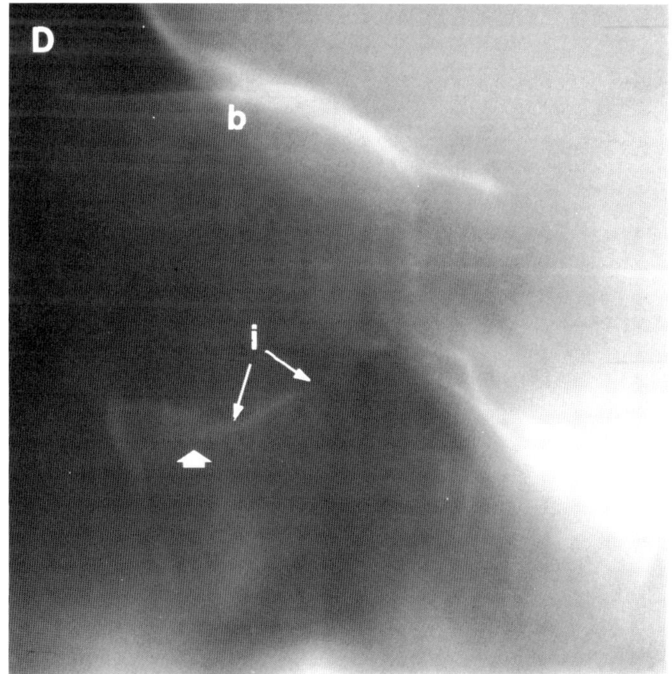

Figure 3.7 *C & D.*

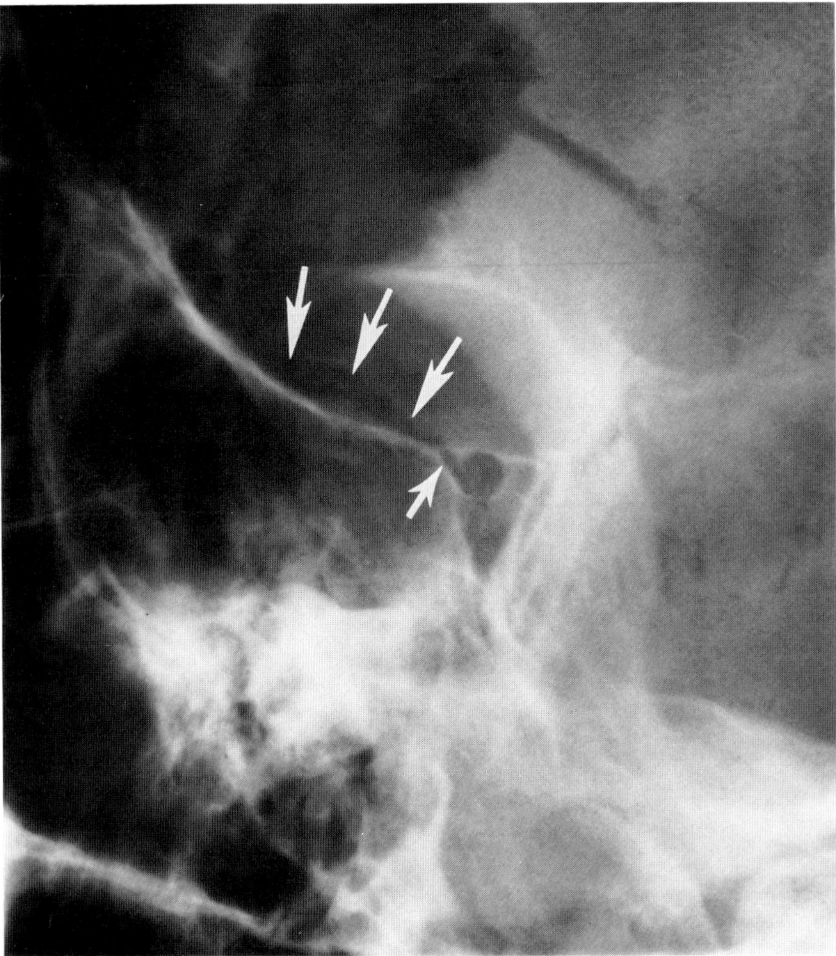

Figure 3.8.   Oblique apical projection of the right optic canal. Note the large fracture extending through the orbital roof (*large arrows*) and across the root of the lesser wing of the sphenoid into the superomedial portion of the optic canal (*small arrow*).

enhancement of orbital tissues. The exception to this is in certain cases of orbital tumors, such as meningiomas, which enhance markedly with contrast medium. In examining CT scans, by convention, axial cuts are viewed as if seen from below, and coronal cuts as if seen from in front.

In evaluating the orbit in the axial plane, the collimated x-ray beam is angled at 0 to −15° to Reid's baseline. Low axial cuts just below the orbit give excellent visualization of the maxillary sinuses, nasopharynx, zygomatic arches, and periorbital soft tissues (Fig. 3.9). The nasolacrimal canal shows as a lucency within the frontal process of the maxillary bone. Sections through the inferior orbit show the globe, inferior orbital fissure, temporalis fossa, nasal cavity, the lateral wall of the orbit, and sphenoid sinus. The orbital floor is not seen well because it lies parallel to the scanning plane, and because of partial volume averaging of the thin floor with adjacent sinus and orbital structures.

The inferior rectus muscle usually cannot be seen along its entire length because of its oblique orientation to the scanning plane. Rather it is seen in oblique cross-section on several adjacent cuts through the inferior orbit (Fig. 3.9).

Optimal midlevel scans through the orbit should include the globes and crystalline lens, medial and lateral orbital walls and rectus muscles, superior orbital fissures, and portions of the optic nerves, all in a single cut (Fig. 3.10). Fine bony septae are often seen within the ethmoid sinuses. The optic nerve is usually not seen in its entirety because of its sinuous course. The best CT projection for its demonstration is at an axial plane of −20° with the patient looking upward at 40° to the horizontal. In this scan the optic nerves are most stretched and nearly straight, and lie parallel to the CT plane (Unsold, et al, 1980). Just above the optic nerve, and coursing medially across it, is the superior ophthalmic vein (Fig. 3.11).

The optic canal is not usually seen completely because of its upward inclination. On midlevel scans the orbital portion is seen just medial to the superior orbital fissure and adjacent to the body of the sphenoid. On higher sections, the cranial portion becomes visible just medial to the anterior clinoid processes (Fig. 3.10).

Behind the clinoids the cavernous sinus and suprasellar cistern are visible, and the optic chiasm can be seen as a U-

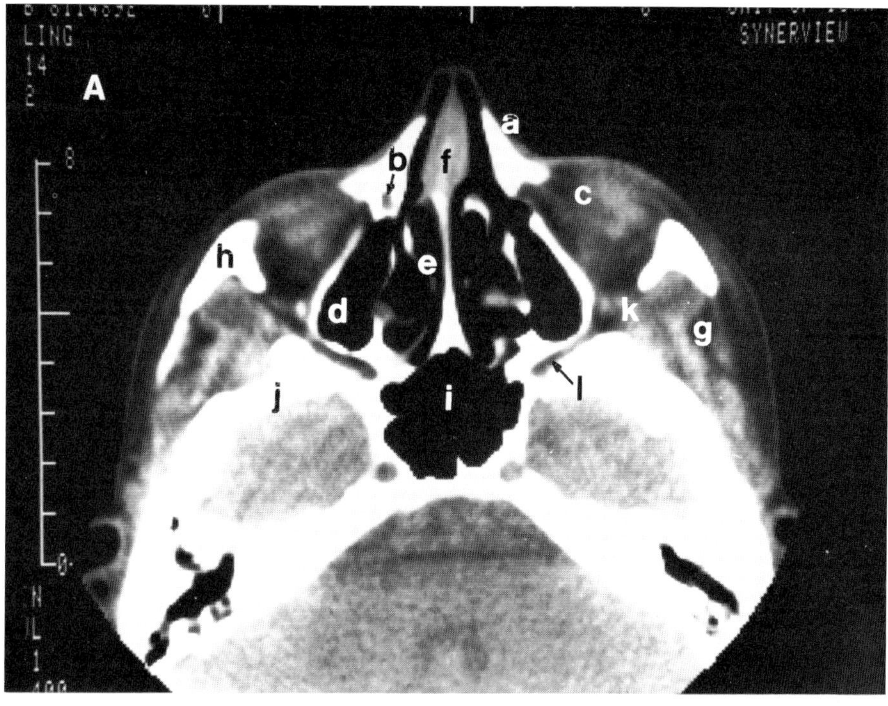

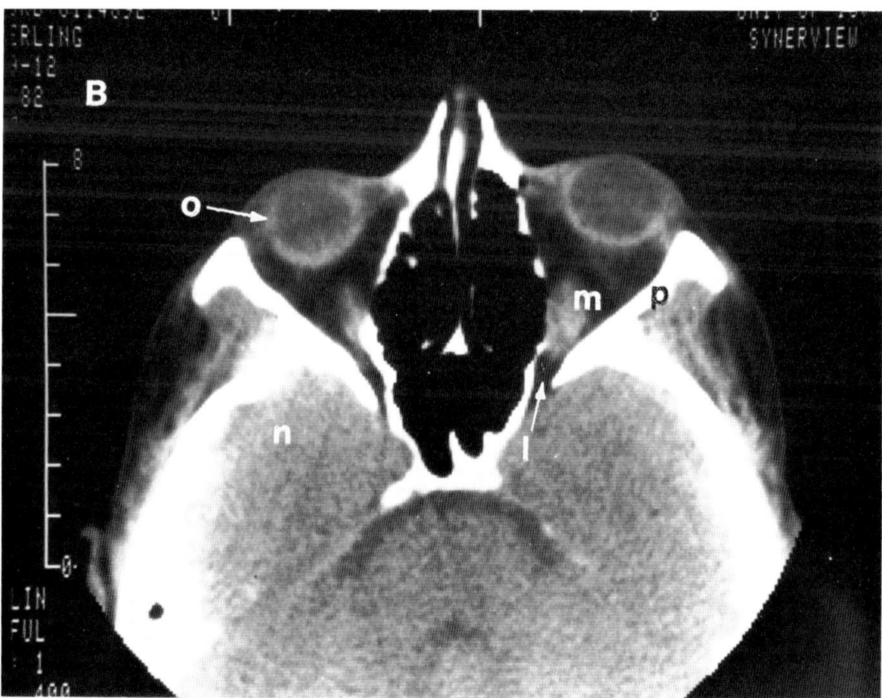

Figure 3.9. Examination of the orbit by CT scanning in the axial projection. Each section represents a high-resolution scan at 2 mm thickness. (*A*) is a low section through the inferior portion of the orbit tangential to the lowermost coats of the globe. (*B*) is a slightly higher section through the inferior rectus muscles. (Abbreviations: *a*, orbital process of the maxillary bone; *b*, nasolacrimal canal; *c*, orbit; *d*, maxillary sinus; *e*, nasal cavity; *f*, nasal septum; *g*, temporalis muscle; *h*, zygomatic bone; *i*, sphenoid sinus; *j*, greater wing of the sphenoid bone forming the anterior wall of the middle cranial fossa; *k*, inferior orbital fissure; *l*, superior orbital fissure; *m*, inferior rectus muscle; *n*, middle cranial fossa and temporal lobe; *o*, sclera and choroid of globe; *p*, lateral orbital wall.)

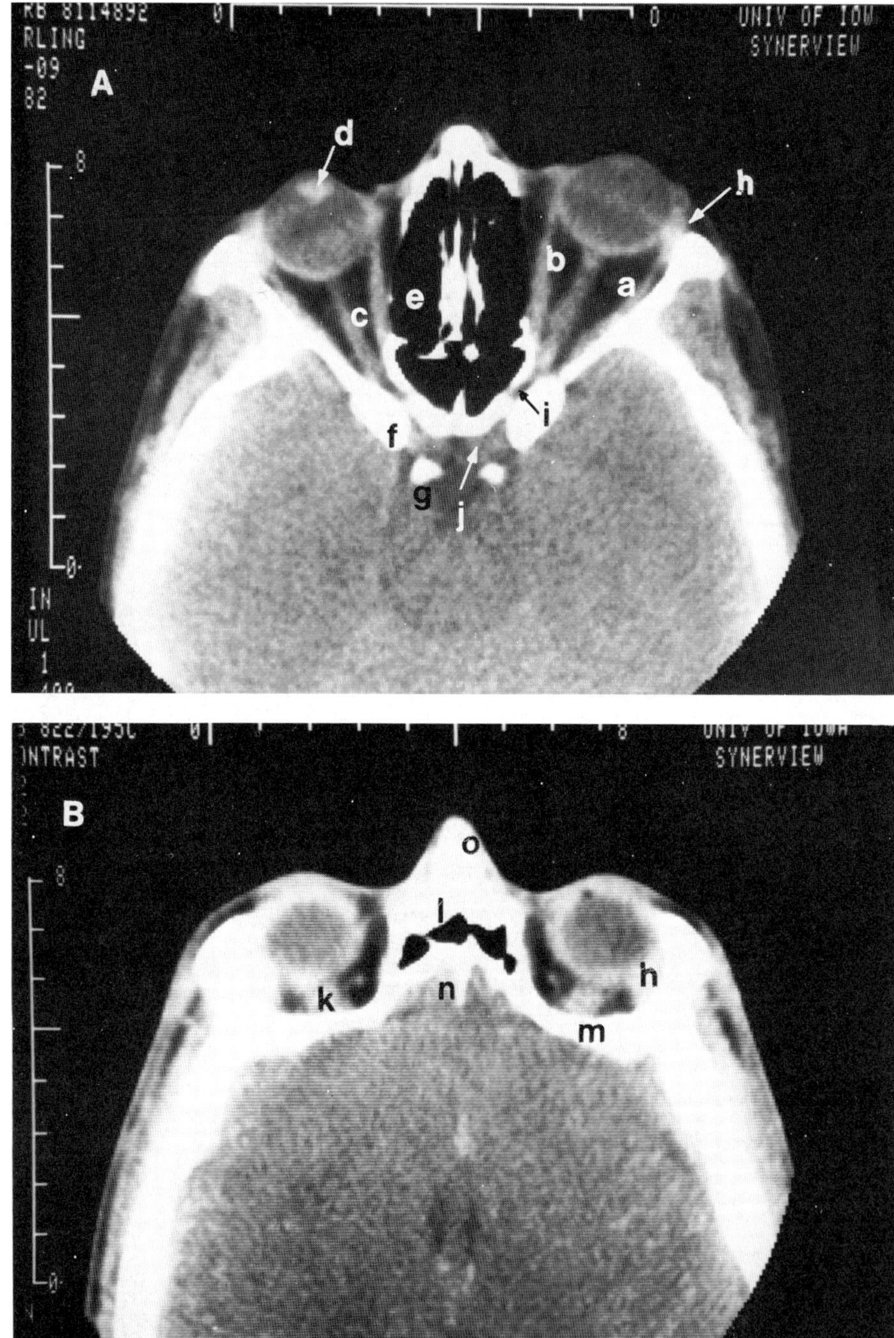

Figure 3.10. Examination of the orbit by CT scanning in the axial projection. (*A*) is a section through the midlevel of the orbit which includes the lens, rectus muscles, and optic nerve. (*B*) is a section through the superior portion of the orbit. (Abbreviations: *a*, lateral rectus muscle; *b*, medial rectus muscle; *c*, optic nerve; *d*, crystalline lens; *e*, ethmoid sinus air cells; *f*, anterior clinoid process; *g*, dorsum sellae; *h*, lacrimal gland; *i*, optic canal; *j*, optic chiasm; *k*, superior rectus-levator palpebrae superioris muscle complex; *l*, frontal sinus; *m*, orbital roof; *n*, falx cerebri; *o*, nasal bone.)

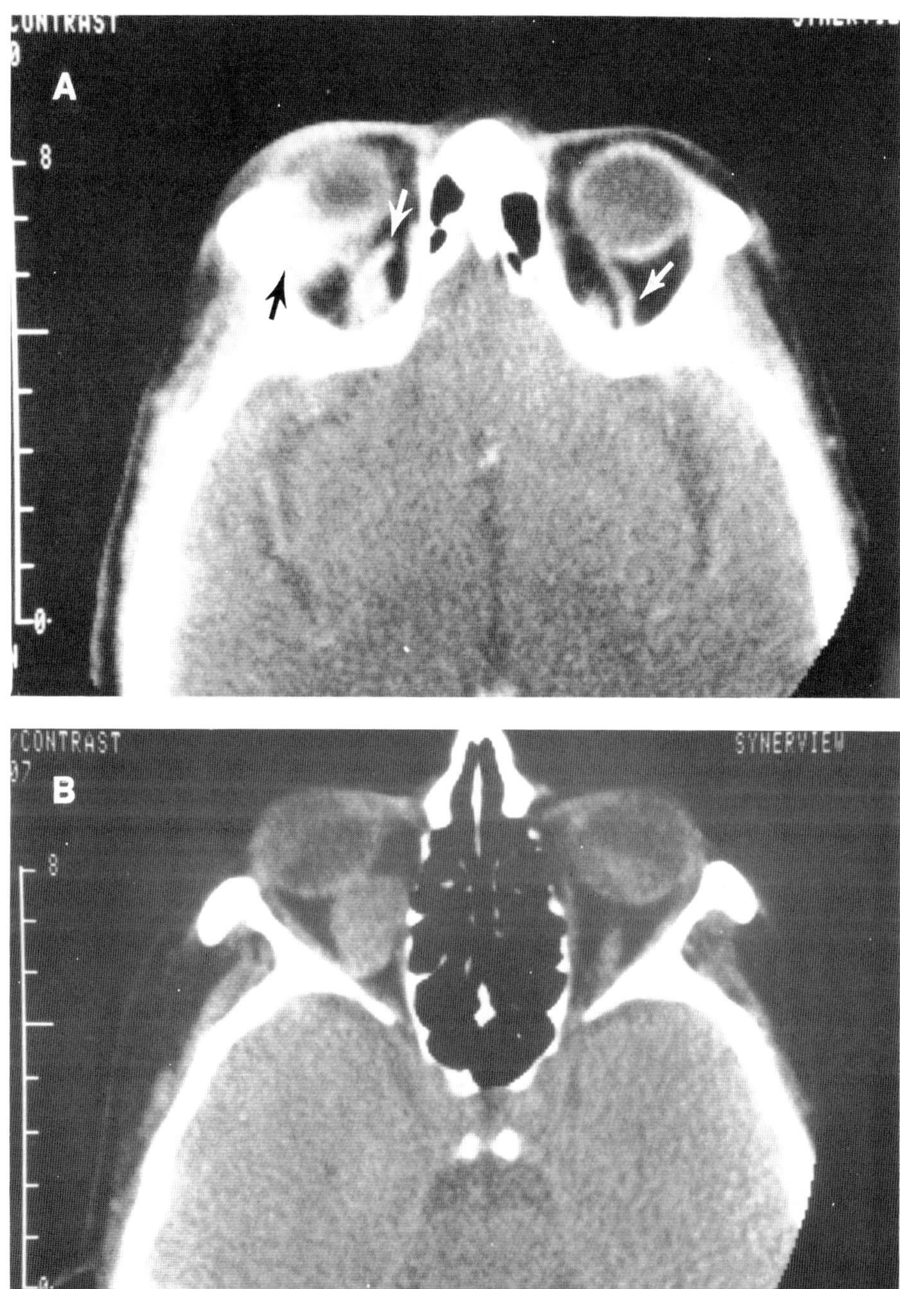

Figure 3.11. (*A*), Axial CT scan of a patient with lymphoma in the superolateral portion of the right orbit (*small black arrow*). Note the superior ophthalmic veins crossing the orbit from lateral to medial on both sides (*white arrows*); (*B*), CT scan of a patient with a well-circumscribed cavernous hemangioma within the muscle cone of the right orbit.

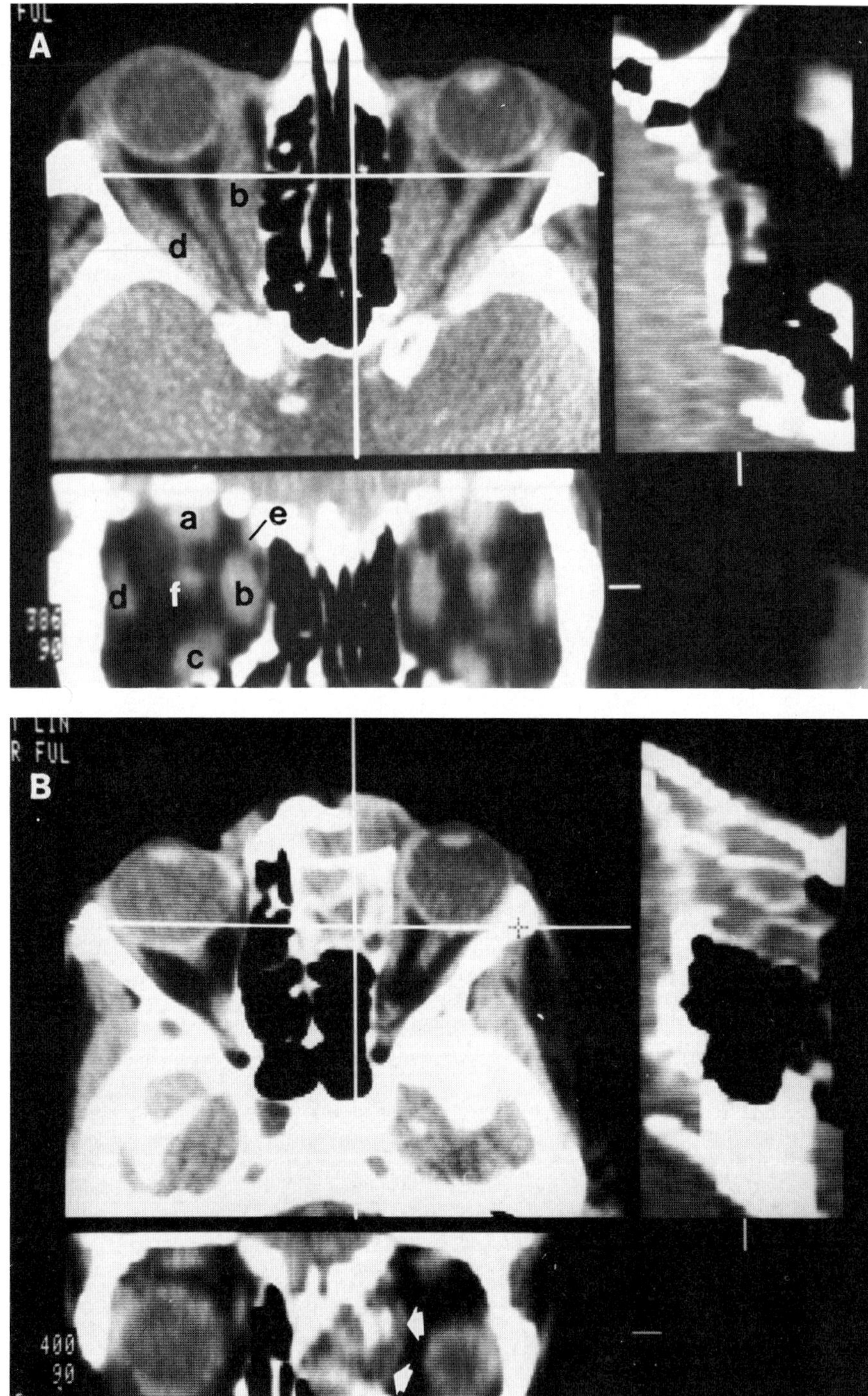

Figure 3.12. (*A*), Orbital examination by computerized tomography of a patient with thyroid ophthalmopathy showing greatly enlarged extraocular muscles and posterior compression of the optic nerve. The direct scan was made in the axial plane at a midorbital level. Computer reconstructed coronal (*below*) and sagittal (*right*) sections provide multiplanar views of orbital structures. (*B*), Patient with a left ethmoid sinus mucocele demonstrated on axial CT with coronal and sagittal reconstructions. The white cross-lines indicate the planes of reconstructed sections. Note the opacified ethmoid sinus and erosion of the lamina papyracea with extension of the mucocele into the orbit (*solid arrow heads*). (Abbreviations: *a*, superior rectus muscle; *b*, medial rectus muscle; *c*, inferior rectus muscle; *d*, lateral rectus muscle; *e*, superior oblique muscle; *f*, optic nerve.)

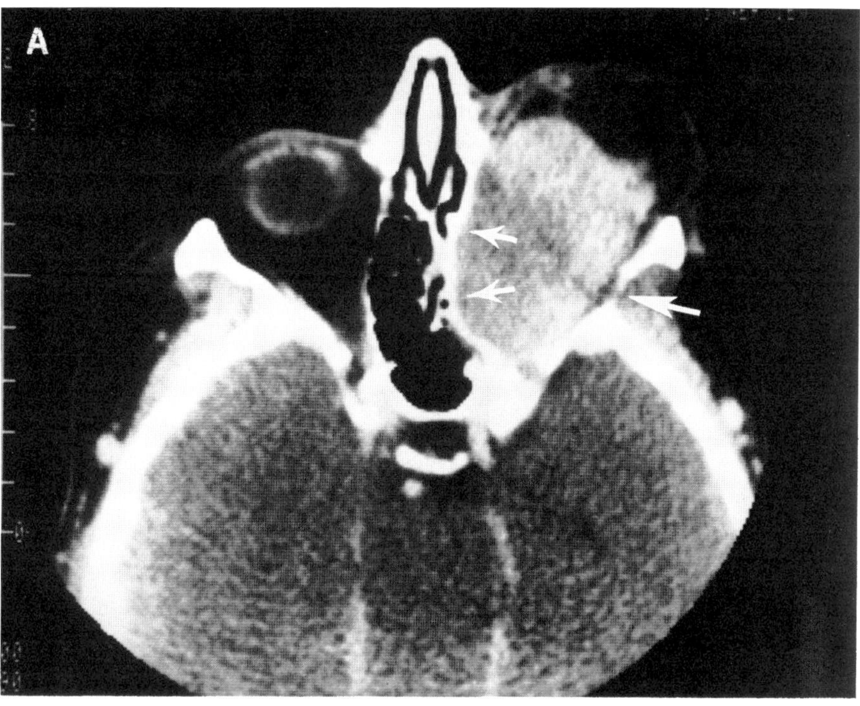

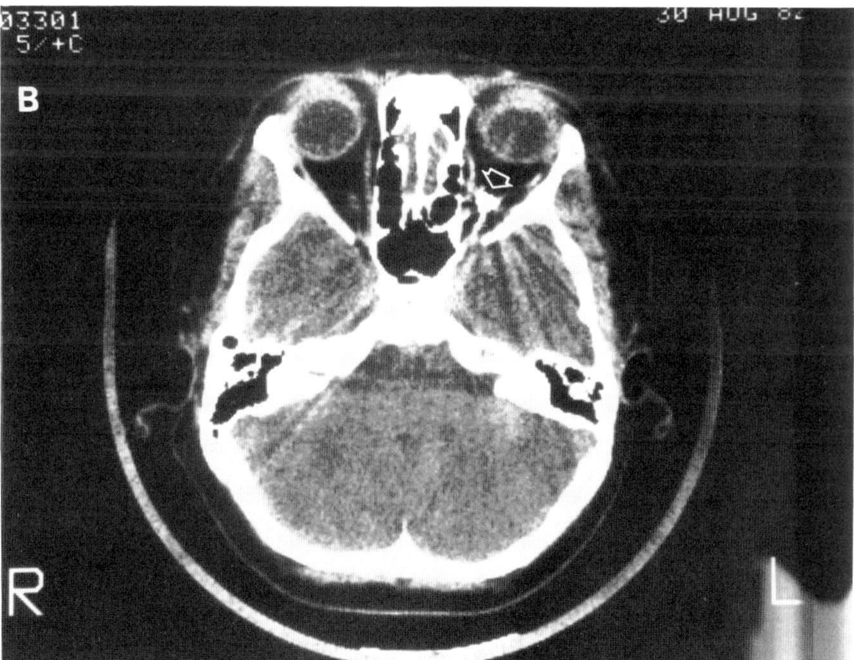

Figure 3.13. (*A*), CT scan in the axial plane of a patient with a large optic nerve sheath meningioma filling the orbit and causing massive proptosis of the left globe. Note erosion of the lateral orbital wall (*large arrow*), and inward displacement of the medial orbital wall (*small arrows*); (*B*), Axial CT scan of a patient with a metallic foreign body near the apex of the left orbit (*open arrow head*). Radiating lines of varying densities extending from the foreign body represent artifacts caused by shadowing of the x-ray beam.

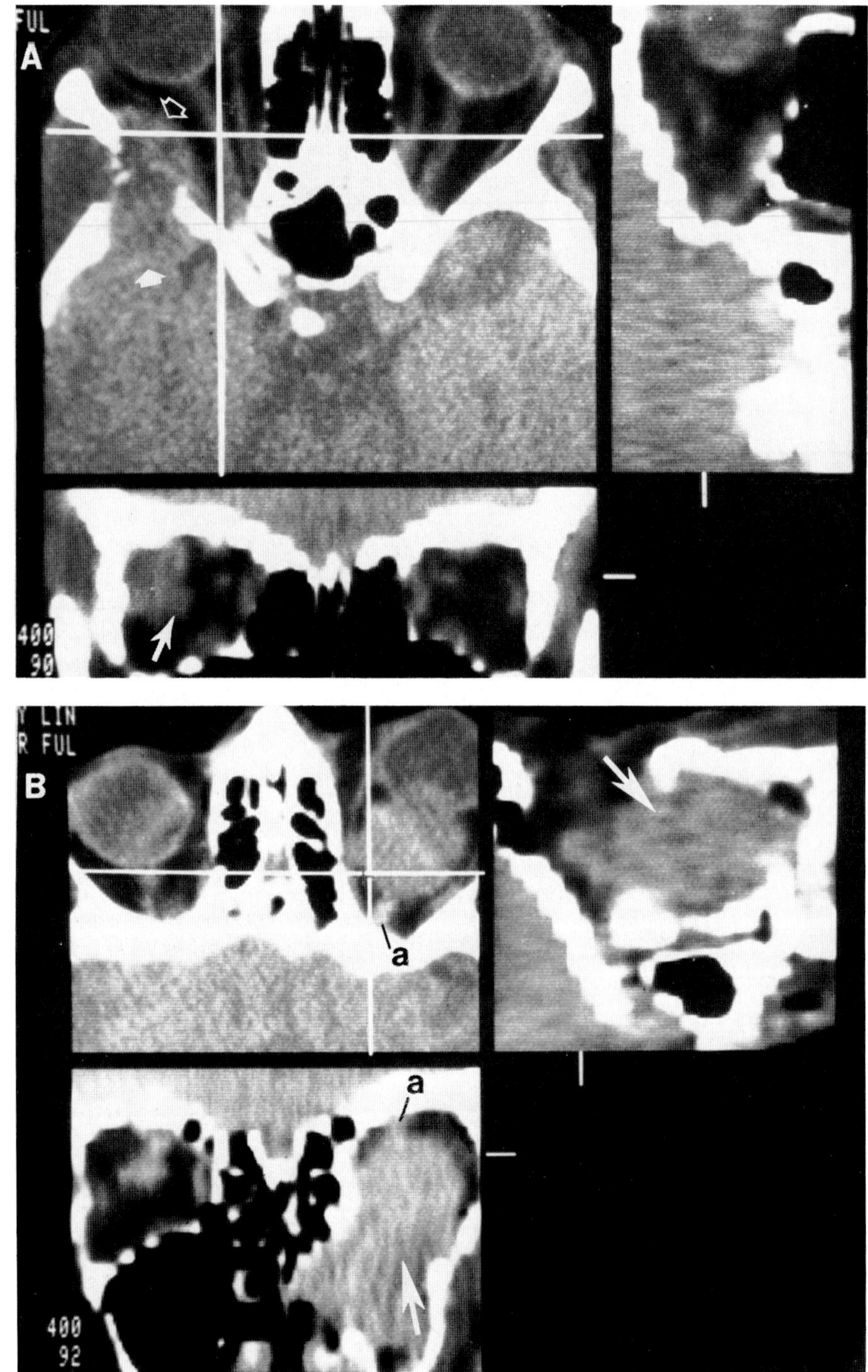

Figure 3.14. (A), A patient with metastatic thyroid carcinoma to the greater wing of the sphenoid bone seen on axial CT with coronal and sagittal reconstructions. The tumor is seen to be eroding into the orbit (*open arrow head*) and into the middle cranial fossa (*solid arrow head*). On the coronal section (*below*) the lateral rectus muscle is seen displaced medially by the expanding lesion (*arrow*); (B), CT examination of a patient with a maxillary sinus tumor that has eroded through the orbital floor (*arrows*) into the left orbit. Note the displacement of the optic nerve upward against the orbital roof (*a*).

shaped density anterior to the dorsum sellae and infundibulum (Dutton, et al, 1982).

In higher sections the globe is cut tangentially through its superior walls, appearing as a more uniformly dense mass in the central orbit. The superior rectus/levator complex forms a broad ribbon of tissue density extending back to the apex, and the orbital roof now appears in the posterior orbit. The upper eyelid density is seen anteriorly. The lacrimal gland is represented as a well circumscribed density in the anterolateral orbit.

Coronal sections of the orbit offer additional information that is essential to a complete evaluation. They offer more accurate determination of spatial relationships within the orbit. Direct coronal sections give the highest resolution images, but most current machines rely on reconstruction from serial axial scans, a technique that exposes the patient to less total radiation. True coronal cuts are seldom employed, but rather the collimated beam is angled back so as to pass through the orbits and in front of the teeth to avoid artifactual image degradation from metallic dental work.

Coronal sections provide superb views of the extraocular muscles and optic nerves in cross-section, and allow an evaluation of their relative sizes (Fig. 3.12). The superior oblique muscle is seen in the superomedial orbit, and the superior ophthalmic vein may be visualized above and medial to the nerve. The orbital floor, which cannot be evaluated on axial cuts, is seen well on coronal cuts.

Sections through the anterior orbit give excellent views of the globes, the frontal sinuses and nasal cavity, and the anterior orbital roof. The lacrimal gland may be seen in the superolateral quadrant. Insertions of the rectus muscles appear as thin densities against the globes. On midlevel sections, the maxillary and ethmoid sinuses are clearly seen separated from the orbit by a well-defined orbital floor and medial wall. The extraocular muscles are seen in cross-section within the lucent orbital fat. On more posterior sections behind the globes, the muscles can be followed into the orbital apex with the optic nerve centrally placed between them, and the superior ophthalmic vein above it. Further posteriorly, the superior and inferior orbital fissures appear as dehiscences in the orbital bones, and the optic nerve may be followed into the optic canal. Just medial to the canal is the sphenoid sinus, and just above is the anterior clinoid process.

ORBITAL PATHOLOGY ON CT SCANS

Orbital CT scanning has proven to be of exceptional value in the diagnosis and localization of foreign bodies (Grove, 1980) (Fig. 3.13). Both axial and coronal views are necessary for accurate three-dimensional placement of foreign material. In cases of orbital trauma and fractures, CT provides evaluation of soft-tissue entrapment and retrobulbar hematomas (Grove, 1979). Coronal views are needed for floor fractures, but axial scans are better for the medial wall.

Enlargement of extraocular muscles, particularly the inferior and medial recti, along with increased edema of retroorbital fat, is a characteristic finding in Grave's ophthalmopathy. CT evaluation of these and associated structures, such as distension of the superior ophthalmic vein and the

pattern of muscle enlargement, may aid in differentiating thickened muscles due to thyroid disease (Fig. 3.12) from those related to carotid-cavernous fistulae, inflammatory myositis, or arteriovenous malformations (Trokel and Hilal, 1979).

Various inflammatory lesions often have distinguishing features on CT. Thus, an irregular density draping the globe, often with thickening of subtenons space, is seen in lymphoid hyperplasia, enlarged muscles in myositis, enlarged lacrimal gland with a typical "ring sign" of edema around the globe in dacryoadenitis, and a more diffuse orbital infiltration with idiopathic orbital pseudotumor (Trokel and Hilal, 1982).

Evaluation of tumors of the orbit causing unilateral or bilateral proptosis is best achieved with CT which often contributes significantly to the correct diagnosis (Gyldensted, et al, 1977; Wende et al, 1977; Price and Danziger, 1978; Guibert-Trainier, et al, 1979; Lloyd, 1979; Alker, et al, 1981). Scans in both the axial and coronal planes can accurately determine the size and position of such lesions within the orbit. They will show if the mass is continuous with the optic nerve, as with gliomas and meningiomas, (Fig. 3.13) or separate from it. Localization of the tumor within the muscle cone, as with typical hemangiomas (Fig. 3.11) and schwannomas may be distinguished from masses outside the cone, as with lacrimal gland tumors. Relationships with other soft tissues such as the globe, periosteum, and vascular structures, and with the paranasal sinuses and intracranial fossae will also be readily determined (Fig. 3.12). Hyperostosis and lytic lesions of bone (Fig. 3.14), although easily seen on CT, are best evaluated with conventional hypocycloidal tomography. The distinction between cystic masses and solid tumors is easily made from their differential attenuation values. Irregular infiltrative lesions are distinguishable from more circumscribed ones, and the presence of a capsule is often evident.

References

Alfano JE, White H: Ocular significance of intracranial calcium deposits. *Arch Ophthalmol* 54:77, 1955.

Alker GJ, Banna M, Rudin S, Oh YS: Computed tomography of the orbit. *CRC Crit Rev Diagn Immaging* 15:27, 1981.

Alper MG: Computed tomography in planning and evaluating orbital surgery. *Ophthalmol* 87:418, 1980.

Anderson RL, Panje WR, Gross CE: Optic nerve blindness following blunt forehead trauma. *Ophthalmol* 89:445, 1982.

Arger PH: Fractures of the orbit. In Arger PH (ed): *Orbit Roentgenology*. New York, J Wiley and Sons, 1977.

Blakeslee GA: Eye manifestations of fracture of the skull. *Arch Ophthalmol* 2:566, 1929.

Caldwell EW: Skiagraphy of the accessory sinuses of the nose. *Am J Roentgen* 5:569, 1918.

Converse JM, Smith B: Enophthalmos and diplopia in fractures of the orbital floor. *Br J Plast Surg* 9:265, 1957.

Converse JM, Smith B: Blow-out fracture of the floor of orbit. *Trans Am Acad Ophthalmol Otol* 64:676, 1960.

Dodick JM, Galin MA, Kwitko M: Medial wall fracture of the orbit. *Can J Ophthalmol* 4:377, 1969.

Dutton JJ, Klingele TG, Burde RM, Gado M: Evaluation of the suprasellar cistern by computed tomography. *Ophthalmol* 89:1220, 1982.

Emery JM, von Noorden GK, Schlermitzauer DA: Orbital floor fractures. Long term follow-up of cases with and without surgical repair. *Trans Am Acad Ophthalmol Otol* 75:802, 1971.

Evans RA, Schwartz JF, Chutorian AM: Radiologic diagnosis in pediatric ophthalmology. *Radiol Clin North Am* 1:459, 1963.

Fradkin FI: Orbital floor fractures and ocular complications. *Am J Ophthalmol* 72:600, 1971.

Fueger GF, Milauskas A: Two new radiographic projections of orbitography in the diagnosis of orbital blow-out injuries in the fracture of the orbit. In Bleeker GM, Lyle IK (eds): *Symposium on Orbital Fractures.* New York, Excerpta Medica, 1970, p 45.

Fukado Y: Results in 400 cases of surgical decompression of the optic nerve. *Mod Prob Ophthalmol* 14:474, 1975.

Goalwin HA: One thousand optic canals. Clinical, anatomic and roentgenologic study. *JAMA* 89:1745, 1922.

Gould HR, Titus CO: Internal orbital fractures: Value of laminography in diagnosis. *Am J Roentgenol* 97:618, 1966.

Grove AS Jr: Orbital trauma evaluation by computed tomography. *Comput Tomogr* 3:267, 1979.

Grove AS Jr: Orbital trauma and computed tomography. *Ophthalmol* 87:403, 1980.

Guibert-Trainier F, Piton J, Calobet A, Caille JM: Orbital syndromes—CT analysis of 100 cases. *Comput Tomogr* 3:241, 1979.

Gyldensted C, Lester J, Fledelius H: Computed tomography of orbital lesions. *Neuroradiology* 13:141, 1977.

Hartman E, Gilles E: *Roentgenologic Diagnosis in Ophthalmology.* Philadelphia, JB Lippincott, 1959.

Hughes B: Indirect injury of the optic nerves and chiasm. *Bull Johns Hopkins Hosp* 111:98, 1962.

Jurkiewicz MJ, Nickell WB: Fractures of the skeleton of the face—a study of diagnosis and treatment based on twelve years' experience in the treatment of over 600 major fractures of the facial skeleton. *J Trauma* 11:947, 1971.

Kennedy RE: The effect of early enucleation on the orbit of animals and humans. *Am J Ophthalmol* 60:277, 1965.

Kieffer SA: Superior orbital fissure. In Newton TH, Potts DG (eds): *Radiology of the Skull and Brain. The Skull,* Vol I, Book 2. St. Louis, CV Mosby, 1971.

Kier EL: Embryology of the normal optic canal and its anomalies. *Invest Radiol* 1:346, 1966.

Lerman S: Blowout fracture of the orbit. Diagnosis and treatment. *Br J Ophthalmol* 54:90, 1970.

Lloyd GAS: Orbital emphysema. *Br J Radiol* 39:933, 1966.

Lloyd GAS: Radiology of the optic nerve in the orbit. *Trans Ophthalmol Soc UK* 96:382, 1976.

Lloyd GAS: CT scanning in the diagnosis of orbital disease. *Comput Tomogr* 3:227, 1979.

Lombardi G: *Radiology in Neuro-ophthalmology.* Baltimore, Williams & Wilkins, 1967.

Merrell RA, Yanagisawa E: Radiographic anatomy of the paranasal sinuses. I. Waters' view. *Arch Otolaryngol* 87:184, 1968.

Milauskas AT, Fueger GF: Serious ocular complications associated with blow-out fractures of the orbit. *Am J Ophthalmol* 62:670, 1966.

Newton TH, Potts DG (eds): *Radiology of the Skull and Brain. The Skull,* Vol I, Book 2. St. Louis, CV Mosby, 1971.

Niloskelainen E, Enzmann DR, Sogg RL, Rosenthal AR: Computerized tomography of the orbits. *Acta Ophthalmol* 55:885, 1977.

Pettinoti S: L'indagine radiologica nella fratture dell' orbita. *Rass Ital Ottol* 24:294, 1955.

Pfeiffer RL: Traumatic enophthalmos. *Arch Ophthalmol* 30:718, 1943.

Potter GD: Radiologic examination of the orbit. *CRC Crit Rev Radiol Sci* 2:145, 1971.

Potter GD: Radiologic diagnosis of tumors of the orbit. In Aston AS, Hornblass SH, Meltzer MA, Rees TD (eds): *Third Intl Symp Plast Reconstr Surg Eye and Adnexa.* Baltimore, Williams & Wilkins, 1982.

Potter GD, Trokel SL: Optic canal. In Newton TH, Potts DG, (eds): *Radiology of the Skull and Brain. The Skull,* Vol I, Book 2. St. Louis, CV Mosby, 1971.

Price HI, Danziger A: The computerized tomographic findings in paediatric orbital tumors. *Clin Radiol* 30:435, 1978.

Pringle JH: Atrophy of the optic nerve following diffused violence to the skull. *Br Med J* 2:1156, 1922.

Reese AB: *Tumors of the Eye,* ed 2. New York, Harper & Row, 1963.

Rischbieth RHC: Significance of enlargement of the superior orbital fissure. *Br J Radiol* 31:125, 1958.

Russell WK: Injury to cranial nerves including the optic nerve and chiasm. In Brock S (ed): *Injuries of the Skull, Brain and Spinal Cord.* London, Bailliere, 1940.

Shapiro R, Robinson F: Alterations of the sphenoid fissure produced by local and systemic processes. *Am J Roentgenol* 101:814, 1967.

Shultz RC: Supraorbital and glabellar fractures. *Plast Reconstr Surg* 45:227, 1970.

Smith B, Reagan WFJ: Blow-out fractures of the orbit: Mechanism and correction of inferior orbital fracture. *Am J Ophthalmol* 44:733, 1957.

Sugita S, Sugita Y, Kawake Y: Die Sehstorung nach Schädeltrauma und ihre operative Behandlung. *Klin Monatsbl Augenheilk* 147:720, 1965.

Trokel SL, Hilal SK: Recognition and differential diagnosis of enlarged extraocular muscles in computed tomography. *Am J Ophthalmol* 87:503, 1979.

Trokel SL, Hilal SK: Submillimeter resolution CT scanning in the analysis of orbital disease. Aston AS, Hornblass SH, Meltzer MA, Rees TD (eds): *Third Intl Symp Plast Reconstr Surg Eye and Adnexa.* Baltimore, Williams & Wilkins, 1982.

Unsold R, Newton TH, Hoyt WF: CT examination technique of the optic nerve. *J Comput Assist Tomogr* 4:560, 1980.

ViGario GD: Blow-out fractures of the orbit. *Br J Radiol* 39:939, 1966.

Waters CA, Waldron CW: Roentgenology of the accessory nasal sinuses describing a modification of the occipito-frontal position. *Am J Roentgenol* 2:633, 1915.

Wende S, Aulich A, Nover A, Lankesch W, Steinhoff H, Meese W, Lange S, Grumme T: Computed tomography of orbital lesions. A comparative study of 210 cases. *Neuroradiol* 13:123, 1977.

Whyte DK: Blow-out fractures of the orbit. *Br J Ophthalmol* 52:721, 1968.

Yanagisawa E, Smith HW: Radiographic anatomy of the paranasal sinuses. IV. Caldwell view. *Arch Otolaryngol* 87:109, 1968.

Yanagisawa E, Smith HW, Thaler S: Radiographic anatomy of the paranasal sinuses. II. Lateral view. *Arch Otolaryngol* 87:196, 1968.

Zizmor J, Raskind RH: Orbital calcification. In Newton TH, Potts DG (eds): *Radiology of the Skull and Brain. The Skull.* Vol I, Book 2. St. Louis, CV Mosby, 1971.

Zizmor J, Lombardi G: *Atlas of Orbital Radiography.* Birmingham, Aesculapius, 1973.

Zizmor J, Noyek AM: Orbital trauma. In Newton TH, Potts DG (eds): *Radiology of the Skull and Brain. The Skull.* Vol I, Book 2. St. Louis, CV Mosby, 1971.

CHAPTER 4

Eyebrows, Eyelids, and Anterior Orbit

The eyelids and eyebrows are functionally and cosmetically very important structures. The eyelids protect the globe from exposure, yet are adequately mobile to prevent visual obstruction. As such, the eyelids have anatomic attributes which enhance their functional role as specialized mobile structures. The same anatomic features which enhance the function of the eyelids also contribute to a variety of pathologic conditions of eyelids.

Classically, the orbital tissues are defined as those of the eyelid or orbit. Theoretically, the anterior boundary of the orbit is defined by the orbital septum. Those structures anterior to the septum are a part of the eyelid, while those structures posterior to the septum constitute the orbit. However, this is an arbitrary distinction since structures such as the aponeurosis of the upper and lower lids originate in the orbit, but extend anterior to their fusion with the orbital septum into the eyelid.

Initially, the eyebrows will be reviewed. The eyelids will then be analyzed in a systematic manner, with the surface topography of the eyelid being reviewed first. This will be followed by evaluation of the eyelid margin, caruncle and plica semilunaris, upper and lower eyelids, and glands of the eyelid.

EYEBROW

The eyebrows extend to the superior orbital margin and are formed by thick skin supported by strong muscle fibers. They have a very characteristic distribution of hair. The eyebrows are actually cranial appendages rather than extensions of facial tissue. Morphologically, they are considered a part of the scalp, and are best described as specialized sliding muscular extensions of the scalp (Fig. 4.1). The mobility of the brow is an important component of facial expression.

For descriptive purposes, one may consider the eyebrow to have a head, body, and tail. The head or medial extension of the eyebrows are separated by an "intersuperciliary region," the glabella, a word derived from the Latin, *glabu,* meaning smooth or hairless. The eyebrows overlie the frontal sinuses nasally, the frontal bone centrally, and extend to approximately the zygomaticofrontal suture temporally. The eyebrow varies in width, being widest nasally and narrowest

laterally. The brow is normally at the level of the superior orbital rim in males and above the superior orbital rim in females.

The eyebrows are composed of four layers: skin, muscle, fat, and aponeurosis. The eyebrows act as a functional unit, but each layer will be discussed separately. The skin of the eyebrow is thick and rich in sebaceous and sweat glands. The skin contains large hair follicles with hairs which are embryologically the first to develop in utero. The hairs are more numerous and more irregularly arranged in males than in females.

The muscular layer of the eyebrow is a composite of four muscle groups: frontalis, procerus, corrugator superciliaris, and orbicularis oculi. These muscle fibers intermingle and are difficult to separate at their cutaneous insertions. The vertically oriented frontalis muscle is the extension of the anterior belly of the epicranius, and is connected to the occipital muscle by the galea aponeurotica (Fig. 4.1). The frontalis muscle has no bony attachments; it intermingles with the other forehead muscles, and has a cutaneous insertion at the level of the eyebrow. The medial portions of the frontalis extend inferiorly to attach to nasal bones to form the procerus muscle (Fig. 4.2). The procerus muscle is structurally and functionally a distinct entity from the frontalis muscle. Since the procerus muscle has a bony attachment, its action will effectively pull the medial portion of the brow inferiorly and has been termed the muscle of aggression.

The obliquely oriented corrugator superciliaris lies beneath the frontalis and orbicularis muscles. The muscle forms a definite coarse band of fibers originating from the frontal bone near the superior medial orbital margin. It extends 2–3 cm laterally, to blend with the orbicularis oculi and frontalis muscles. Action of the corrugator superciliaris muscle pulls the eyebrow medially and inferiorly and produces vertical glabellar furrows.

The peripheral fibers of the concentrically oriented orbital orbicularis muscle interdigitate with the inferior extension of the frontalis (Fig. 4.2). The skin is easily dissected from the underlying orbicularis except at the eyebrow, where the orbicularis interdigitates with the subcutaneous tissues.

The aponeurosis of the brow is formed by the central aponeurotic layer of the scalp, the galea aponeurotica, which is continuous with the occipital muscle. The galea splits to

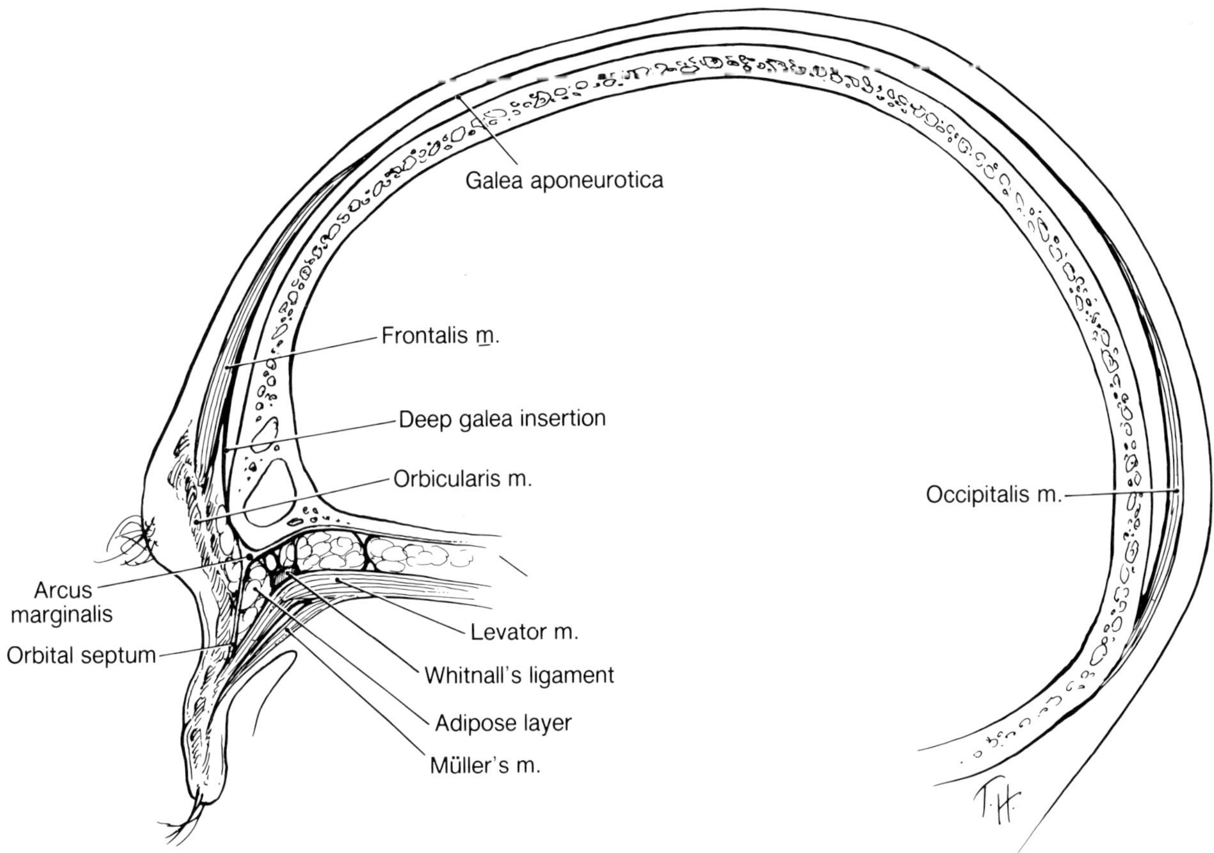

Figure 4.1.   Anatomy of the brow. The frontalis muscle is connected to the occipitalis muscle by the galea aponeurotica.

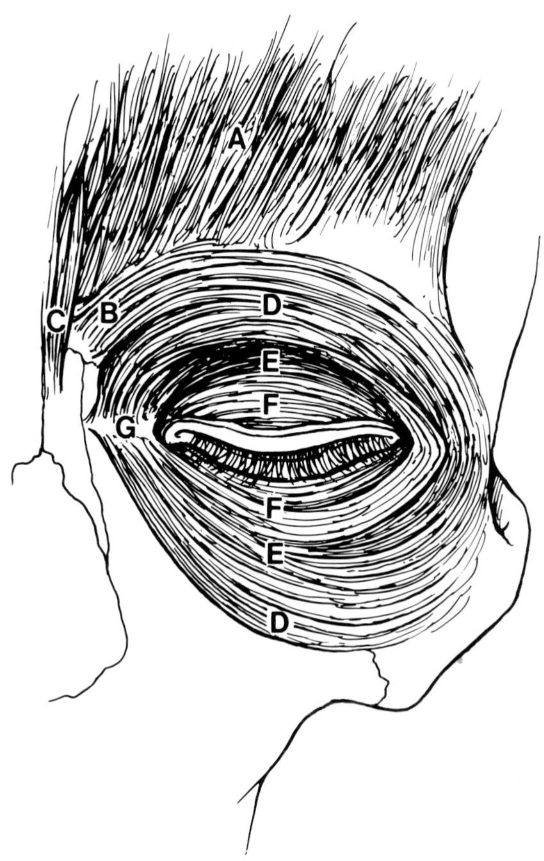

form a thin, superficial layer over the frontalis muscle and a deeper, well-defined layer passing beneath the muscles where it is firmly attached to the supraorbital margin, contributing to the arcus marginalis.

The fat layer of the eyebrows separates the muscles of the eyebrow from the aponeurotic layer. The adipose layer is most abundant in the lower portion of the eyebrow at the superior orbital margin. The presence of the adipose tissue spaced between the periosteum and the muscular layer of the brow and forehead enhances the mobility of the brow (Lemke and Stasior, 1982). The adipose layer extends inferiorly into the upper eyelid, forming a small fat pad beneath the superior portion of the orbital orbicularis muscle.

The firm attachment of the superficial muscle plane to adipose and aponeurotic layer helps suspend the brow. This attachment is most pronounced in the medial two-thirds of the brow and corresponds to the supraorbital ridge (Fig. 4.3). The muscular layer has little significant attachment in the lateral one-third of the brow (Fig. 4.4), accounting for brow ptosis occurring most often in this region.

CLINICAL NOTE

Eyebrow ptosis will be evident in those patients with facial nerve palsy and involutional changes. Pa-

Figure 4.2.   Musculature of the brow and eyelids: *A*, frontalis muscle; *B*, corrugator superciliaris muscle; *C*, procerus muscle; *D*, orbital orbicularis muscle; *E*, preseptal orbicularis muscle; *F*, pretarsal orbicularis muscle; *G*, medial canthal tendon.

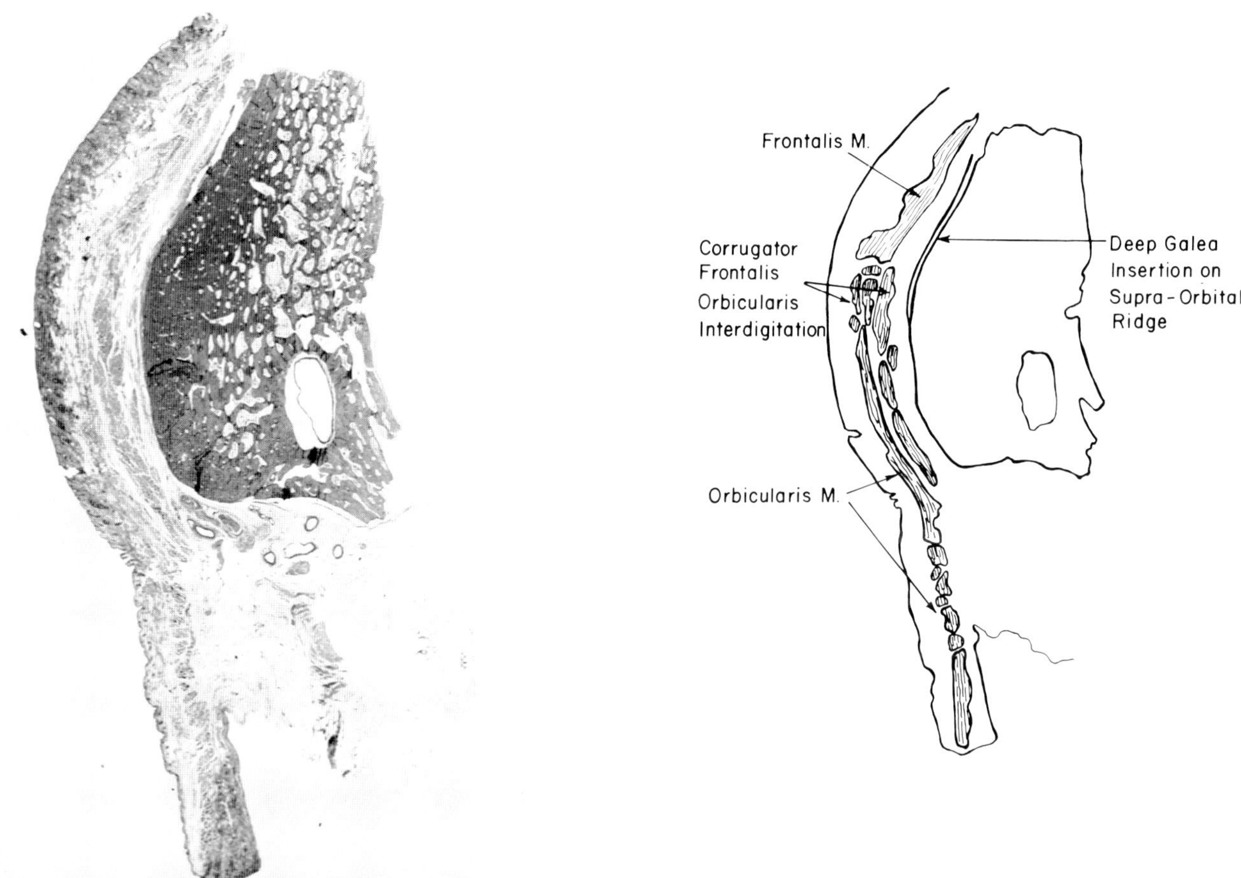

Figure 4.3. Parasagittal section through the medial portion of the brow. Muscle groups overlap at the supraorbital ridge and the deep galea aponeurotica inserts on the supraorbital ridge. (Reproduced from Lemke B, Stasior OG: *Archives of Ophthalmology* 100:981–986. 1982, American Medical Association.)

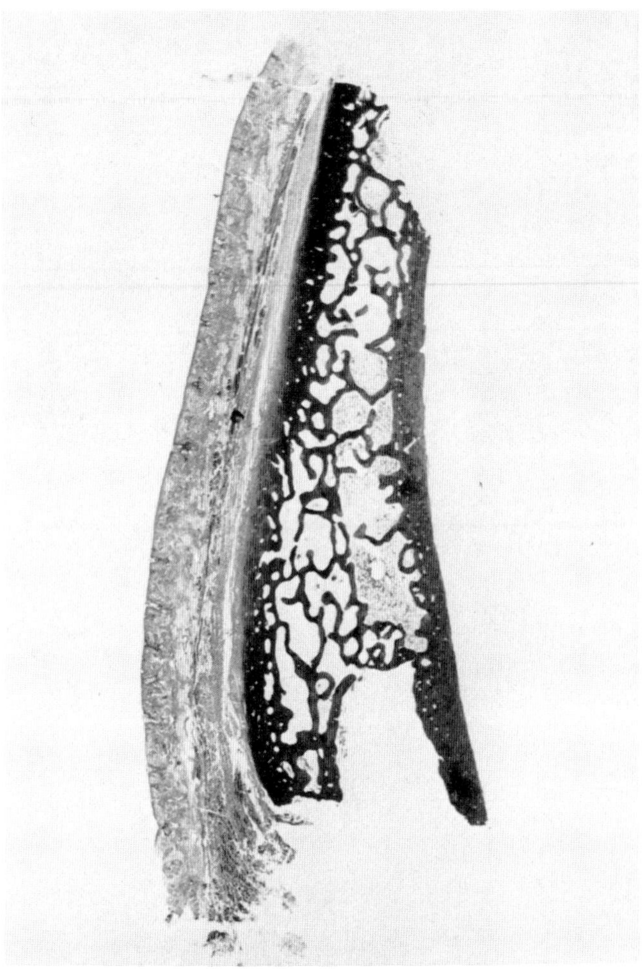

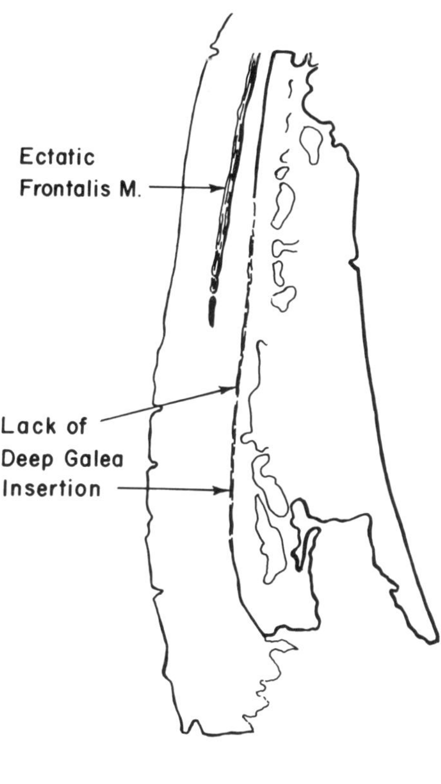

Figure 4.4. Parasagittal section through the lateral portion of the brow. The frontalis muscle becomes thinned and attenuated prior to reaching the supraorbital rim and the galea aponeurotica does not insert into the periorbita. (Reproduced from Lemke B, Stasior OG: *Archives of Ophthalmology* 100:981–986. 1982, American Medical Association.)

tients with a facial palsy will have eyebrow ptosis secondary to the gravitational effects of the denervated forehead. Due to the lack of periosteal attachments, the forehead slides into the upper eyelid. The ptotic brow and secondary dermatochalasis may produce a significant superior lateral visual field loss (Fig. 4.5).

Involutional brow ptosis results from a relaxation of the scalp and forehead structures. In addition, atrophy of brow fat pad enhances brow mobility due to a weakening of deep attachments of the frontalis muscle to the underlying periosteum. Due to the relative lack of deep attachments laterally, eyebrow ptosis is exaggerated in this portion of the eyebrow. Attempts to elevate the brow may result in deep, horizontally oriented forehead furrows.

Appreciation of eyebrow ptosis is important in the evaluation of upper lid blepharoplasty. Lateral brow ptosis will exaggerate upper eyelid dermatochalasis, producing a functional loss of the superior lateral visual field of gaze. An upper eyelid blepharoplasty will act to exaggerate the brow ptosis, actually pulling

the brow into the upper eyelid, due to excessive eyelid skin removal (Fig. 4.6 A). A brow elevation will restore brow position and prevent a compromised upper eyelid blepharoplasty (Fig. 4.6 B).

Brow mobility can be employed in the management of ptotic patients with a severe restriction of levator function. Congenitally ptotic patients with very poor or absent levator function are best managed with a frontalis brow suspension. However, aponeurotic ptosis surgery has refined the management of ptosis and patients with as little as 3–4 mm of function have been effectively treated with modifications of aponeurotic ptosis procedures (Anderson and Dixon, 1979; Doxanas and Dryden, 1981; Anderson, 1982). At the time of brow suspension, the upper eyelid should be suspended to the frontalis muscle to enhance its mobility. Novice ptosis surgeons may suspend the eyelid to the periosteum of the superior orbital margin which restricts upper eyelid mobility (Dryden, et al, 1982).

The connection of the frontalis and the occipital muscles by the galea aponeurotica may help explain the occurrence of occipital headaches. Sustained con-

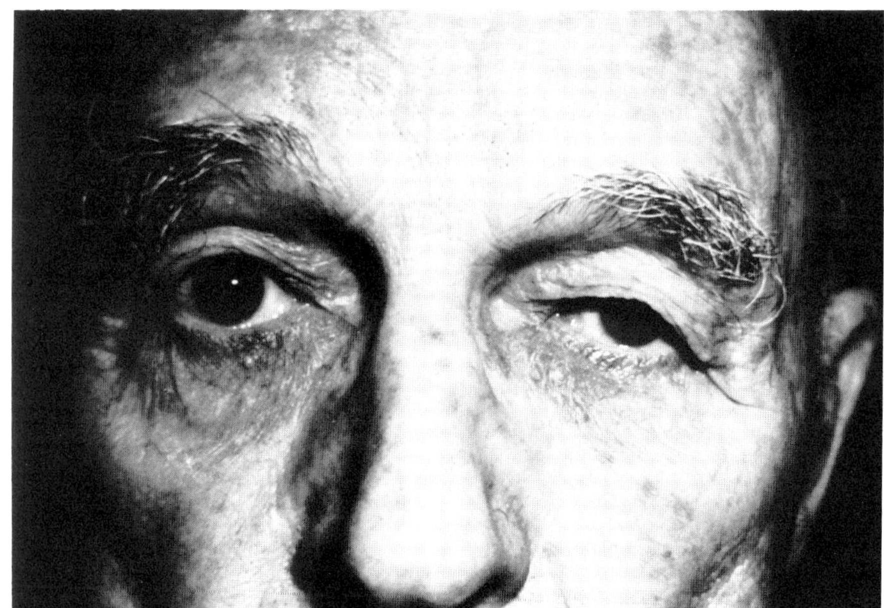

Figure 4.5. Brow ptosis following facial nerve palsy. Brow ptosis accentuates upper eyelid dermatochalasis.

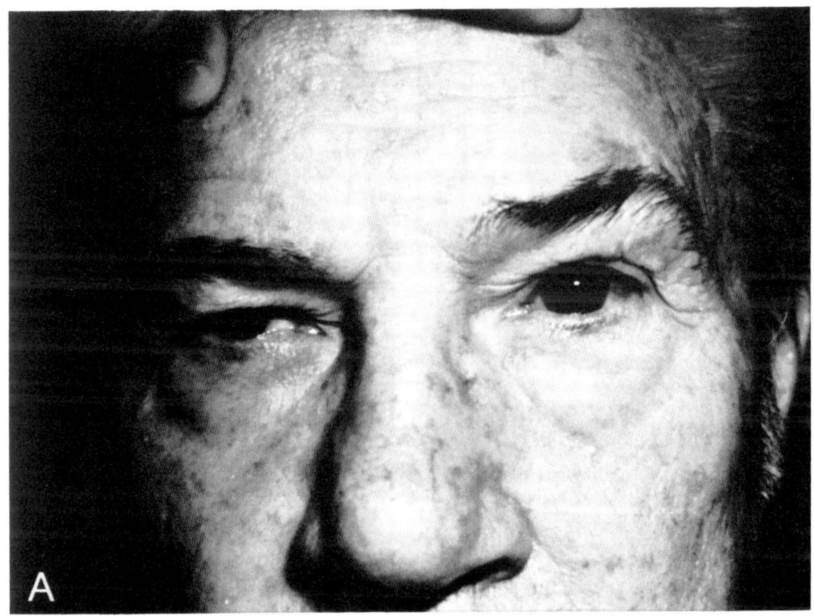

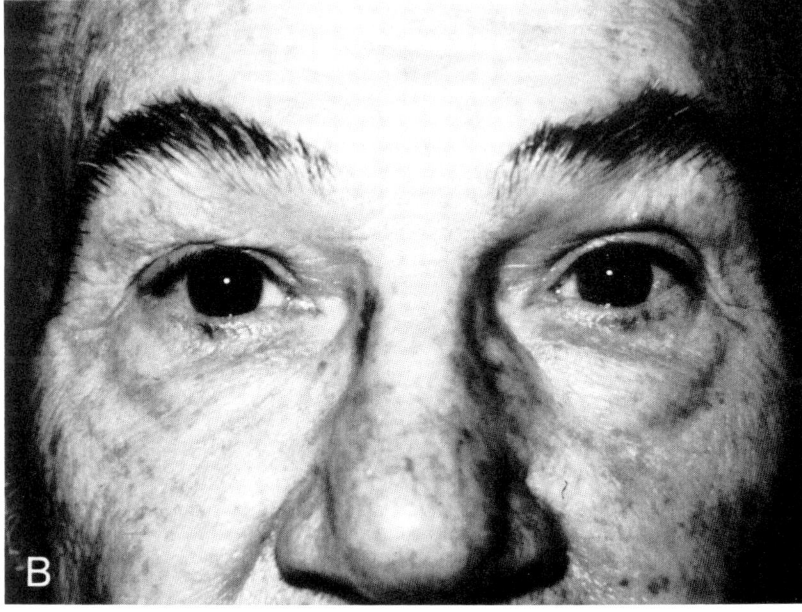

Figure 4.6. Brow ptosis and blepharoplasty. (*A*), Patient 1 year following an upper eyelid blepharoplasty. Brow ptosis results in a continued superior and lateral visual field defect. (*B*), Following supraciliary brow elevation, a better functional and cosmetic result is obtained.

traction of the frontalis muscle, in response to brow ptosis, accommodative effort, or in reaction to a strong light may be manifested as a frontaloccipital headache.

EYELIDS—TOPOGRAPHICAL ANATOMY

The eyelids blend imperceptibly with surrounding facial structures. However, in pathological states, such as eyelid edema, the extent of the lids can be delineated. Eyelid skin is thin with a loose connective tissue devoid of fat, permitting distention with tissue edema. The skin of the brow and temple is thicker, with a dense connective tissue which helps restrict involvement in edematous conditions. The upper eyelid extends to the eyebrow and lateral orbital rim, while the lower eyelid extends beyond the inferior orbital rim to the nasojugal and malar skin folds (Fig. 4.7).

The eyelids have a characteristic surface anatomy. The upper eyelid is divided into an orbital (preseptal) and a tarsal portion. The orbital portion lies between the orbital rim and the superior border of the tarsus. The tarsal portion of the eyelid contains the tarsal plate. The upper eyelid crease, approximately 8–11 mm above the eyelid margin, is formed by the attachment of the levator aponeurosis in the subcutaneous tissues below its fusion with the orbital septum. The skin in the preseptal or orbital portion of the eyelid is loose-textured and has no attachments of the apo-

neurosis in the subcutaneous tissue. It, therefore, may overhang the eyelid crease to produce the eyelid fold. The skin beneath the eyelid crease is more adherent to deeper tissues by the distal insertions of the levator aponeurosis into the subcutaneous tissues.

A lid crease is frequently apparent in the lower eyelid in children. This crease may be lost with age. It is nearer the eyelid margin than is the upper eyelid crease. It generally extends to within 2–3 mm of the eyelid margin medially and is approximately 5–6 mm below the eyelid margin laterally. Two other folds may be present around the lower eyelid. The nasojugal fold runs inferiorly and laterally from the medial canthal region along the side of the nose. A malar fold may run from the lateral side of the lower eyelid toward the inferior portion of the nasojugal fold. These folds are produced at the junction of skin bound by loose connective tissue in the eyelid, and denser connective tissue in the cheek.

Orientals have an exaggerated upper lid skin fold and a poor lid crease (Fig. 4.8 *A*). The fold essentially obliterates the eyelid crease. The skin may fold over the roots of the eyelashes to obscure the upper eyelid margin (Fig. 4.8 *A*). The anatomical basis for the appearance of the oriental or Mongolian eyelid results from the relationship between the levator aponeurosis and the orbital septum (Fig. 4.8 *B*). In Orientals the orbital septum inserts lower on the aponeu-

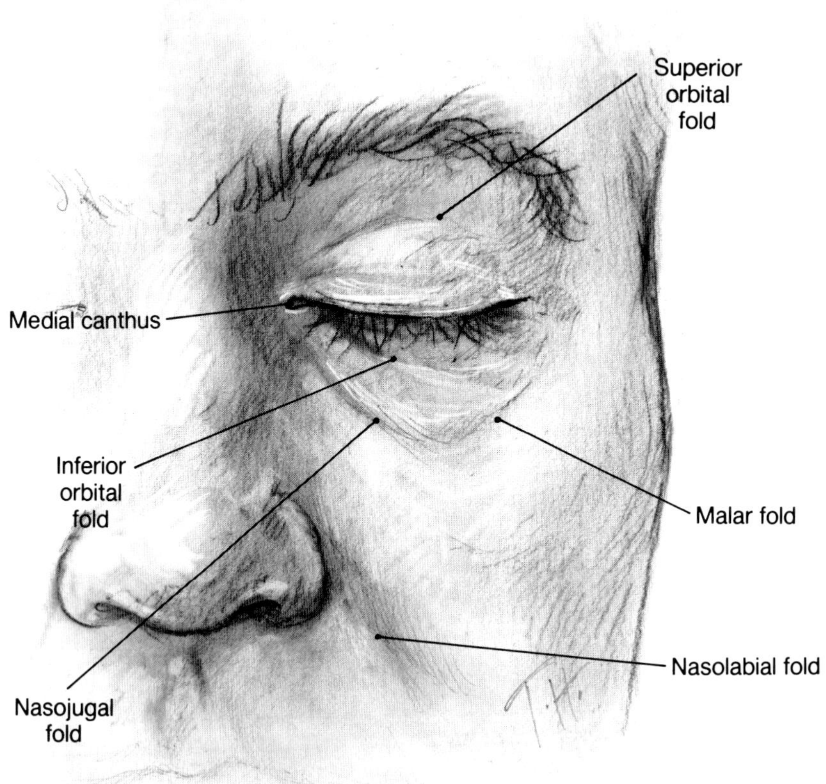

Figure 4.7.   Schematic demonstrating eyelid folds.

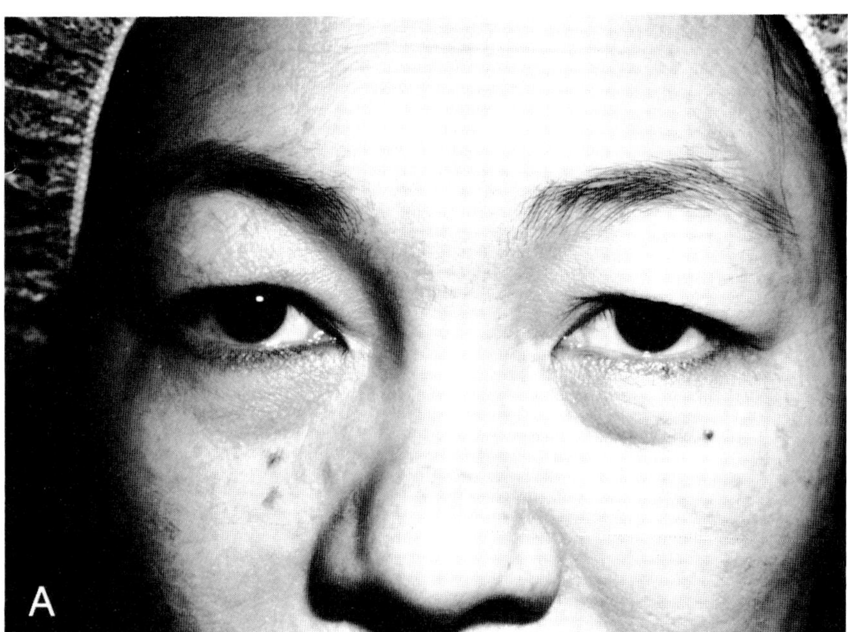

Figure 4.8. Oriental eyelids. (*A*), Clinical appearance. (*B*), Schematic of eyelid anatomy. In the Oriental eyelid the orbital septum fuses with the levator aponeurosis below the superior tarsal border.

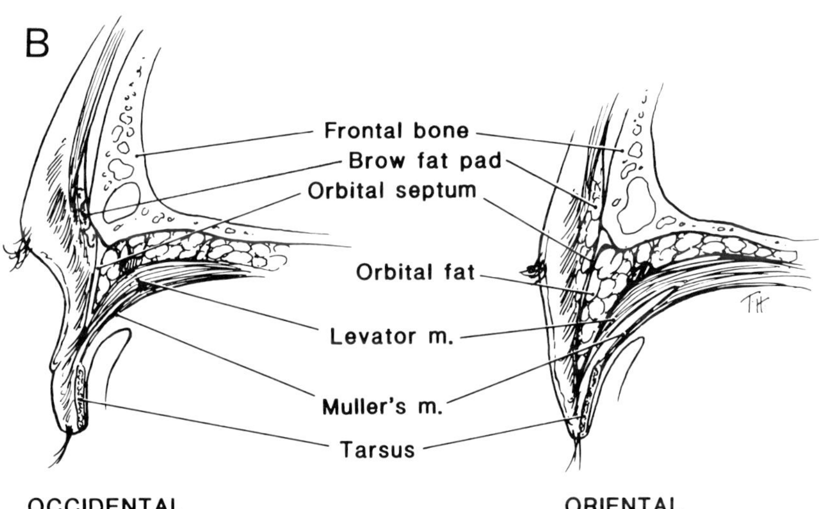

rosis to allow a more inferior extension of the orbital fat. The inferior extension of the orbital septum also blocks the superficial course of the anterior levator aponeurosis, which normally creates the eyelid crease. In some Orientals there also seems to be fatty tissues anterior to the levator aponeurosis which may block the subcutaneous attachments of the aponeurosis. These findings contribute to the characteristic Mongolian or Oriental eyelid.

CLINICAL NOTE

Epicanthus is a fold of skin in the medial canthal region. Johnson (1956, 1968, 1978) has recognized various types of epicanthal folds based upon the origin and termination of the fold. The four types of epicanthal folds are: supraciliaris, palpebralis, tarsalis, and inversus (Fig. 4.9 *A–D*). Epicanthus supraciliaris is formed by the fold arising from the region of the eyebrow and terminating over the lacrimal sac (Fig. 4.9 *A*). Epicanthus palpebralis is formed by the fold arising in the upper eyelid and extending to the lower eyelid (Fig. 4.9 *B*). Epicanthus tarsalis arises from the tarsal fold and extends to the inner canthus (Fig 4.9 *C*). Epicanthus tarsalis is frequently present in Orientals. Epicanthus inversus is a small fold arising in the lower eyelid and extending upward, partially obscuring the medial canthus (Fig. 4.9 *D*). Epicanthus inversus is associated with the blepharophimosis syndrome, which also includes ptosis, telecanthus, and a short palpebral fissure. Epicanthal folds usually diminish with the development of the nasal bridge in adolescence (Bergin and La Piana, 1981).

Epiblepharon is characterized by the presence of a fold of skin that overlaps the eyelid margin, which rolls the lashes against the globe (Fig. 4.10). The eyelid margin is not rolled against the globe. Epiblepharon is frequently observed in Oriental patients. Epiblepharon is commonly confused with congenital entropion, which is an extremely rare condition (Fox, 1956; Tse and Anderson, 1983). The eyelid margin is rolled

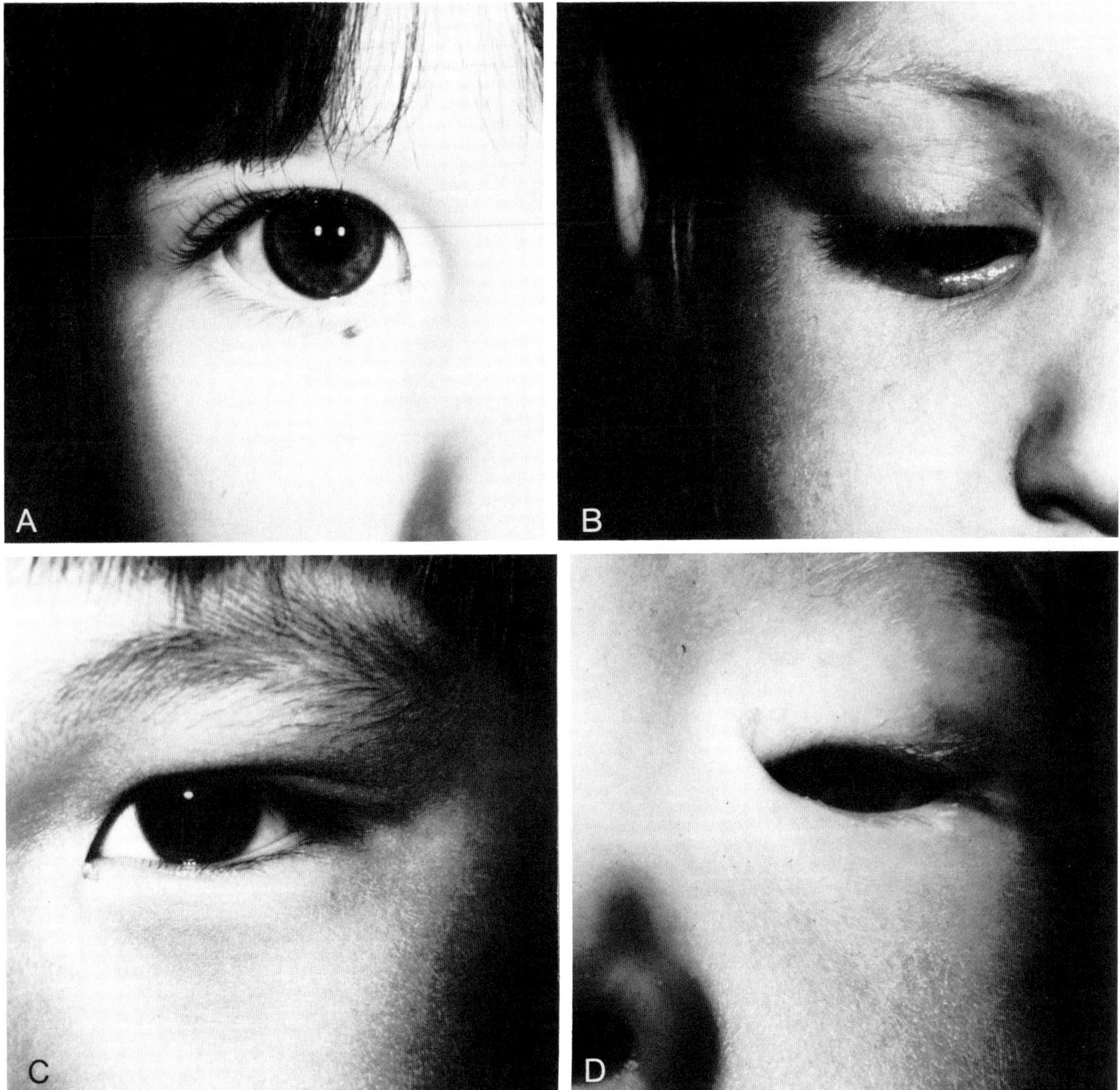

Figure 4.9. Epicanthal folds. (*A*), Epicanthus supraciliaris—the fold arises near the brow to terminate over the lacrimal sac. (*B*), Epicanthus palpebralis—the fold arises in the upper lid and terminates in the lower eyelid. (*C*), Epicanthus tarsalis—the fold arises from the tarsal fold and extends to the medial canthus. (*D*), Epicanthus inversus—the fold arises in the lower eyelid to obscure the medial canthus.

against the globe in patients with congenital entropion, distinguishing it from epiblepharon. Despite the fact that eyelashes are rolled against the cornea, epiblepharon may be well-tolerated. With time and development of the eyelid, epiblepharon may diminish or disappear. If epiblepharon persists, producing corneal irritation or photosensitivity, surgical intervention is indicated (Johnson, 1968; McCord, 1980).

EYELID MARGIN

The eyelid margin provides a transitional zone between the skin of the eyelid and the palpebral conjunctiva. The eyelid skin has a keratinized stratified squamous epithelium, while the eyelid margin and conjunctiva have a nonkeratinizing stratified squamous and stratified columnar epithelium, respectively. The transition to a nonkeratinized epithelium provides the eyelid margin a smooth, nonabrasive surface.

The eyelid margin can be divided in its horizontal dimensions at the junction of the medial 1/6 and the lateral 5/6. This junction is defined by the papilla lacrimalis, which contains the punctum lacrimalis at its summit. This nipple-like elevation of the eyelid margin is directed slightly inwards to

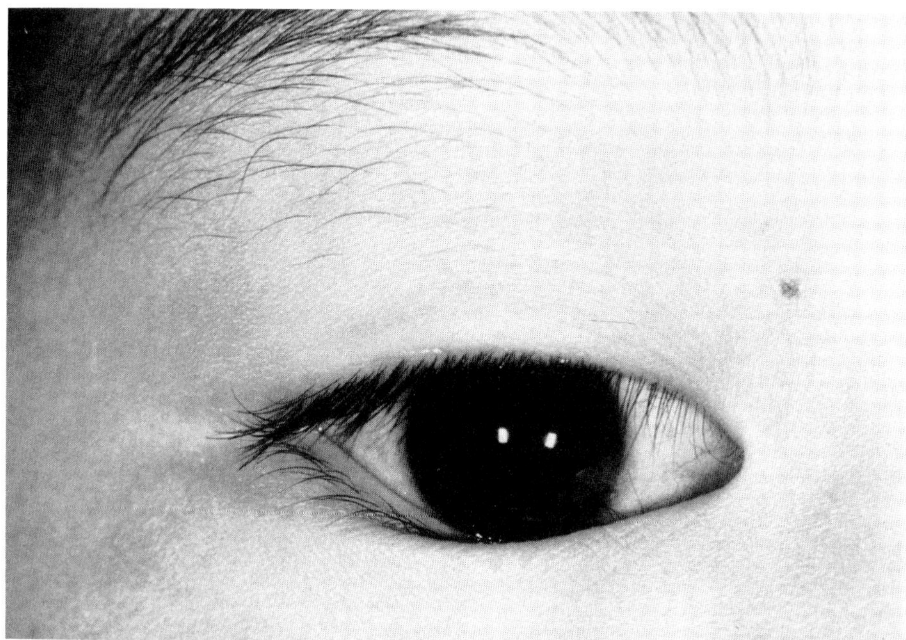

Figure 4.10. Epiblepharon. A fold of skin overlaps the eyelid margin and rolls the lashes against the globe.

approximate the globe. The medial ⅙ of the eyelid margin is the lacrimal portion reflecting its close association to the lacrimal excretory system. This area extends approximately 5–8 mm, forming the medial canthus. The sulcus in the posterior medial canthal region is occupied by the caruncle, which approximates the plica semilunaris. The area within the medial canthus provides a place for tears to pool and forms the lacus lacrimalus. The lateral ⅚ of the eyelid margin forms the bulbar or ciliary portion. This portion of the eyelid contains tarsus and is thicker and better defined.

The ciliary portion of the eyelid margin can be defined by distinct areas. A cross section of the eyelid margin (Fig. 4.11) reveals two portions: anterior-ciliary and posterior-tarsal. The ciliary margin contains the cilia and orbicularis muscle. The pilar apparatus marks the transition of the surface epithelium from a keratinizing to a nonkeratinizing stratified squamous. The cilia form two or three irregularly arranged rows and are more numerous in the upper eyelid than in the lower eyelid. The pilosebaceous apparati are embedded in a dense connective tissue extending to the anterior surface of the tarsus. The sebaceous glands contiguous with the pilar units are the glands of Zeis. In addition, the secretory nodules of the apocrine glands of Moll open into the ciliary follicles, into the duct of a Zeis gland, or directly onto the eyelid margin. A small strip of orbicularis muscle may lie posterior to the pilosebaceous apparati, yet anterior to the tarsus, forming the muscle of Riolan.

The tarsal portion of the eyelid is separated from the ciliary portion by the sulcus intermarginalis of Graefe, more commonly called the greyline. The intermarginal sulcus is a fine groove which is best appreciated in the upper eyelid (Fig. 4.12). The Meibomian gland orifices are centered in the tarsal portion of the eyelid margin.

UPPER EYELID

The upper and lower eyelids are analogous structures, their main differences existing in the eyelid retractors. In

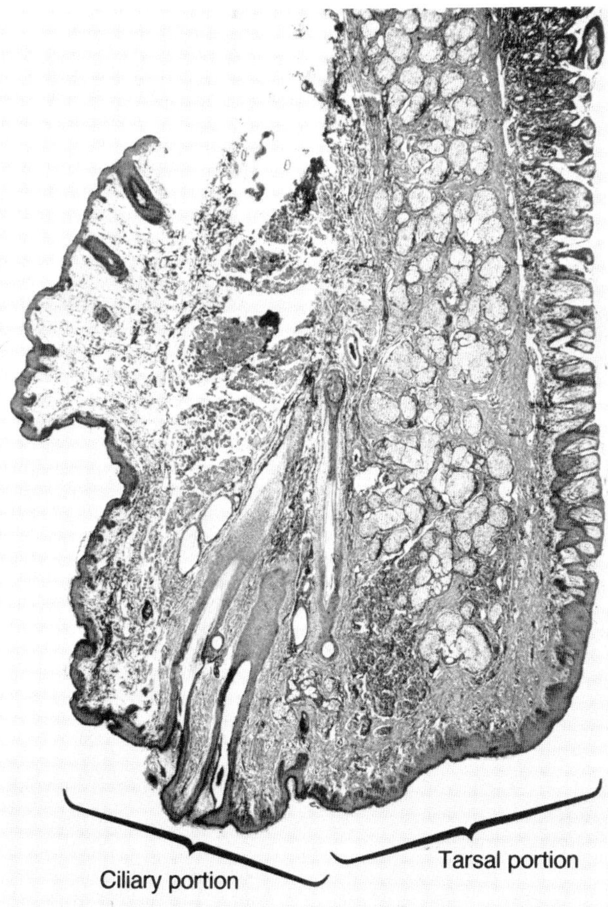

Figure 4.11. Cross section of the eyelid margin. The anterior eyelid margin forms the ciliary portion while the posterior eyelid margin produces the tarsal portion of the eyelid margin. (Reproduced with permission from W. R. Green.)

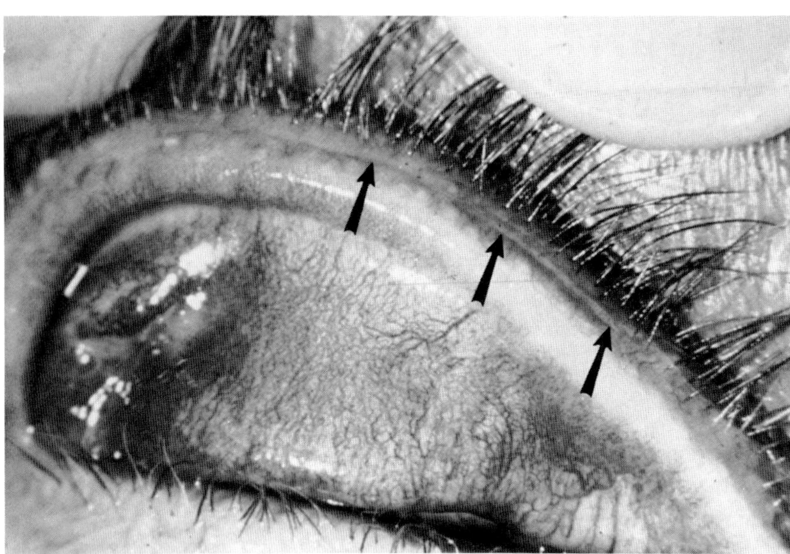

Figure 4.12. Everted upper eyelid demonstrates the greyline (intermarginal sulcus of von Graefe) (*arrows*). (Reproduced with permission from W. R. Green.)

the upper eyelid, the levator muscle and its aponeurosis are a distinct entity, which has evolved from the superior rectus muscle as the skeletal retractor. The lower eyelid retractor is a fascial extension of the inferior rectus (capsulopalpebral fascia, or the lower eyelid aponeurosis). Phylogenetically, it seems that as mammals were no longer aquatic and became upright, greater mobility and elevation of the upper eyelid was required. Lower aquatic mammals still have a capsulopalpebral head of the superior rectus rather than a levator muscle. Thus, the differences in the upper and the lower eyelids seem to have an evolutionary basis.

The eyelid is generally considered to have six layers: skin, orbicularis muscle, levator aponeurosis, Müller's muscle, tarsal plate, and palpebral conjunctiva. However, such a six-layer portrayal of the eyelid results in an inadequate, if not incorrect, description. One must first define the level of the eyelid being evaluated before discussing layers (Fig. 4.13). Near the eyelid margin (marginal portion) there are only four major layers (Fig 4.13 *A*). These are skin, orbicularis muscle, tarsus, and conjunctiva. Just above the marginal portion of the eyelid (approximately 3 mm above the margin), the levator aponeurosis has a firm attachment on the tarsal plate. The levator aponeurosis, therefore, forms a fifth layer on the anterior tarsal portion of the eyelid (Fig. 4.13 *B*). Müller's muscle inserts on the superior border of the tarsal plate, thus at the upper tarsal border there remains five layers, with the addition of Müller's muscle and the loss of the tarsus (Fig. 4.13 *C*). Higher in the eyelid (septal portion) there are an additional two layers, the orbital septum and the underlying preaponeurotic fat. In the septal portion of the eyelid, there are seven major layers: skin, orbicularis muscle, orbital septum, orbital fat, levator aponeurosis, Müller's muscle, and conjunctiva.

CLINICAL NOTE

It is useful to consider the tarsal portion of the lids as being composed of an anterior and a posterior lamella. The anterior lamella consists of the skin and orbicularis muscle, while the posterior lamella consists

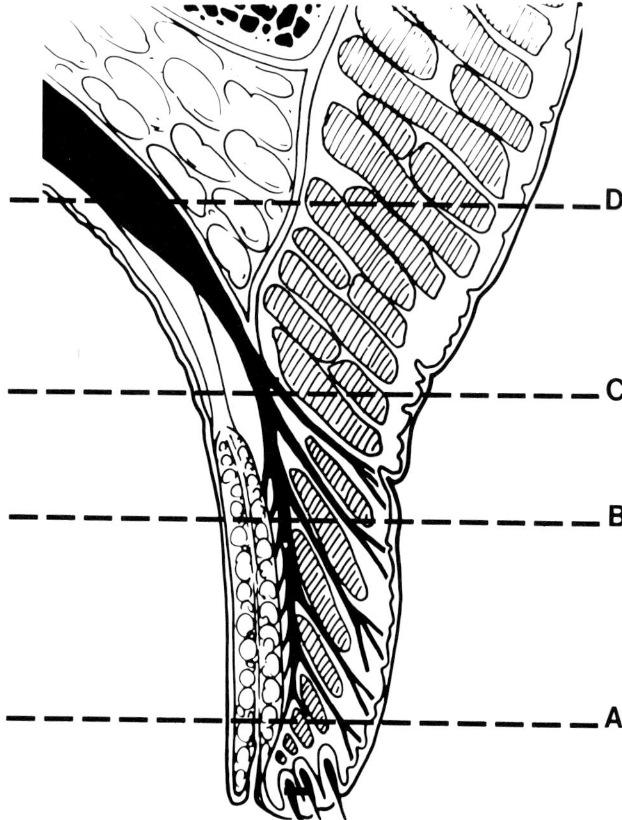

Figure 4.13. Schematic cross section of the eyelid. Layers of the eyelid at various levels. (*A*), near lid margin; (*B*), tarsal; (*C*), supratarsal; (*D*), septal portions of the eyelids.

of the tarsus, eyelid retractors, and conjunctiva. This concept is particularly helpful in evaluating abnormalities of the eyelid position and contour, such as entropion and ectropion. In addition, eyelid reconstruction following tumor excision or trauma should reconstitute both anterior and posterior lamellae.

The skin of the eyelid is the thinnest in the body. Histologically, this skin may be only 3–4 cell layers thick. The skin contains very fine, downy hairs which grow from rather underdeveloped hair follicles. The dermal sweat and sebaceous glands are also less well-developed than elsewhere in the body. The pretarsal skin is attached to the underlying orbicularis muscle by a fine connective tissue which is devoid of fat. The thinness of this skin and the fine underlying connective tissue permits the tissue to be quite distensible, as evidenced by patients with periorbital cellulitis and ecchymosis of the eyelids.

CLINICAL NOTE

A distinction should be made between dermatochalasis and blepharochalasis. There is a frequent confusion in the literature regarding these two conditions. Dermatochalasis is a nonspecific aging change and consists of stretching or senile elastosis of the upper eyelid skin to produce an exaggerated eyelid fold (Fig. 4.14). Although ptosis does not generally accompany dermatochalasis, Dryden and Leibsohn (1978) recognized aponeurotic defects in many patients. Stretching and weakness in the orbital septum may allow prolapsing orbital fat into the eyelids.

Blepharochalasis, meaning "relaxed eyelid," is fortunately a rare distinct condition produced by recurrent attacks of eyelid edema, typically beginning in adolescence (Fig. 4.15). It tends to be a familial condition with recurrent unpredictable attacks of idiopathic edema of the eyelid which thin, stretch, and wrinkle the eyelid skin, and make the blood vessels of the eyelid more prominent. As the disease process increases in severity, the deeper eyelid layers will be involved, allowing prolapse of preaponeurotic fat and lacrimal gland and ptosis secondary to levator aponeurotic defects. In some cases, the lower eyelids are also affected (Jones and Wobig, 1976).

The etiology of recurrent eyelid edema is unknown. The onset of blepharochalasis is in adolescence. The attacks of eyelid edema last for several days, then subside, only to return again in a matter of weeks to months. The severity of each episode of eyelid edema may increase with each recurrent attack. As one advances in age, the recurrent attacks of eyelid edema subside. Surgical management of blepharochalasis requires excision of redundant skin and prolapsed orbital fat and repair of aponeurotic defects. In addition, prolapsed lacrimal glands may be sutured back into position (Stieglitz and Crawford, 1974; Collin, et al, 1979; Brazin, et al, 1979).

The thinness of the eyelid skin creates problems in grafting in this region. Donor sites for eyelid skin include upper eyelid skin and the retroauricular, supraclavicular, preauricular, and inner arm regions. When possible, the contralateral upper eyelid is recommended as the donor site in upper lid grafts.

The thin epidermis of the eyelid skin may also predispose this region to benign and malignant epithelial tumors. Despite their negligible body surface area, the eyelids account for a significant percentage of epithelial tumors.

The orbicularis muscle is concentrically arranged around the palpebral opening and is divided into three basic portions: orbital, palpebral, and tarsal (Fig. 4.2). These portions are confluent and act in unison to form the protractors of the eyelids. Innervation is by the facial nerve (CN VII).

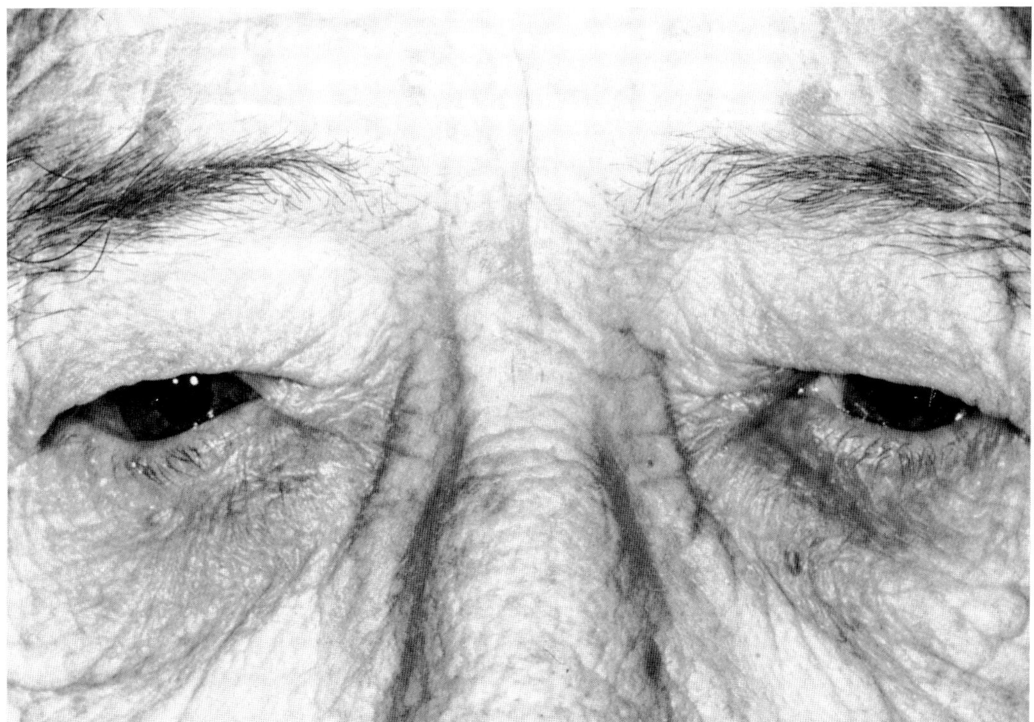

Figure 4.14.   Dermatochalasis. Redundant skin of the upper eyelid producing an exaggerated eyelid fold.

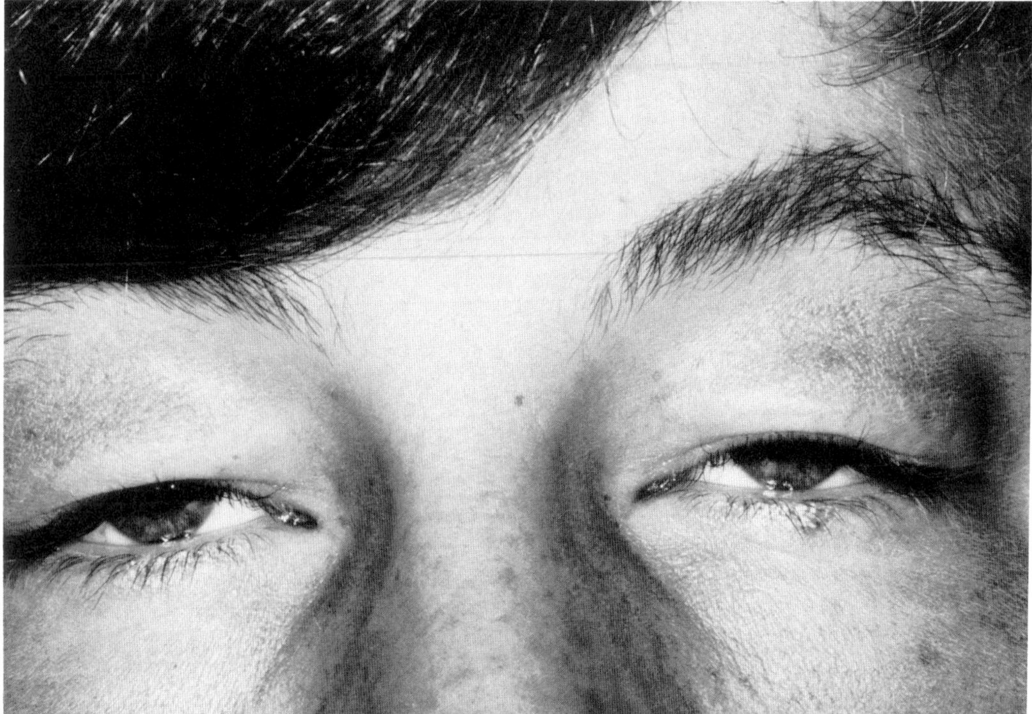

Figure 4.15.   Blepharochalasis. Recurrent eyelid edema stretches all tissues of the eyelid, allowing prolapse of preaponeurotic fat and lacrimal gland, ptosis, and redundant skin of the upper eyelid.

The orbital portion of the orbicularis muscle overlies the orbital rims. The muscle extends into the region of the eyebrow to intermix with the frontalis, procerus, and corrugator supercilia muscles. Laterally, the orbital portion of the orbicularis lies over the anterior portion of the temporalis muscle and extends into the cheek, where it covers the origin of the lip elevator muscles. The bony attachment for the orbital orbicularis fibers are the medial orbital margin, frontal process of the maxilla, and frontal bones. The bony attachment of the orbital orbicularis extends from the supraorbital notch to near the infraorbital foramen. Fibers adhere to the medial canthal tendon, which interrupts the medial bony attachment. Fibers originating from the medial bony attachment superiorly sweep uninterrupted around the lateral orbital rim to attach to the orbital margin below the medial canthal tendon in a horseshoe fashion.

The palpebral portion of the orbicularis muscle is subdivided into the preseptal and pretarsal portions. The palpebral orbicularis is composed of half ellipses, which extend from the medial orbital margin to the lateral palpebral raphe. The lateral palpebral raphe is formed by interlacing orbicularis fibers and does not represent a termination of the tarsal plates (Fig. 4.16).

The preseptal part of the orbicularis muscle overlies the orbital septum. It has similar parts in both upper and lower lids that arise from the medial canthal tendon and from deeper attachments to the lacrimal crest and diaphragm. The upper and lower lid portions join at the lateral raphe overlying the lateral canthal tendon and the lateral orbital rim. A portion of the preseptal orbicularis extends into the

fascia overlying the lacrimal sac. Jones (1960) described this muscle and its importance in the lacrimal pump mechanism. Thus, the deep heads of the preseptal muscles are referred to as Jones muscle.

The pretarsal part of the orbicularis muscle lies anterior to the tarsus. It has two heads: one forms the anterior limb of the medial canthal tendon (MCT), and the other, deeper head, is the tensor tarsi muscle of Horner which forms the posterior limb of the MCT. The anterior head lies anterior to the canaliculus, while the latter passes posteriorly to it and inserts into the lacrimal fascia and the posterior lacrimal crest. The upper and lower pretarsal muscles join to form the lateral canthal tendon (LCT), which inserts onto the lateral orbital tubercle (Fig. 4.17). This tendon is not as well-defined as the MCT. A small strip of orbicularis located at the eyelid margin is called the muscle of Riolan, which was once thought to be a major factor in "spastic entropion."

The medial canthal tendon has two heads: one deep and one superficial (Fig. 4.18). The deep portion is the thinnest and runs behind the lacrimal fossa to attach to the posterior lacrimal crest. The superficial portion is much thicker and lies anterior to the lacrimal fossa. It is formed by fusion of a superior crus from the upper eyelid orbicularis and inferior crus from the lower eyelid orbicularis. This common tendon adheres to the frontal process of the maxillary bone. Directly beneath the anterior portion of the medial canthal tendon is the lacrimal canaliculus. Thus, injuries to the medial canthal tendon may involve the canaliculus.

The medial canthal tendon also has a superior supporting branch which unites the tendon to the frontal bone (Ander-

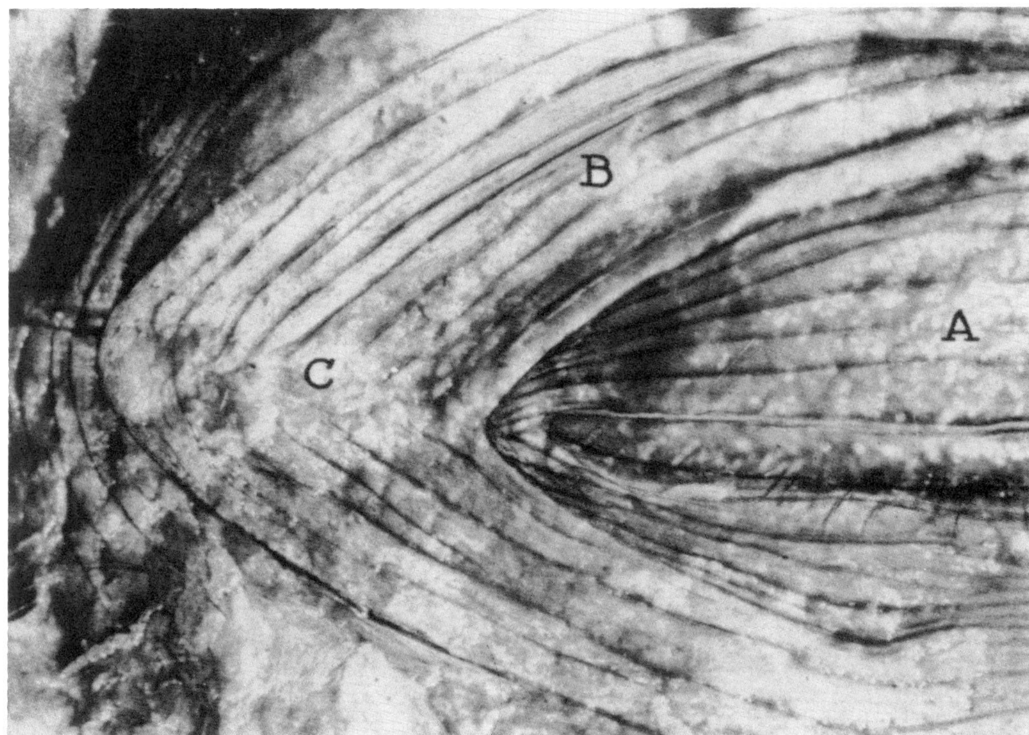

Figure 4.16. Lateral palpebral raphe. (*A*), pretarsal orbicularis; (*B*), preseptal orbicularis; (*C*), lateral palpebral raphe. (Reproduced from Jones LT, Wobig JL: *Surgery of the Eyelids and Lacrimal System.* 1976, Birmingham, Aesculapius Publishing Co.)

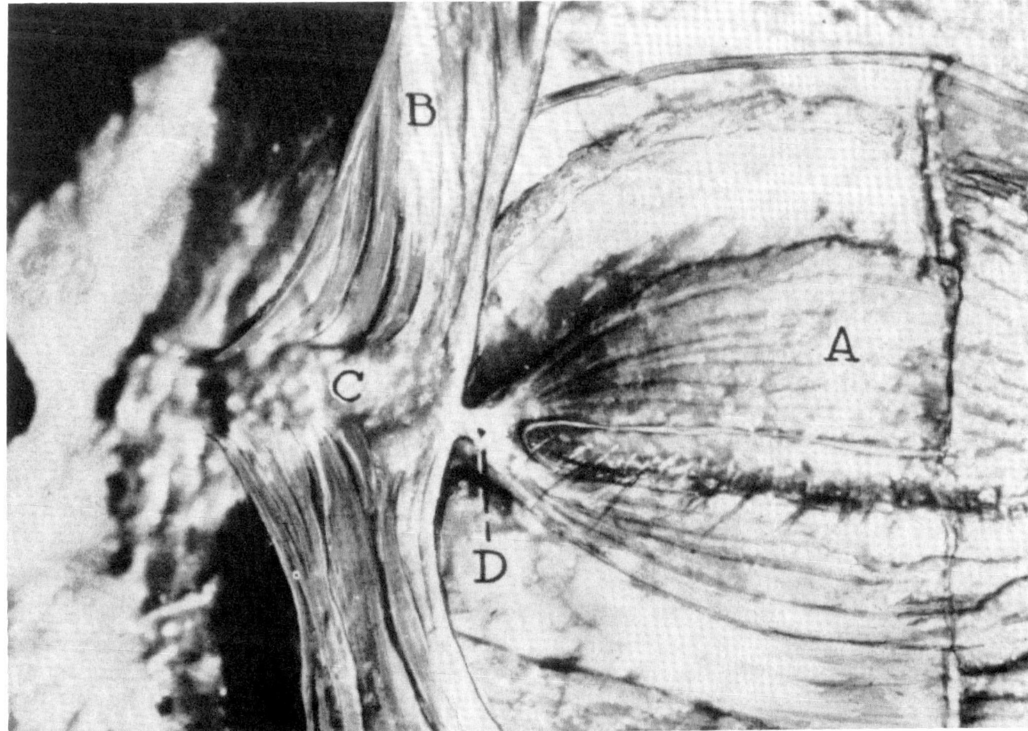

Figure 4.17. Lateral canthal tendon. The tendon is produced by the fusion of the pretarsal orbicularis muscle from the upper and lower eyelids. (*A*), pretarsal orbicularis; (*B*), reflected preseptal orbicularis; (*C*), lateral orbital tubercle; and, (*D*), lateral canthal tendon. (Reproduced from Jones LT, Wobig JL: *Surgery of the Eyelids and Lacrimal System.* 1976, Birmingham, Aesculapius Publishing Co.)

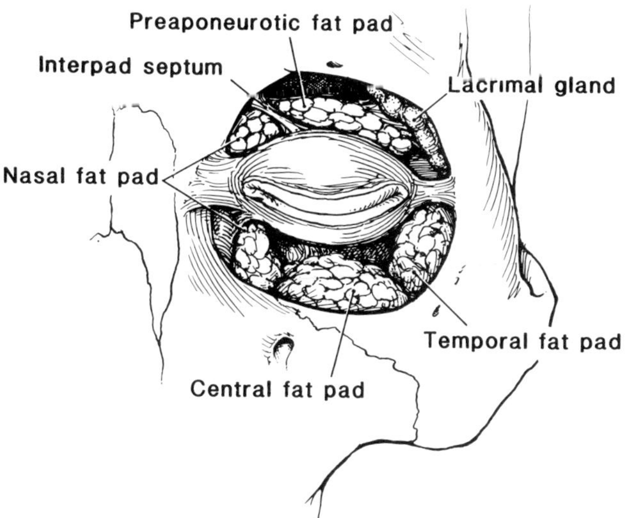

Figure 4.20. Orbital fat pockets.

bral fascia and the orbital septum, usually at a point between the lateral ⅓ and central ⅔ of the lid. The nasal fat pad, which is the larger, is further divided by the inferior oblique muscle and its fascial attachments into two compartments.

CLINICAL NOTE

The orbital septum is thinnest in the medial aspect of the upper and lower eyelid. In addition, the septum is perforated by nerves and vessels, especially in the medial aspect of the upper eyelid. The inherent weakness of the septum in these areas presumably contributes to the fact that orbital fat is more likely to herniate or distend the orbital septum in this region. This presents clinically as fat pouches which have been termed steatoblepharon.

There has been some debate concerning the number of fat pockets in the eyelids. Classically, there are two fat pockets of the upper eyelid (central and medial), while three fat pockets (medial, central, and lateral) are described in the lower eyelid (Castanares, 1951). Dye injection failed to reveal distinct fat pockets and anatomical barriers in cadaver studies (Hugo and Stone, 1974). However, Barker (1977) did show compartmentalization of dye in patients undergoing blepharoplasty. Removal of orbital fat is an important consideration in patients undergoing cosmetic blepharoplasty and it is useful to consider the fat pockets as different compartments which must be individually considered preoperatively and operatively.

The levator palpebrae superioris is the main upper eyelid retractor. This muscle originates from the annulus of Zinn at the orbital apex and proceeds anteriorly in the superior aspect of the orbit (Fig. 4.21 A and B). The levator muscle proceeds anteriorly adjacent to the periorbita of the orbital plate of the frontal bone. Approximately 15–20 mm above the superior border of the tarsus, horizontally oriented fibers serve as a suspensory ligament for the upper eyelid (Anderson, 1979) (Fig. 4.22). This ligament has been called the superior transverse ligament or Whitnall's ligament. At

Whitnall's ligament, the levator muscle is reoriented from an anterior-posterior to a superior-inferior direction. The levator muscle will gradually become aponeurotic at the superior transverse ligament, producing a white, shining aponeurotic sheet 10–12 mm above the border of the tarsus (Fig. 4.23).

At this level, 10–12 mm above the superior border of the tarsus, the retractor of the upper eyelid has developed an anterior and posterior lamella. The anterior lamella is the levator aponeurosis and the posterior lamella is the sympathetic muscle of Müller. The aponeurosis has diffuse terminal attachments, while Müller's muscle has a tendinous insertion on the superior border of the tarsus.

The insertion of the upper eyelid retractors is an important consideration. The superior tarsal muscle is densely adherent to the underlying conjunctiva and has a loose connective tissue attachment to the overlying levator aponeurosis. This accounts for the plane of separation appreciated while performing aponeurotic ptosis or lid retraction surgery (Anderson and Dixon, 1979; Harvey and Anderson, 1981). The levator aponeurosis inserts widely (Figs. 4.24 and 4.25). The aponeurosis inserts onto the anterior aspect of the tarsus, the fibrous septa of the orbicularis, and to the subcutaneous tissues at the eyelid crease and below (Anderson and Beard, 1977; Collin, et al, 1978).

The orbital septum fuses with the levator aponeurosis prior to reaching the superior border of the tarsus (Figs. 4.1 and 4.13). At the inferior edge of this fusion, the aponeurosis divides into a fan-like arrangement of fibers. The first anterior fibers can be traced through the orbicularis muscle to insert in the subcutaneous tissues at the level of the eyelid crease. The upper eyelid fold is produced by loose, redundant skin draping over the eyelid crease. The remaining anterior fibers of the levator aponeurosis further branch to form the septa around the pretarsal orbicularis and pass through the pretarsal orbicularis to terminate as delicate branching fibers into the subcutaneous tissues below the eyelid crease (Fig. 4.25). The posterior portion of the aponeurosis firmly attaches to the lower 7–8 mm tarsus. The aponeurosis has very loose attachments to the superior 2–3 mm of tarsus and is most firmly adherent to the tarsus about 3 mm above the eyelid margin (Anderson and Beard, 1977).

The levator aponeurosis fans both medially and laterally, creating its "horns." The lateral horn of the aponeurosis is a well-defined structure which passes through the lacrimal gland between the orbital and palpebral lobes. The lateral horn then attaches to the periosteum of the lateral orbital tubercle (Whitnall's tubercle), contributing to lateral canthal tendon and lateral retinaculum. The medial horn of the levator aponeurosis is not as clearly defined. It passes inferiorly to contribute to the posterior portion of the medial canthal tendon.

The superior transverse ligament of Whitnall consists of a condensation of fascial sheaths around the levator muscle (Fig. 4.26). It inserts medially at the trochlear region. The ligament extends laterally through the lacrimal gland to insert on the superior portion of the lateral orbital wall. It also sends a small branch to the superior portion of the lateral retinaculum.

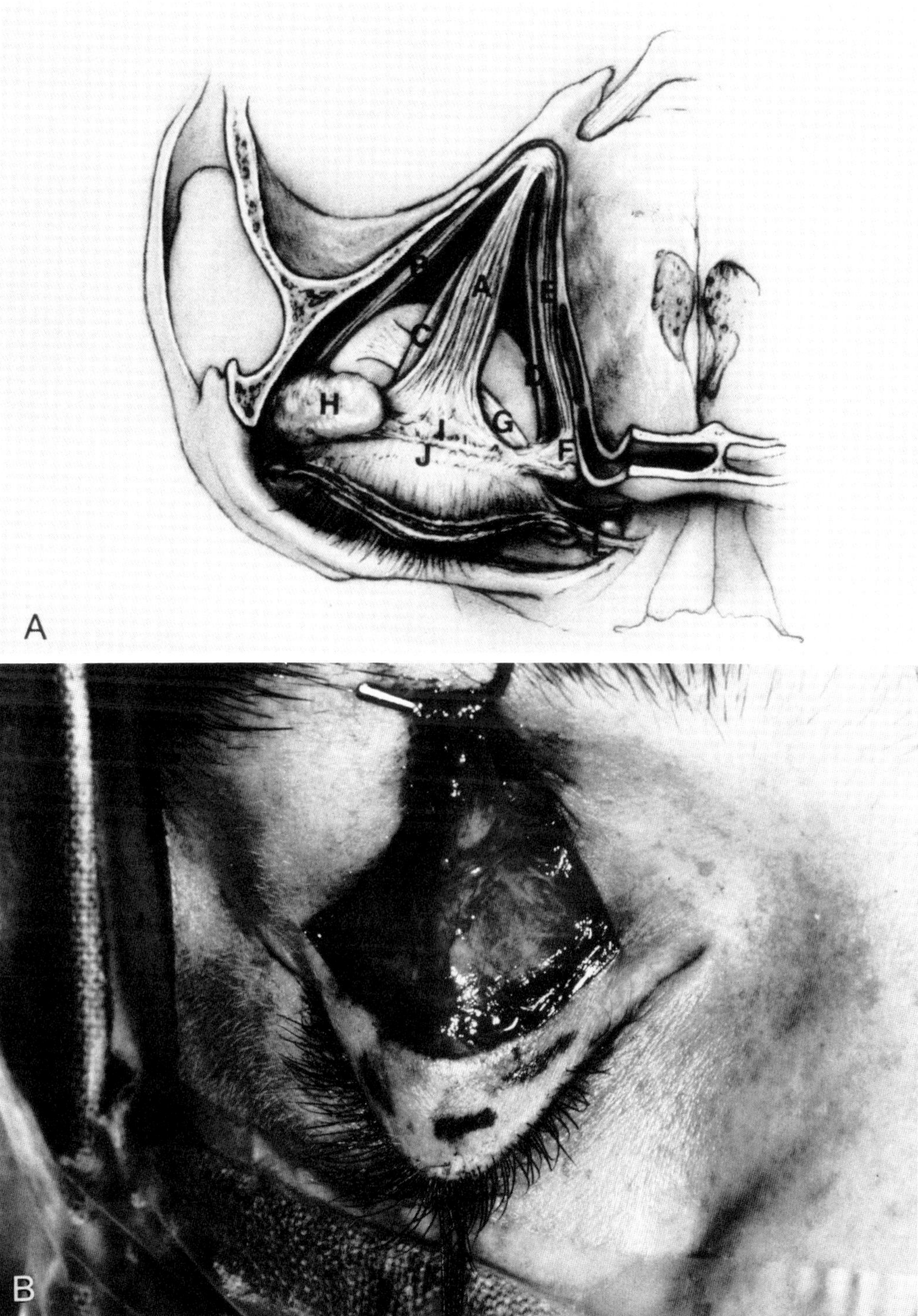

Figure 4.21. Schematic of the superior view of the orbit. (*A*): *A*, Levator muscle; *B*, lateral rectus muscle; *C*, superior rectus muscle; *D*, medial rectus muscle; *E*, superior oblique muscle; *F*, trochlea; *G*, superior oblique tendon; *H*, lacrimal gland; *I*, Whitnall's ligament; *J*, levator aponeurosis; *K*, lateral canthal tendon; *L*, medial canthal tendon. (*B*): Clinical photo demonstrating levator muscle posterior (on hook) to Whitnall's ligament. ((*A*) reproduced from Beard C: *Ptosis* (ed 3). 1981, St. Louis, C.V. Mosby Co.)

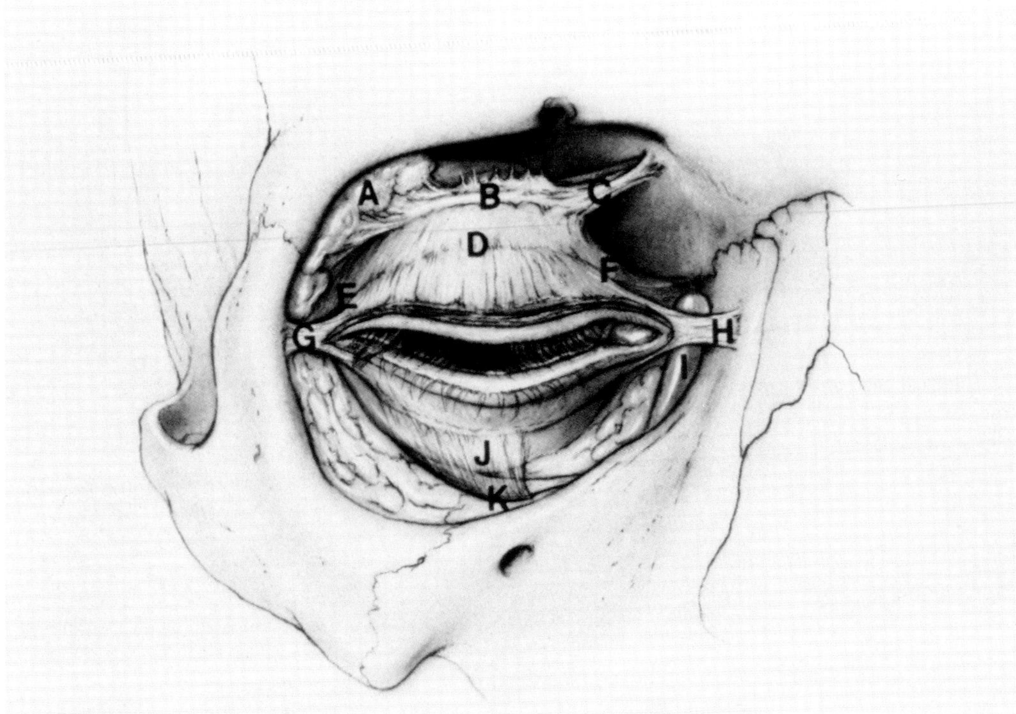

Figure 4.22.   Whitnall's ligament and anterior orbital structures: *A*, lacrimal gland; *B*, Whitnall's ligament; *C*, superior oblique tendon sheath; *D*, levator aponeurosis; *E*, lateral horn of the levator aponeurosis; *F*, medial horn of the levator aponeurosis; *G*, lateral canthal tendon; *H*, medial canthal tendon; *I*, lacrimal sac; *J*, lower eyelid retractors; and *K*, inferior oblique muscle. (Reproduced from Beard C: *Ptosis* (ed 3). 1981, St. Louis, C.V. Mosby Co.)

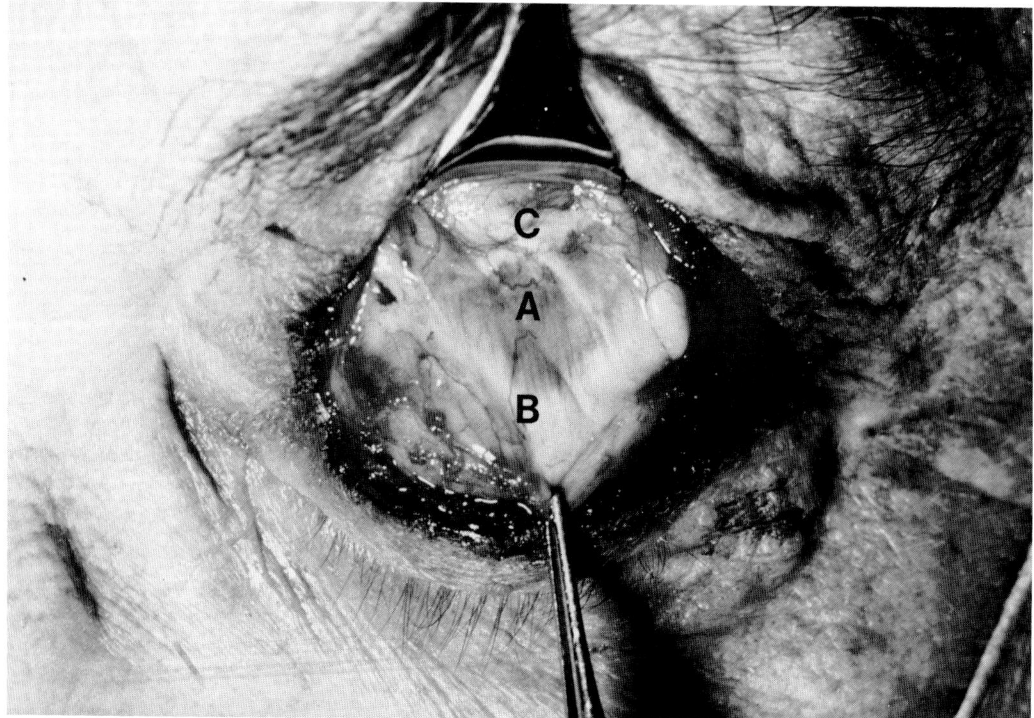

Figure 4.23.   Levator aponeurosis. The levator is muscular *A* at the level of Whitnall's ligament *C*, and gradually transforms to a white aponeurotic structure *B* prior to inserting in the anterior surface of the tarsus.

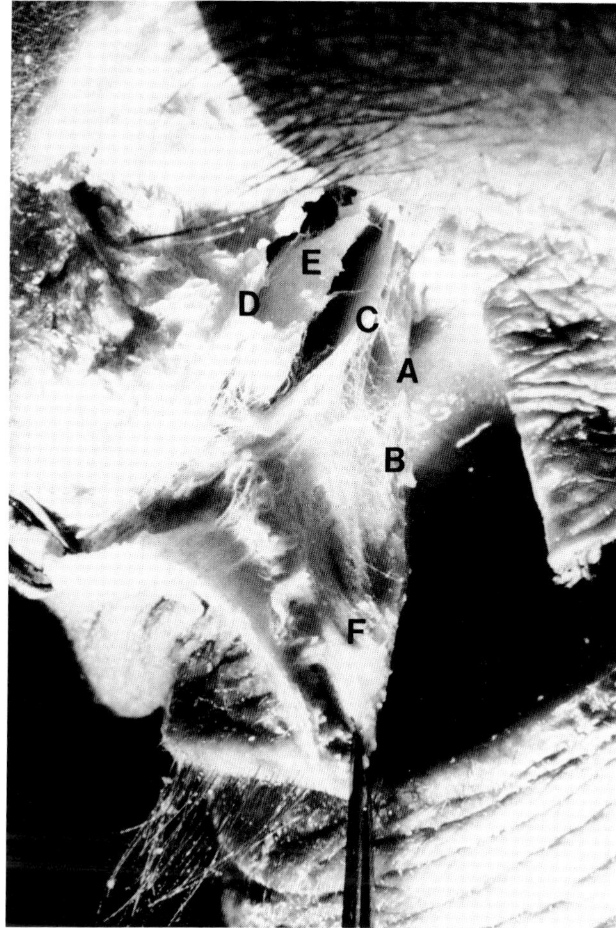

Figure 4.24. Sagittal section of the eyelid demonstrating the diffuse attachments of the levator aponeurosis. *A*, conjunctiva; *B*, Müller's muscle; *C*, levator aponeurosis; *D*, orbital septum; *E*, preaponeurotic fat; *F*, firm attachment of the levator aponeurosis onto the anterior surface of the lower tarsal plate.

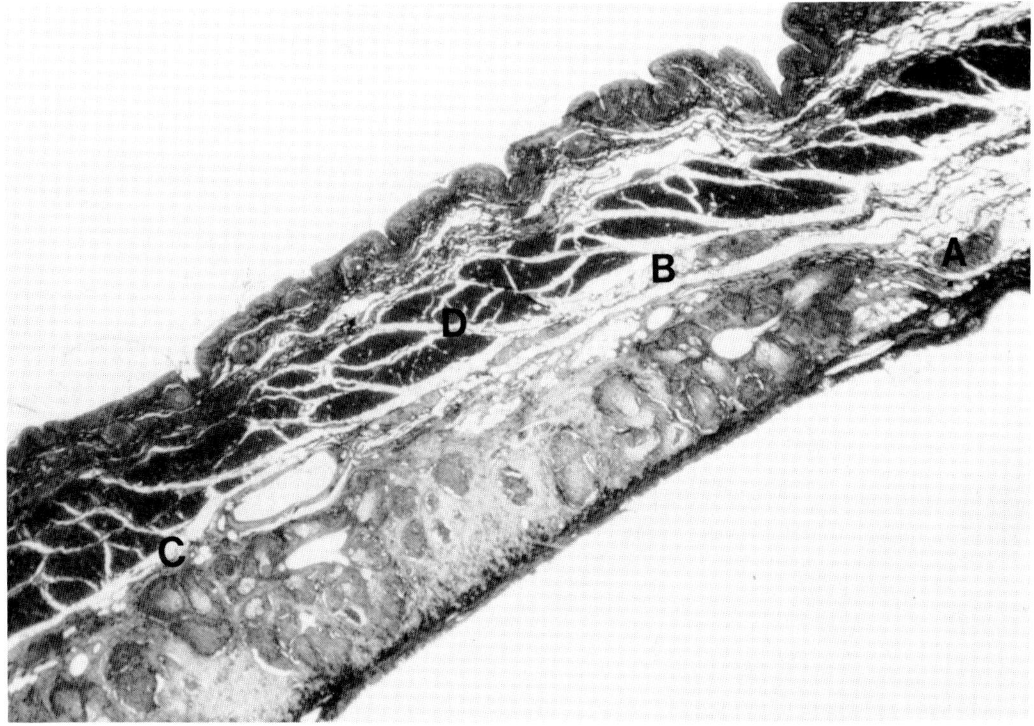

Figure 4.25. The tarsal portion of the eyelid. Müller's muscle inserts at the superior border of the tarsus (*A*). The levator aponeurosis displays loose attachments to the superior aspect of the tarsus (*B*). The strongest attachments of the levator aponeurosis are present 2–3 mm above the eyelid margin (*C*). Superficial extensions of the levator aponeurosis extend through the orbicularis muscle (*D*) and insert on the subcutaneous tissues of the skin (magnification ×4).

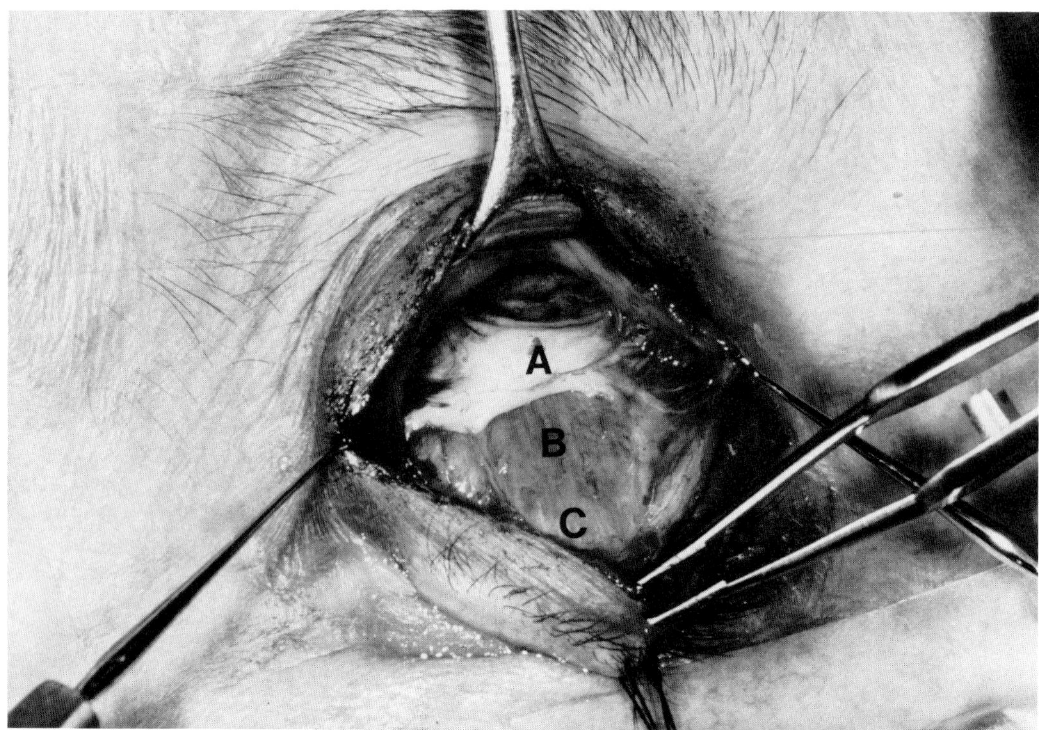

Figure 4.26. Whitnall's ligament. Abnormally prominent Whitnall's ligament. *A*, Whitnall's ligament; *B*, levator aponeurosis with muscle fibers; *C*, levator aponeurosis. (Reproduced from Anderson RL, Dixon RS: *Archives of Ophthalmology* 97:705–707. 1979, American Medical Association.)

While Whitnall correctly described the anatomic position of this structure, he misrepresented its function to be a check ligament of the levator. This was unfortunate, for this misrepresentation of function has resulted in inappropriate cutting of this structure at ptosis surgery.

Functionally, Whitnall's ligament acts as a pulley or fulcrum for the upper eyelid, changing the contraction force of the levator muscle from an anterior-posterior to a superior-inferior direction (Fig. 4.27 *A* and *B*) (Anderson and Dixon, 1979). Whitnall's ligament is also, unfortunately, mistakenly identified as the levator horns and is cut at ptosis surgery by a maneuver referred to as "cutting the levator horns" in many descriptions of standard levator resections. If this occurs much more levator resection is required because of the loss of the pulley effect creating the inferior-superior force vector and effective lengthening of the levator complex by its loss of superior support. Cutting Whitnall's ligament may also result in temporal fullness and droop of the upper eyelid due to prolapse of the lacrimal gland, as it provides support to the lacrimal gland as well as the superior orbit. Whitnall's ligament is analogous to Lockwood's ligament in the lower lid and each are supportive structures of the orbit and should be preserved.

CLINICAL NOTE

The appreciation of the levator muscle and its aponeurosis has revolutionized ptosis surgery. Jones and co-workers (1975) first reported aponeurotic defects in acquired ptosis and introduced a procedure to iden-tify and correct these defects. Since 1975, numerous reports have emphasized the beneficial aspects of aponeurotic surgery (Older, 1978; Collin, 1979; Anderson and Dixon 1979; Dortzbach and Sutula, 1980).

Aponeurotic defects have been identified in the majority of patients with acquired ptosis with good levator function. While much less common, aponeurotic defects may also occur in congenital ptosis cases with good function (Anderson and Gordy, 1979; Doxanas and Dryden, 1981). Aponeurotic ptosis repairs identify and eliminate the anatomical defect while correcting the ptosis.

A thorough appreciation of eyelid anatomy is essential prior to performing aponeurotic ptosis surgery. Occasional ptosis surgeons have expressed difficulty in identifying the levator aponeurosis while performing aponeurotic ptosis procedures. Identification of the levator and its aponeurosis is simplified by a stepwise dissection. The skin and orbicularis muscle are incised. Dissection is carried superiorly to identify the orbital septum. The orbital septum is incised across its entire horizontal dimension. The preaponeurotic or fat pad is then reflected to consistently expose the levator aponeurosis.

Anderson and Beard (1977) have described three types of aponeurotic defects: rarification, dehiscence, and disinsertion. The aponeurosis may be rarified, where it is thinned but no defects are present. A dehiscence is a circumscribed defect of the aponeurosis with the surrounding aponeurosis intact. A true disinsertion is where a distinct, white, sometimes rolled edge of the aponeurosis has separated from the tarsal

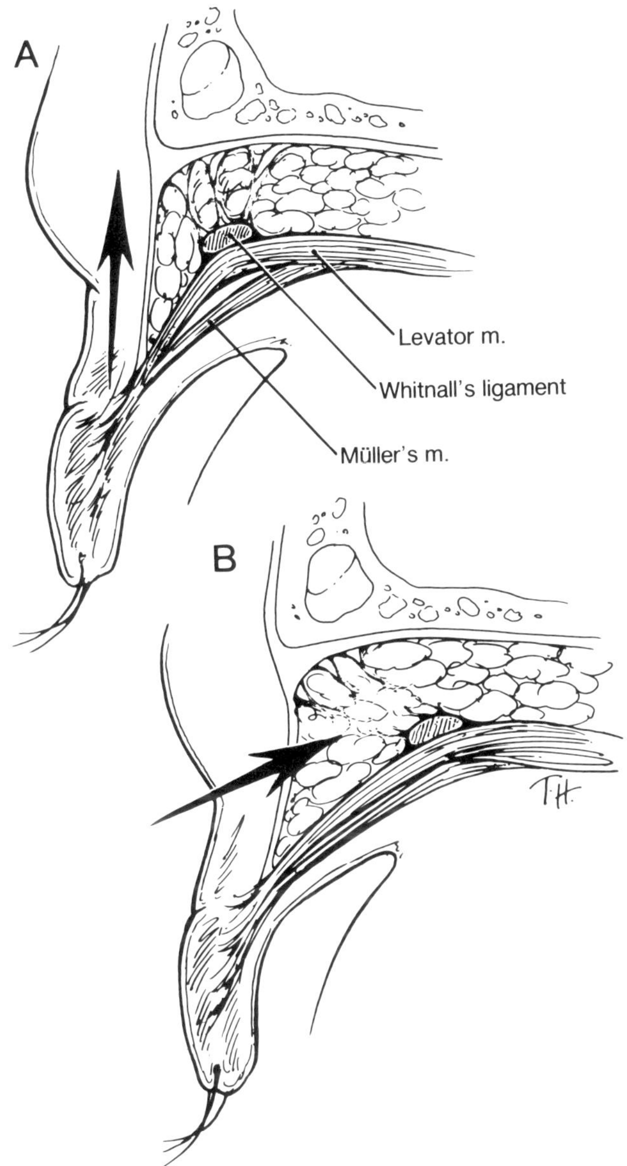

Figure 4.27. Schematic demonstrating the function of Whitnall's ligament. Whitnall's ligament acts as a pulley, changing the direction of the levator action to a vertical direction (*A*). If Whitnall's ligament is cut (*B*), it no longer acts as a pulley and no longer supports the levator muscle/aponeurotic complex with the resultant contractile force directed more posteriorly.

plate and is found above the superior tarsal border (Fig. 4.28 *A* and *B* and Fig. 4.29). Using this classification we find rarifactions and disinsertions to be the most common aponeurotic defects.

The superior tarsal muscle of Müller originates from the under surface of the levator at about the muscle-aponeurotic junction. As the striated muscles disappear, nonstriated or smooth muscle fibers are noted at the posterior aspect of the levator. So abrupt is the origin of smooth muscle that the smooth and striated muscle seems to intermingle for a

few millimeters (Kuwabara, et al, 1975). At approximately 10 mm above the superior tarsus, where the transition of the levator to its aponeurosis is clinically complete, a well-defined superior tarsal muscle is present (Figs. 4.29 and 4.30).

In the upper eyelid, Müller's muscle (superior tarsal muscle) inserts in the superior border of the tarsus (Fig. 4.25). In the lower eyelid the sympathetic muscle (inferior tarsal muscle) is also thought to insert on the inferior border of the tarsus, along with the capsulopalpebral fascia. However, recent histologic studies by Hawes and Dortzbach (1982) have failed to identify the inferior tarsal muscle inserting onto the tarsus. The first identifiable smooth muscle strands of the inferior tarsal muscle averaged 2–5 mm from the inferior tarsal border.

The superior and inferior tarsal muscles are innervated by sympathetic nerve fibers; however, their course is poorly understood. Collin and co-workers (1979) postulated four possible courses of sympathetic fibers to Müller's muscle: (1) via the levator muscle; (2) from the perivascular plexus around the marginal artery at the upper border of the tarsus; (3) from the perivascular plexus around other arterioles supplying the muscle; and (4) via the sensory nerves.

CLINICAL NOTE

Interruption of the sympathetic nervous system from its origin in the hypothalamus to its termination in the eye will produce the characteristic features of Horner's syndrome. The classic triad of Horner's syndrome is ptosis, miosis, and apparent enophthalmos (Fig. 4.31). The apparent enophthalmos is secondary to the mild ptosis of the upper eyelid and elevation of the lower eyelid from the loss of sympathetic tone to the respective superior and inferior tarsal muscles. Anisocoria, produced by miosis of the affected pupil, is most pronounced in dim illumination. With the complete syndrome, the involved side of the face is also anhidrotic, warm, and hyperemic due to denervation of sweat and vasoconstrictor fibers. In a congenital Horner's syndrome hypochromia of the involved iris is present.

Eyelid retraction is a common manifestation of Graves' disease, accounting for significant functional and cosmetic problems (see Fig. 7.11). Sympathetic muscle contraction due to adrenergic stimulation and later fibrosis appears to be a contributing factor to upper and lower eyelid retraction. In the upper eyelid additional factors producing eyelid retraction are infiltration of the levator muscle by glycoprotein and mucopolysaccharide and overaction of the levator palpebrae superioris-superior rectus complex in response to a fibrotic inferior rectus muscle. In the lower eyelid, fibrosis of the inferior rectus muscle may directly produce lower eyelid retraction via the capsulopalpebral head (aponeurosis) of the lower lid.

Surgical management of eyelid retraction has been directed toward recession of the eyelid retractors with or without the placement of spacers (Callahan, 1965; Quickert and Dryden, 1971; Flanagan, 1974; Dryden and Soll, 1977; Doxanas and Dryden, 1981). Chalfin and Putterman (1979), and Harvey and Anderson (1981) recommend recession of the aponeurosis with

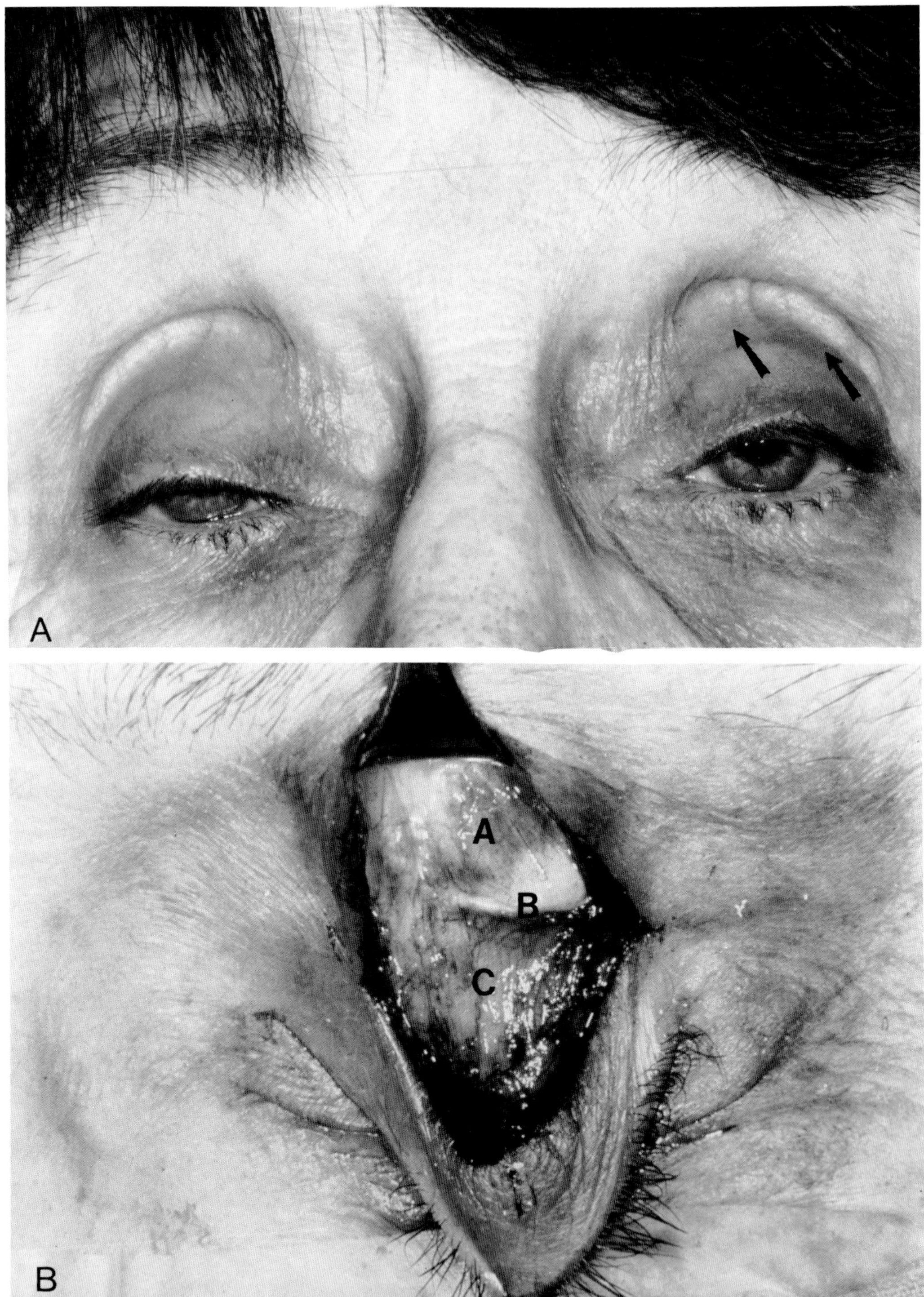

Figure 4.28. Clinical and anatomical features of ptosis. (*A*), Clinical appearance of bilateral ptosis secondary to levator aponeurotic disinsertion. There is a loss of the eyelid crease, sunken superior sulcus, and a ridge in the midaspect of the upper eyelid (*arrow*). At surgery (*B*) a levator aponeurotic disinsertion was present and the clinically apparent ridge in the eyelid was secondary to the rolled edge of the disinserted levator aponeurosis. *A*, Levator aponeurosis; *B*, rolled dehisced edge of the levator aponeurosis; and *C*, Müller's muscle. (Reproduced from Anderson RL, Dixon RS: *Archives of Ophthalmology* 97:1129–1131. 1979, American Medical Association.)

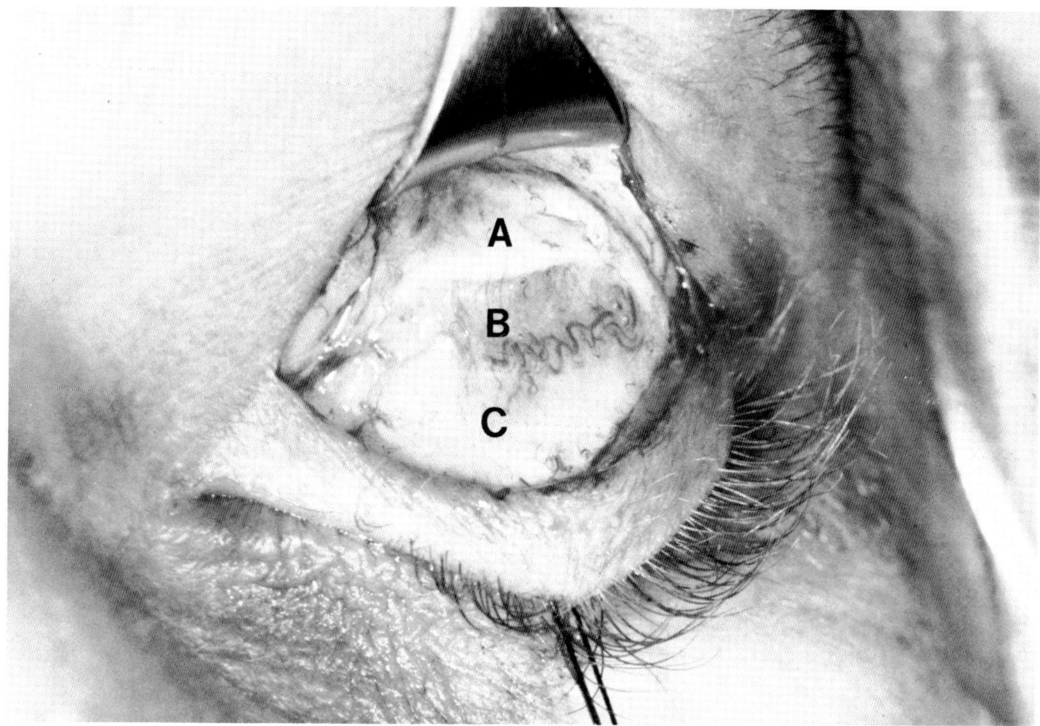

Figure 4.29. Müller's muscle. Levator aponeurotic defect exposes Müller's muscle. *A*, Levator aponeurosis; *B*, Müller's muscle; and *C*, tarsus.

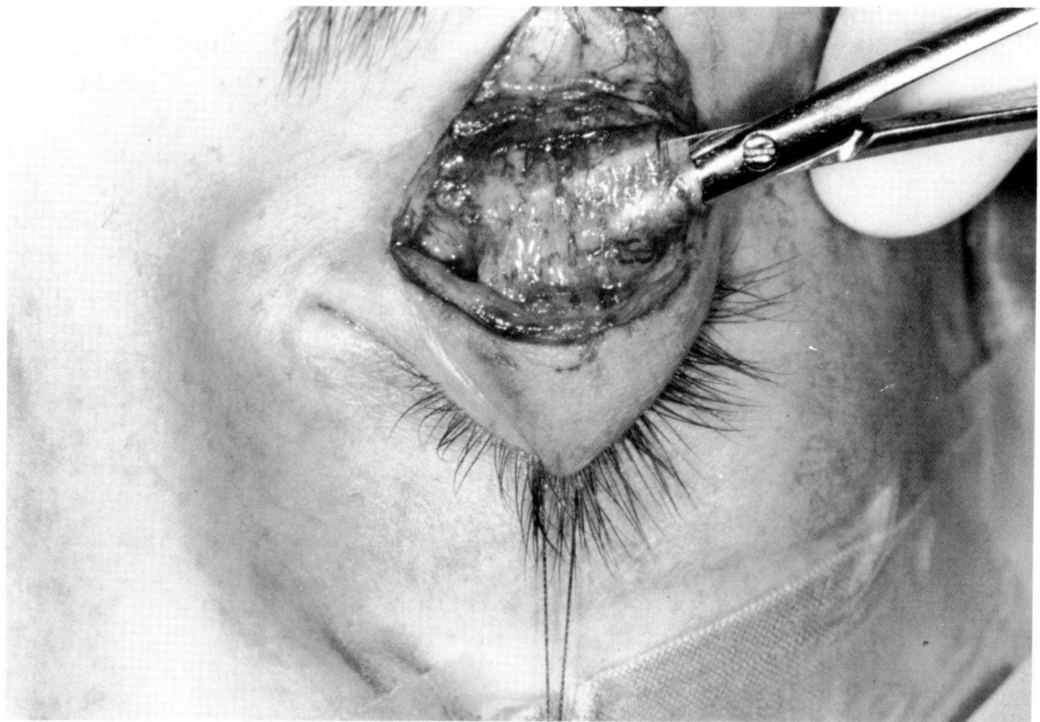

Figure 4.30. Müller's muscle. Dissection of Müller's muscle from conjunctiva demonstrates its thin appearance. (Reproduced from Harvey JT, Anderson RL: *Ophthalmology* 88:513, 1981.)

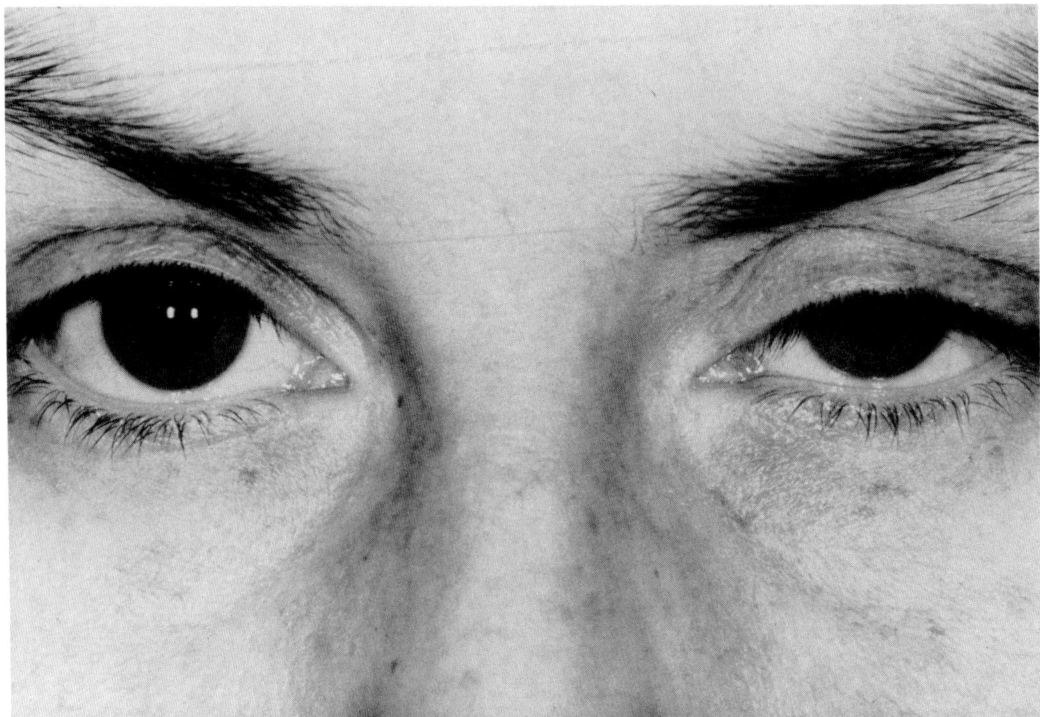

Figure 4.31.   Horner's syndrome.

excision of the sympathetic muscles to effectively treat the retraction without the use of spacers. Removal of the sympathetic muscles in the latter procedures eliminates one of the underlying pathophysiologic mechanisms responsible for the marked variability in cases of eyelid retraction.

LOWER EYELID

The lower eyelid retractor, analogous to the levator muscle and its aponeurosis, is the capsulopalpebral fascia or lower eyelid aponeurosis (Fig. 4.32 *A* and *B*). This dense, fibrous sheath is a direct extension of the inferior rectus muscle and contributes to Lockwood's ligament, which is analogous to Whitnall's ligament. After originating from the inferior rectus muscle, the capsulopalpebral fascia splits into two sections to envelop the inferior oblique muscle (Fig. 4.33). The layer superior to the inferior oblique is thicker than the layer beneath the inferior oblique muscle. Anterior to the inferior oblique muscle the two portions of the capsulopalpebral head reunite to help form Lockwood's ligament (Lockwood, 1886; Jones, 1968). Anterior to Lockwood's ligament, the lower lid retractor is termed the capsulopalpebral fascia or aponeurosis. The bulk of the capsulopalpebral fascia inserts near the inferior border of the lower lid tarsus. However, some strands extend through the orbital fat to the orbital septum and help divide the orbital fat into pockets. In addition, some fibers extend from the inferior fornix to contribute to Tenon's capsule.

There has been some discrepancy or points of disagreement in several anatomic concepts of the lower eyelid. There is general agreement in the literature that the capsulopalpebral fascia inserts at the lower tarsal border (Whitnall,

1921; Jones, 1960; Hargiss, 1973; Dryden, et al, 1978). However, some disagreement exists as to the insertion of the inferior tarsal muscle. Most previously published reports identify the inferior tarsal muscle inserting to the inferior tarsal border (Jones, 1960; Hargiss, 1974; Dryden, et al, 1978; Beard and Quickert, 1977; Warwick, 1976). However, histologic evaluation of the lower eyelid (Whitnall, 1932) recognized that the inferior tarsal muscle did not insert on the inferior tarsal border. This observation was recently confirmed by Hawes and Dortzbach (1982), who identified the bulk of smooth muscle to be near the inferior fornix. In younger eyelids the inferior tarsal muscle was closer to the inferior border of the tarsus than in older eyelids.

In addition, confusion has existed regarding the termination of the orbital septum of the lower eyelid. Some authors (Jones, 1974; Hargiss, 1974; Beard and Quickert, 1977) described the orbital septum inserting on the inferior tarsal border. Hawes and Dortzbach (1982) confirmed previous observations (Dryden, et al, 1977; Putterman, 1978) that the orbital septum fused with the capsulopalpebral fascia several mm beneath the inferior tarsal border and did not terminate at the inferior tarsal border.

CLINICAL NOTE

Appreciation of the lower eyelid anatomy has enabled a better understanding of involutional entropion and ectropion. These disorders of lower eyelid position may share some similar pathophysiologic mechanisms. In both conditions there may be horizontal laxity of the lower eyelid and a weakness or frank defect in the lower eyelid aponeurosis (capsulopalpebral fascia) which results in an instability of the eyelid margin

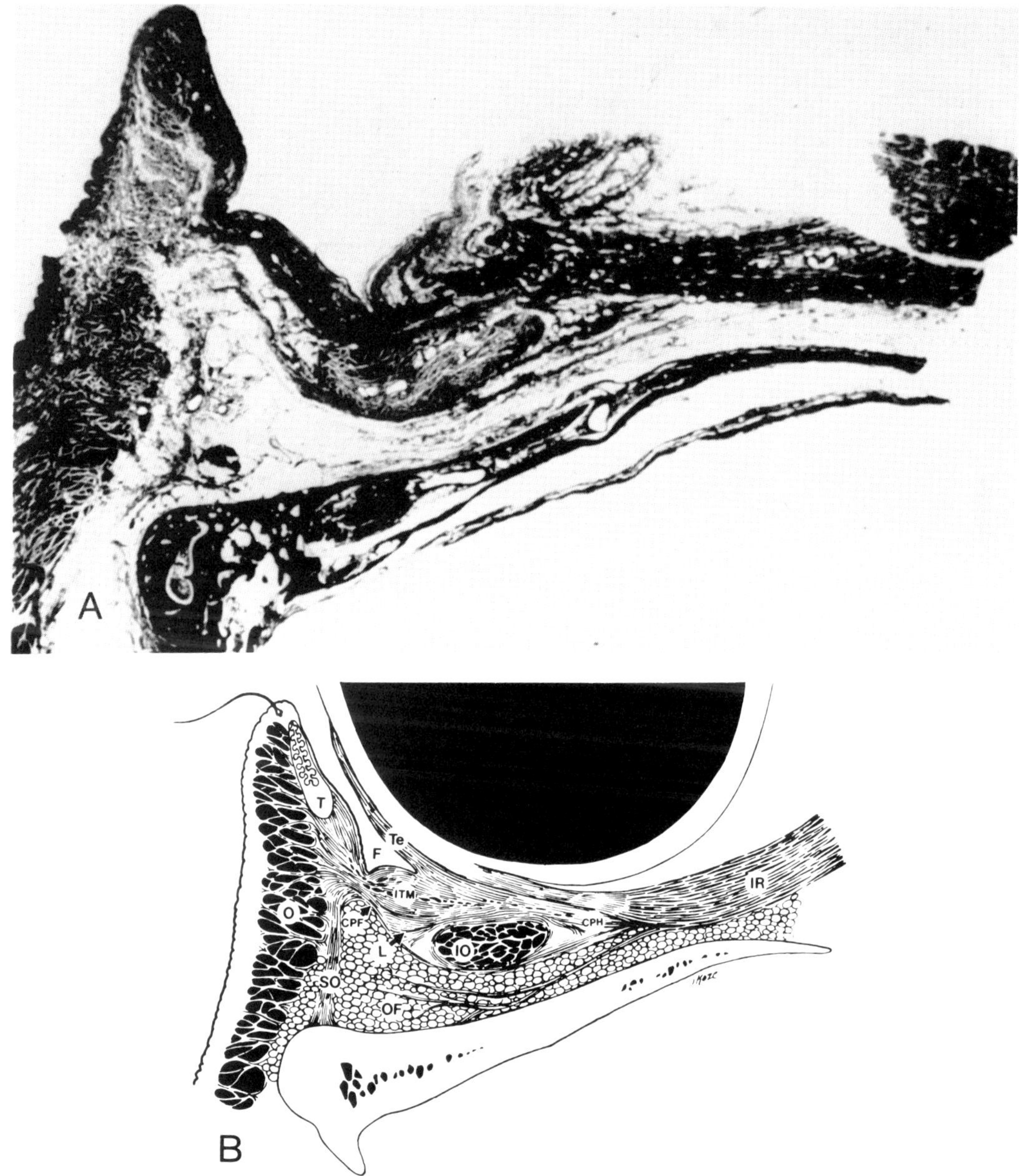

Figure 4.32. Lower eyelid retractors. Histologic (*A*) and schematic (*B*) representation of the lower eyelid retractors. (Abbreviations: *T*, Tarsus; *F*, fornix; *Te*, Tenon's capsule; *ITM*, inferior tarsal muscle; *CPDH*, capsulopalpebral head; *IR*, inferior rectus; *O*, orbicularis oculi muscle; *CPF*, capsulopalpebral fascia; *L*, Lockwood's ligament; *IO*, inferior oblique muscle; *SO*, orbital septum; and *OF*, orbital fat. (Reproduced from Hawes MJ, Dortzbach RK: *Archives of Ophthalmology* 100:1313–1318. 1982, American Medical Association.)

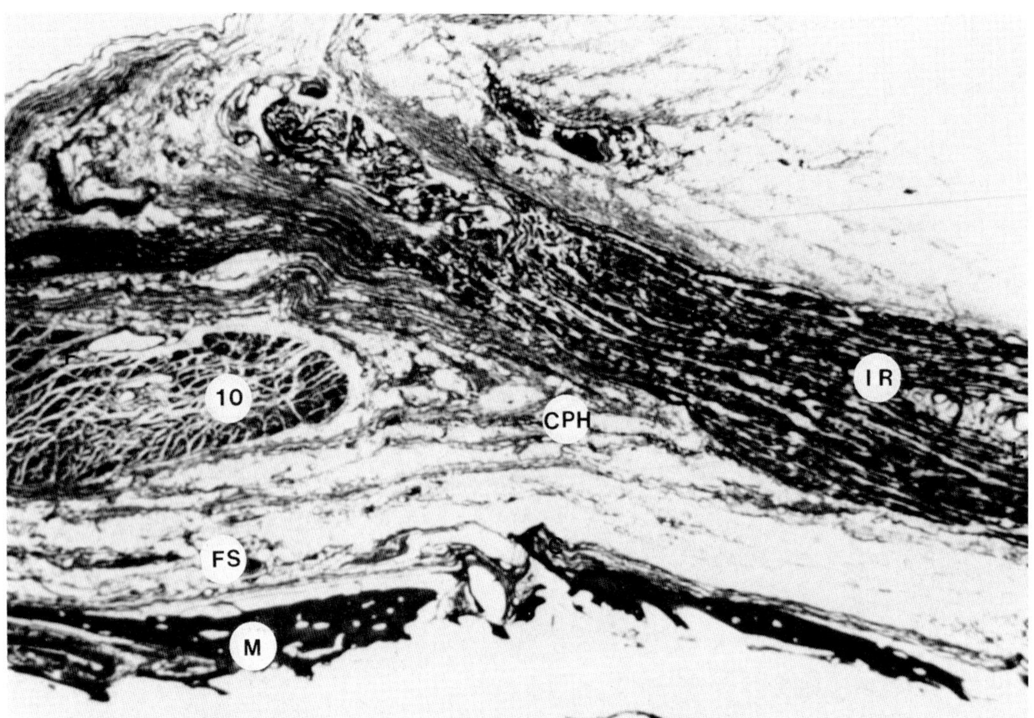

Figure 4.33.   Origin of the capsulopalpebral head (*CPH*) from the inferior rectus muscle (*IR*). The fascial extension of the inferior rectus incapsulates the inferior oblique muscle (*IO*) and has strands extending to the orbital floor (*FS*). (Reproduced from Hawes MJ, Dortzbach RK: *Archives of Ophthalmology* 100:1313–1318. 1982, American Medical Association.)

(Fig. 4.34). The eyelid laxity is probably a result of stretching of the lateral and medial canthal tendons as opposed to actual tarsal lengthening. Differential forces between the anterior (skin and orbicularis muscle) and the posterior lamella (tarsus and lid retractors) help determine if ectropion or entropion will result.

Ousterhout and Weil (1982) measured the horizontal length of the tarsus in advancing age groups and found no elongation of the lower eyelid in the various age groups. The medial and lateral canthal tendons are being recognized as the pathophysiologic mechanisms of eyelid laxity. Surgical procedures have been reported which effectively strengthen the lateral canthus as opposed to performing a block resection of the lower eyelid (Tenzel, et al, 1977; Anderson and Gordy, 1979; Schaefer, 1980). A block resection exaggerates the laxity of the medial and lateral canthal tendon and may also produce a horizontally narrowed palpebral fissure. Procedures which tighten and strengthen the medial canthal tendon and correct medial ectropion have also been shown to be efficacious (Stasior, 1976; Anderson, et al, 1979; Collin and Rathbun, 1978).

Surgical procedures designed to reapproximate the clinically apparent lower eyelid retractor defects are reported to be effective in the management of entropion and even selected cases of ectropion (Wesley, 1982; Putterman, 1978; Dryden, et al, 1978; Wesley and Collins, 1983).

Lockwood's suspensory ligament is composed of the fascia of several structures. It has contributions from the lower eyelid aponeurosis, Tenon's capsule, intermuscular fibrous septae, and the fascia surrounding the inferior oblique and inferior rectus muscles. The inferior oblique muscle crosses the medial portion of the orbit, directed posterior-laterally, through the lower eyelid retractors. The lower eyelid aponeurosis and the inferior oblique fascia are the major contributions to Lockwood's ligament. This structure is difficult to isolate from its component structures, but is anatomically and functionally important in that it helps support the orbital structures, particularly in the absence of the orbital floor.

Other structures that are difficult to isolate from their component parts are the medial and lateral orbital retinaculum. The lateral retinaculum consists of the lateral canthal tendon, lateral horn of the levator aponeurosis, retractor aponeurosis, Lockwood's ligament, the check ligaments of the lateral rectus and a small slip of Whitnall's ligament (although this primarily inserts higher at the superior lateral orbital wall). The medial retinaculum consists of corresponding structures inserting at the level of the medial canthal tendon, except Whitnall's ligament, which has no branch to the medial retinaculum.

The tarsal plate forms the basic support structure of the eyelids. It extends from the lacrimal punctum to the lateral canthus with a tendinous attachment at the medial and lateral canthal tendons. The height of the tarsus of the upper eyelid is approximately 10 mm while the lower tarsus is considerably narrower. Wesley and co-workers measured the vertical height of the lower eyelid tarsus and found it to average 3.7 mm. The tarsus is composed of highly organized

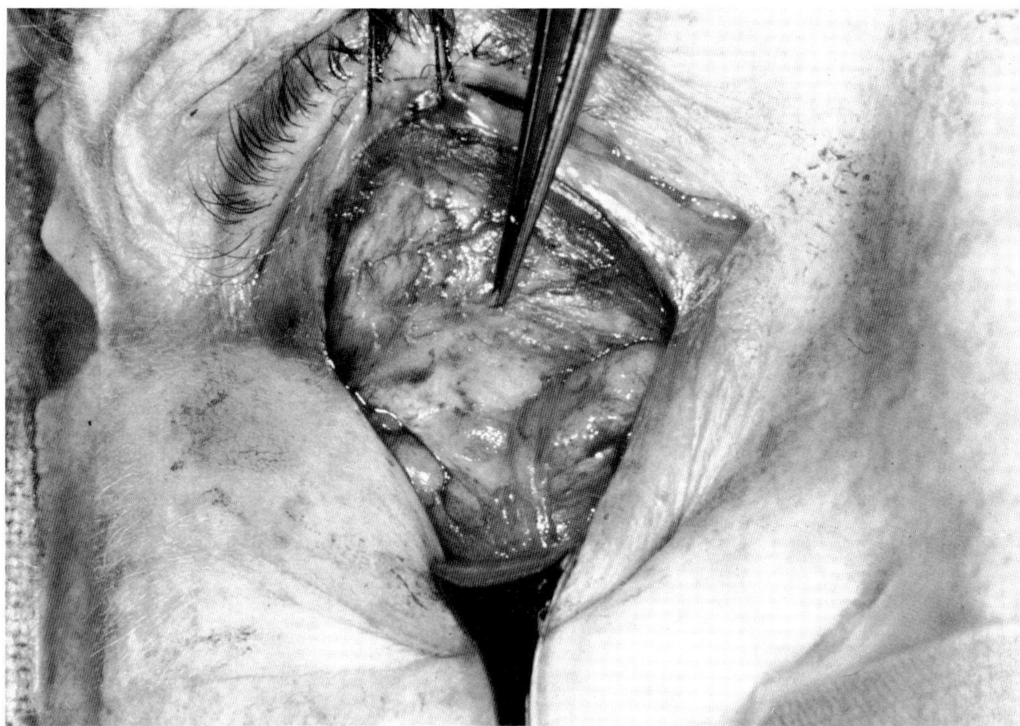

Figure 4.34. Lower lid retractor defect in patient with involutional entropion. Forceps is grasping the upper edge of the disinserted lower eyelid retractor.

collagen fibers. These run in a horizontal and vertical direction to envelop the alveolar units of the Meibomian glands.

GLANDS OF THE EYELIDS

The eyelids contain sebaceous, apocrine, and eccrine glands (Fig. 4.35). The sebaceous glands consist of the Meibomian glands embedded within the tarsus and the glands of Zeis associated with the cilia. The apocrine glands of the eyelid are referred to as the glands of Moll and are also associated with the cilia at the eyelid margin. In addition, eccrine sweat glands are diffusely distributed in the eyelids. The lacrimal glands, the main and accessory glands of Krause and Wolfring, are discussed in Chapter 5, "The Lacrimal System."

Sebaceous glands are characterized by cells exhibiting holocrine secretion, in which the entire cell decomposes (Fig. 4.36). The Meibomian glands are unique sebaceous glands since they are not associated with follicles. These multilobulated glands are embedded within the tarsus, numbering approximately 25 in the upper eyelid and 20 in the lower eyelid. The lobules of sebaceous glands undergo an orderly maturation from an outer basal layer to large foamy vacuolated degenerated cells at the center of each lobule. Secretion is collected by a central duct, whose orifice is in the posterior aspect of the eyelid margin. The lipid-rich secretion of the Meibomian glands produces the outer most (lipid) layer of the tear film.

CLINICAL NOTE

The Meibomian glands are unique since sebaceous glands elsewhere in the body are a portion of a pilo-

sebaceous apparatus. However, in rare conditions the Meibomian glands may become a pilosebaceous unit. A double row of lashes may be present at birth which is called congenital distichiasis (Fig. 4.37). Hoover and Kelley (1971) reported the association of congenital distichiasis and lymphedema as an autosomal dominant syndrome. Lymphedema of the lower extremities presented from birth to early adulthood. Additional abnormalities, such as spinal extradural cysts, may be associated with this syndrome (Robinow, et al, 1970). In addition, Anderson and Harvey (1981) have recognized acquired distichiasis in which abnormal lashes emanate from the region of the Meibomian gland orifices in response to chronic inflammatory conditions such as Steven-Johnson syndrome, ocular pemphigoid, or chemical and physical injuries (Fig. 4.38). This probably represents a regressive metaplasia of a specialized sebaceous unit back to a pilosebaceous unit. These conditions should be differentiated from trichiasis, which implies inturned or misdirected lashes from their normal anatomical position in the eyelid.

Chalazion is the most commonly encountered tumor of the eyelid (Doxanas, et al, 1976; Doxanas and Green, 1982), and represents a lipogranulomatous inflammation of the Meibomian gland. The most accepted theory in the production of chalazion is that of a blockage or obstruction of the Meibomian gland orifice, which causes the gland to become impacted. These impacted secretions of the "constipated" Meibomian gland are then extruded into the surrounding tissues producing an intense lipogranulomatous inflammation.

Gutgesell and co-workers (1982) found abnormali-

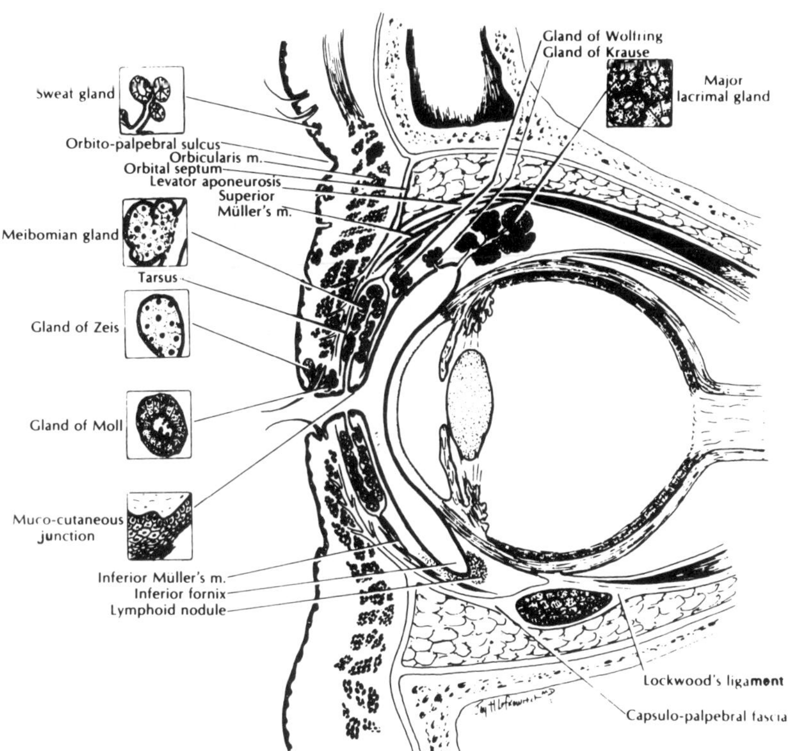

Figure 4.35. Glands of the eyelids. (Reproduced from Jakobiec FA, Iwamoto F. In Duane TD, Jaeger EA (eds): *Biomedical Foundations of Ophthalmology*, Vol. 1, Chapter 28. 1982, Philadelphia, Harper and Row Publishers.)

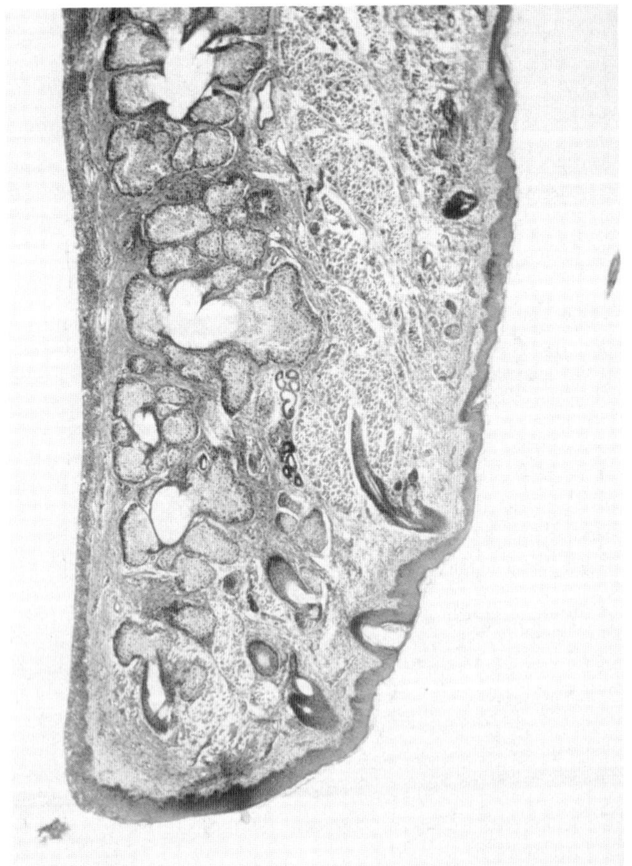

Figure 4.36 Meibomian glands contained within the tarsus of the upper eyelid. (Reproduced with permission from W. R. Green.)

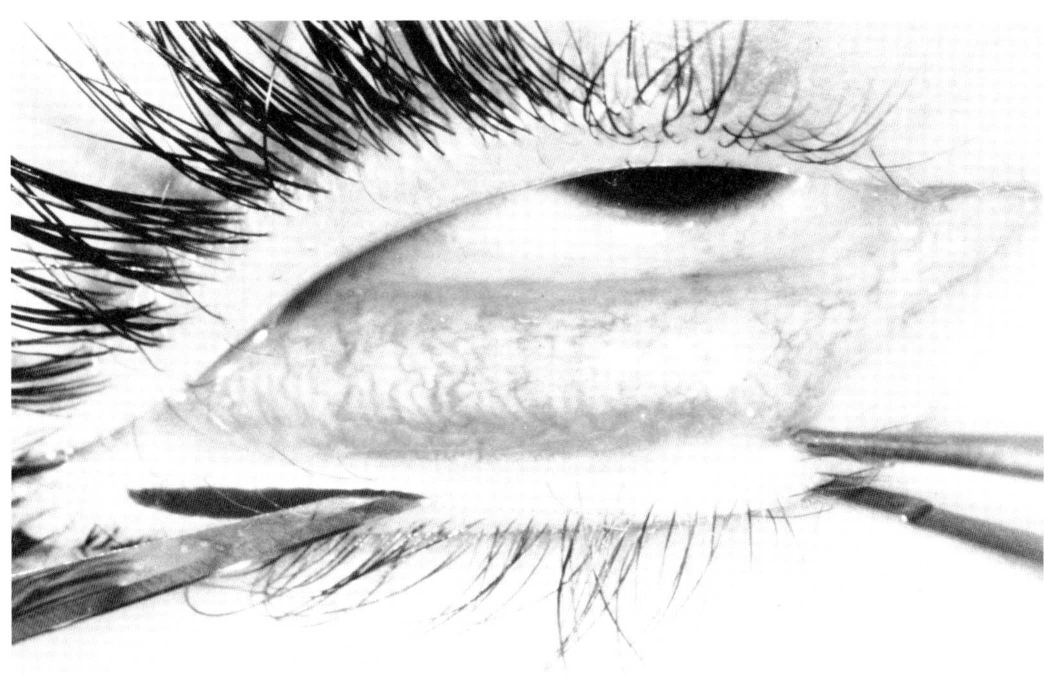

Figure 4.37. Congenital distichiasis. An extra row of eyelashes extending from the Meibomian gland orifices. (Reproduced from Anderson RL, Harvey JT: *Archives Ophthalmology* 99:631–634. 1981, American Medical Association.)

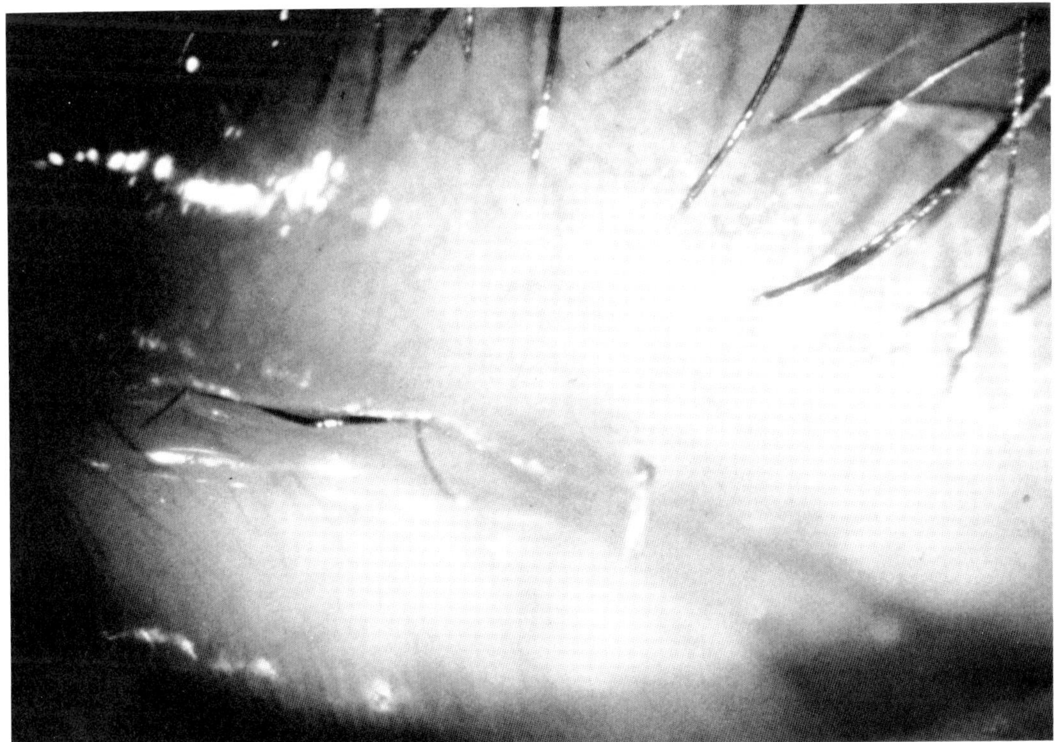

Figure 4.38. Acquired distichiasis. Chronic inflammatory conditions induce a regressive metaplasia of the Meibomian glands to a pilosebaceous apparatus. (Reproduced from Anderson RL, Harvey JT: *Archives of Ophthalmology* 99:631–634. 1981, American Medical Association.)

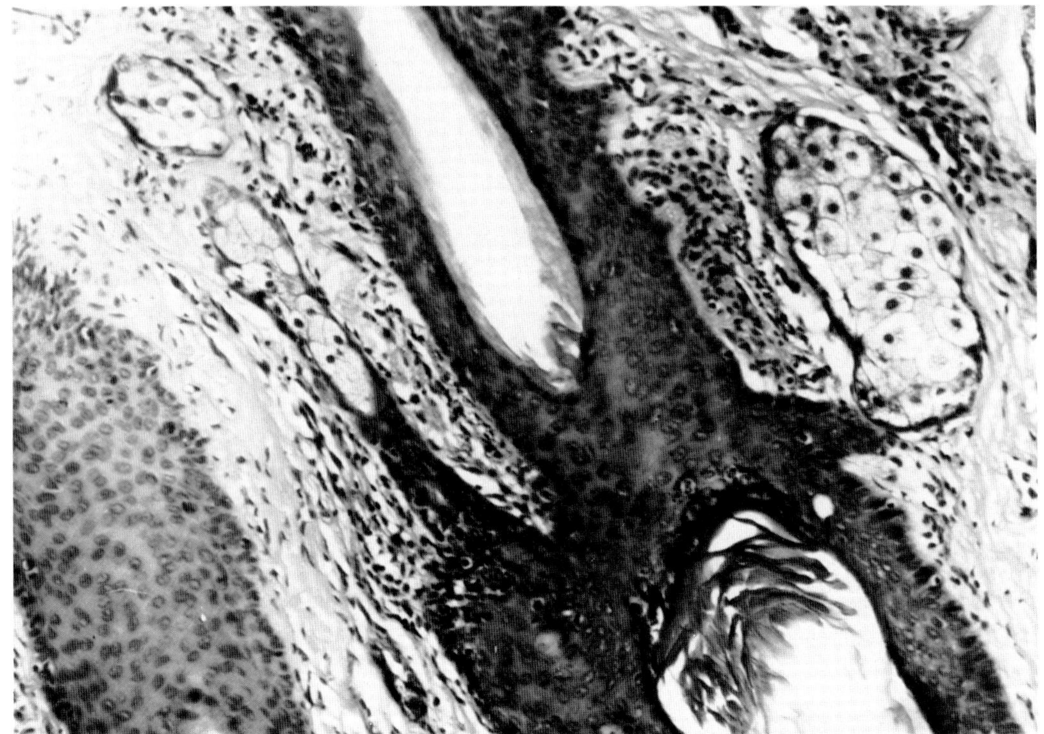

Figure 4.39. Glands of Zeis associated with pilosebaceous apparati.

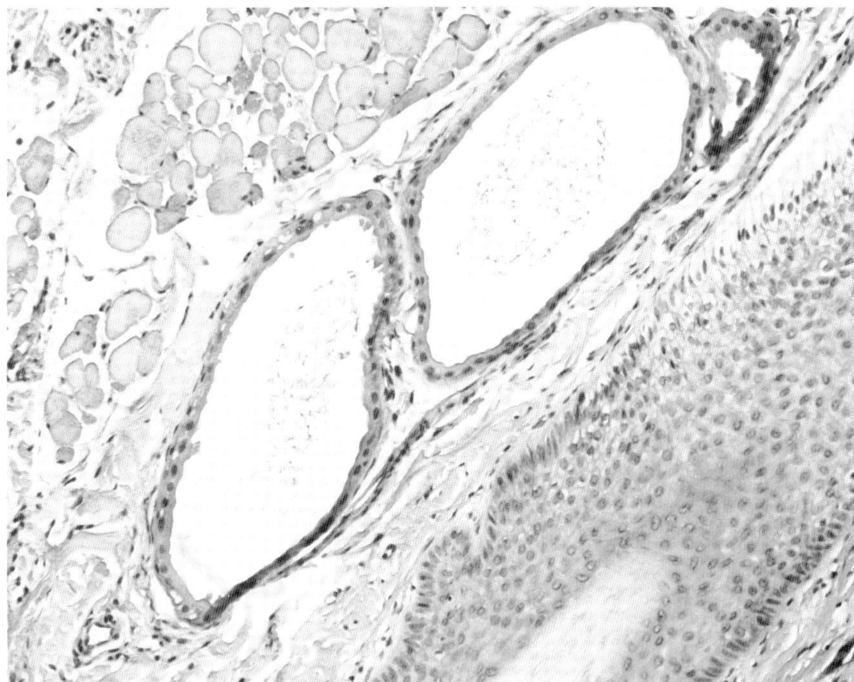

Figure 4.40. Apocrine glands of Moll.

ties of keratinization, including hyperkeratinization, and sloughing of keratin in ducts and ductules, in patients with Meibomian gland dysfunction. These authors speculated that abnormalities of keratinization producing glandular obstruction played an important role in disorders of the Meibomian gland.

The glands of Zeis consist of a single lobule of sebaceous cells which empty their secretions into the follicles of the cilia (Fig. 4.39). The individual lobules of sebaceous cells have a similar appearance and characteristic holocrine secretion as the Meibomian glands. Inflammation of the gland of Zeis is called a hordeolum.

The glands of Moll are apocrine glands, and like the glands of Zeis empty their secretions into the follicles of the cilia. Apocrine glands are formed by two layers of epithelium: an outer, flattened myoepithelial layer, and an inner, low cuboidal layer with ongoing decapitation secretion (Fig. 4.40). The inner cuboidal cells are deeply eosinophilic and toward the central lumen of the gland, apical cytoplasmic processes appear to be pinched off, producing the characteristic decapitation secretion.

The eccrine sweat glands are not associated with the cilia at the lid margin, but are distributed throughout the skin of the eyelids. The eccrine gland consists of a basal coil located at the junction of the reticular dermis and the subcutaneous tissues. The basal coil consists of two rows of cells surrounding a central lumen and embedded within a delicate connective tissue with abundant vessels and nerves. The outer layer is produced by a flattened myoepithelial cell. The inner layer is produced by two types of cells: a clear cell and a dark cell. The clear cell produces the clear component of sweat and the dark cells produce sialomucin. The cells of the inner layer empty their contents into the lumen by exocytosis.

References

Anderson RL: Medial canthal tendon branches out. *Arch Ophthalmol* 95:2051, 1977.

Anderson RL: The aponeurotic approach to ptosis surgery. In Bosniak SL, Smith BC (eds): *Advances in Ophthalmic Plastic and Reconstructive Surgery.* New York, Pergamon Press, 1982, p 183.

Anderson RL, Beard C: The levator aponeurosis. Attachments and their clinical significance. *Arch Ophthalmol* 95:1437, 1977.

Anderson RL, Dixon RS: The role of Whitnall's ligament in ptosis surgery. *Arch Ophthalmol* 97:705, 1979.

Anderson RL, Dixon RS: Neuromyopathic ptosis: A new surgical approach. *Arch Ophthalmol* 97:1129, 1979.

Anderson RL, Dixon RS: Aponeurotic ptosis surgery. *Arch Ophthalmol* 97:1123, 1979.

Anderson RL, Gordy DD: Aponeurotic defects in congenital ptosis. *Ophthalmology* 86:1493, 1979.

Anderson RL, Gordy DD: The tarsal strip procedure. *Arch Ophthalmol* 97:2192, 1979.

Anderson RL, Harvey JT: Lid splitting and posterior lamellar cryosurgery for congenital and acquired distichiasis. *Arch Ophthalmol* 99:631, 1981.

Anderson RL, Hatt MU, Dixon R: Medial ectropion. A new technique. *Arch Ophthalmol* 97:521, 1979.

Barker DE: Dye injection studies of orbital fat compartments. *Plast Reconstr Surg* 59:82, 1977.

Beard C: *Ptosis,* ed 3. St. Louis, CV Mosby, 1981, p 46.

Beard C, Quickert M: *Anatomy of the Orbit.* Birmingham, Aesculapius, 1977, p 14.

Bergin DJ, LaPiana FG: Natural history of the congenital eyelid tetrad (Komoto's syndrome). *Ann Ophthalmol* 13:1145, 1981.

Brazin SA, Stern LJ, Johnson WT: Unilateral blepharochalasis. *Arch Dermatol* 115:479, 1979.

Callahan A: Levator recession with reattachment to the tarsus with collagen film. *Arch Ophthalmol* 73:800, 1965.

Castanares S: Blepharoplasty for herniated intraorbital fat. Anatomical basis for a new approach. *Plast Reconstr Surg* 8:46, 1951.

Chalfin J, Putterman AM: Müller's muscle excision and levator muscle resection in retracted upper lid. Treatment of thyroid-related retraction. *Arch Ophthalmol* 97:1487, 1979.

Collin JRO: A ptosis repair of aponeurotic defects by the posterior approach. *Br J Ophthalmol* 63:586, 1979.

Collin JRO, Beard C, Wood I: Experimental and clinical data on the insertion of the levator palpebrae superioris muscle. *Am J Ophthalmol* 85:792, 1978.

Collin JRO, Beard C, Wood I: Terminal course of nerve supply to Müller's muscle in the rhesus monkey and its clinical significance. *Am J Ophthalmol* 87:234, 1979.

Collin JRO, Beard C, Stern WH, Schoengarth D: Blepharochalasis. *Br J Ophthalmol* 63:542, 1979.

Collin JRO, Rathbun JE: Involutional entropion: A review with evaluation of a procedure. *Arch Ophthalmol* 96:1058, 1978.

Dortzbach RK, Sutula FC: Involutional blepharoptosis: A histopathologic study. *Arch Ophthalmol* 98:2045, 1980.

Doxanas MT, Green WR: Adult lid lesions: A histopathologic review. In preparation.

Doxanas MT, Green WR, Arentsen JJ, Elsas FJ: Lid lesions of childhood: A histopathologic survey at the Wilmer Institute (1923–1974). *J Pediat Ophthalmol* 13:1, 1976.

Doxanas MT, Dryden RM: The use of sclera in the treatment of dysthyroid eyelid retraction. *Ophthalmology* 88:887, 1981.

Doxanas MT, Dryden RM: Levator aponeurotic defects in congenital ptosis. Presented at the Twelfth Annual Symposium of the American Society of Ophthalmic Plastic and Reconstructive Surgery, Atlanta, Nov 6, 1981.

Dryden RM: The role of the aponeurosis in ptosis surgery. Presented at the Eleventh Annual Scientific Symposium of the American Society of Ophthalmic Plastic and Reconstructive Surgery, Chicago, Nov. 7, 1980.

Dryden RM, Fleming JC, Quickert MH: Levator transposition and frontalis sling procedure in severe unilateral ptosis and the paradoxically innervated levator. *Arch Ophthalmol* 100:462, 1982.

Dryden RM, Leibsohn J, Wobig J: Senile entropion: Pathogenesis and treatment. *Arch Ophthalmol* 96:1883, 1978.

Dryden RM, Leibsohn J: The levator aponeurosis in blepharoplasty. *Ophthalmology* 85:718, 1978.

Dryden RM, Soll DB: The use of scleral transplantation in cicatricial entropion and eyelid retraction. *Trans Am Acad Ophthalmol Otolaryng* 83:669, 1977.

Flanagan JC: Eye bank sclera in oculoplastic surgery. *Ophthalmol Surg* 5:45, 1974.

Fox SA: Primary congenital entropion. *Arch Ophthalmol* 56:839, 1956.

Gutgesell VJ, Stern GA, Hood CI: Histopathology of Meibomian gland dysfunction. *Am J Ophthalmol* 94:383, 1982.

Hargiss JL: Inferior aponeurosis vs. orbital septum tucking for senile entropion. *Arch Ophthalmol* 89:210, 1973.

Harvey JT, Anderson RL: The aponeurotic approach to eyelid retraction. *Ophthalmology* 88:513, 1981.

Hawes MJ, Dortzbach RK: The microscopic anatomy of the lower eyelid retractors. *Arch Ophthalmol* 100:1313, 1982.

Hoover RE, Kelley JS: Distichiasis and lymphedema. A hereditary syndrome with possible multiple defects. A report of a family. *Trans Am Ophthalmol Soc* 69:293, 1971.

Hugo NE, Stone E: Anatomy for a blepharoplasty. *Plast Reconstr Surg* 53:381, 1974.

Johnson CC: Operations for epicanthus and blepharophimosis. *Am J Ophthalmol* 41:71, 1956.

Johnson CC: Epiblepharon. *Arch Ophthalmol* 66:1172, 1968.

Johnson CC: Epicanthus and epiblepharon. *Arch Ophthalmol* 96:1030, 1978.

Jones LT: Senile Entropion. In *Surgery of the Orbit and Adnexa.* St. Louis, CV Mosby, 1974, p 210.

Jones LT: The anatomy of the lower eyelid and its relation to the cause and cure of entropion. *Am J Ophthalmol* 49:29, 1960.

Jones LT: A new concept of the orbital fascia and rectus muscle sheaths and its surgical implications. *Trans Am Acad Ophthalmol* 72:755, 1968.

Jones LT, Quickert MH, Wobig JL: The cure of ptosis by aponeurotic repair. *Arch Ophthalmol* 93:629, 1975.

Jones LT, Wobig JL: *Surgery of the Eyelids and Lacrimal System.* Birmingham, Aesculapius, 1976, p 84.

Kuwabara T, Cogan DG, Johnson CC: Structure of the muscles of the upper eyelid. *Arch Ophthalmol* 93:1189, 1975.

Lemke BN, Stasior OG: The anatomy of eyebrow ptosis. *Arch Ophthalmol* 100:981, 1982.

Lockwood CB: The anatomy of the muscles, ligaments, fascia of the orbits, etc. *J Anat* 20:1, 1886.

McCord CD: The correction of telecanthus and epicanthal folds. *Ophthalmol Surg* 11:446, 1980.

Older JJ: Levator aponeurosis disinsertion in the young adult. *Arch Ophthalmol* 96:1857, 1978.

Ousterhout DK, Weil RB: The role of the lateral canthal tendon in lower eyelid laxity. *Plast Reconstr Surg* 69:620, 1982.

Putterman AM, Urist MJ: Surgical anatomy of the orbital septum. *Ann Ophthalmol* 8:319, 1974.

Putterman AM: Ectropion of the lower eyelid secondary to Müller muscle-capsulopalpebral fascia detachment. *Am J Ophthalmol* 85:814, 1978.

Quickert MH, Dryden RM: Lower eyelid advancement. Read before the American Society of Ophthalmic Plastic and Reconstructive Surgery, Las Vegas, 1971.

Robinow M, Johnson GF, Verhagen AD: Distichiasis-lymphedema: A hereditary syndrome of multiple congenital defects. *Am J Dis Child* 119:343, 1970.

Schaefer AJ: Lateral canthal tendon tuck. *Ophthalmology* 86:1879, 1979.

Stasior OG: Complications of ophthalmic surgery and their prevention: The Wendell L. Hughes Lecture. *Trans Am Acad Ophthalmol Otolaryng* 90:543, 1976.

Stieglitz LN, Crawford JS: Blepharochalasis. *Am J Ophthalmol* 77:100, 1974.

Tenzel RR, Buffam FV, Miller GR: The use of the "lateral canthal sling" in ectropion repair. *Can J Ophthalmol* 12:199, 1977.

Tse D, Anderson RL: Aponeurosis disinsertion in congenital entropion. *Arch Ophthalmol* 101:436, 1983.

Warwick R: *Eugene Wolff's Anatomy of the Eye and Orbit,* ed 7. Philadelphia, WB Saunders, 1976, p 188, p 270.

Wesley RE: Tarsal ectropion from detachment of the lower eyelid retractors. *Am J Ophthal* 93:491, 1982.

Wesley RE, Collins JW: Combined procedure for senile entropion. *Ophthalmic Surg* 14:401, 1983.

Wesley RE, McCord CD, Jones NA: Height of the tarsus of the lower eyelid. *Am J Ophthalmol* 90:102, 1980.

Whitnall SE: *The Anatomy of the Human Orbit and Accessory Organs of Vision.* London, Henry Frowde & Hoddes & Stoughton, 1921, p 131, 291.

Whitnall SE: *The Anatomy of the Human Orbit and Accessory Organs of Vision,* Milford H (ed). London, Oxford University Press, 1932, p 146.

CHAPTER 5

The Lacrimal System

The lacrimal system allows for the production, distribution, and drainage of tears. The tear film protects the cornea to facilitate optical integrity. This chapter is divided into three primary sections which correspond to the functions of the lacrimal system: secretory, tear distribution, and excretory (Fig. 5.1).

LACRIMAL SECRETORY SYSTEM

The secretory portion of the lacrimal system primarily consists of the main and accessory lacrimal glands (Fig. 5.2). The main lacrimal gland lies in the superior lateral portion of the anterior orbit. This gland lies just posterior to the orbital rim in a concavity in the frontal bone, called appropriately, the fossa glandular lacrimalis. The lacrimal gland is divided into two lobes by the lateral horn of the levator aponeurosis (Fig. 5.3 A and B). The superior or orbital lobe is larger than the inferior or palpebral lobe. As the names imply, the orbital lobe is located primarily within the superior lateral orbital confines, while the palpebral lobe is more inferiorly located in the eyelid. The accessory lacrimal glands consist of the glands of Wolfring and Krause. The glands of Krause are located in the conjunctival fornix, and the glands of Wolfring are positioned along the border of the tarsus.

The orbital lobe of the lacrimal gland is an oval, bean-shaped mass approximately 20 mm long, 12 mm wide, and 5 mm deep (Whitnall, 1932). The orbital lobe extends inferiorly to the level of the zygomaticofrontal suture. The inferior portion of the orbital lobe lies directly on, and is molded by, the globe. The anterior portion of the lower pole extends to and rests on the lateral horn of the levator aponeurosis.

The inferior, or palpebral lobe of the lacrimal gland is approximately one-third the size of the orbital lobe. The palpebral lobe lies between the levator aponeurosis and the palpebral conjunctiva. The palpebral lobe is firmly attached to the conjunctiva, and the ducts of both lobes of the lacrimal gland exit through the palpebral lobe. If the upper eyelid is everted, the palpebral lobe is seen to extend to within a few millimeters of the superior aspect of the tarsus, laterally. The lacrimal gland has a glandular-appearing surface and is pinkish-grey in color. The color, texture, consistency, and location of the lacrimal gland are important considerations in distinguishing it from orbital fat when prolapsed tissue occurs in the lateral orbital region.

The ducts of the lacrimal gland accumulate secretions which exit on the palpebral conjunctiva. Approximately 12 ducts exit on the conjunctiva; the ducts from the orbital lobe traverse the palpebral lobe to enter the conjunctival sac (Fig. 5.4). Most of the ducts open 4–5 mm from the superior border of the tarsus; however, the ducts' orifices may extend into the lateral canthus. All the ducts from the lacrimal gland extend through the palpebral lobe of the gland; thus, excision of a prolapsed palpebral lobe of the lacrimal gland will destroy all the glandular functions and obviously should be avoided (Smith and Petrelli, 1978).

The lacrimal gland is supported in the superior lateral aspect of the orbit by various attachments. The suspensory "ligament of Soemmering" fixes the superior surface of the gland to the periorbita of the frontal bone. This "ligament" consists of loose connective tissue strands which pass from the stroma of the gland to the periorbita. However, the major supportive structure for the lacrimal gland is Whitnall's ligament and the lateral horn of the levator aponeurosis (Fig. 5.3 A). As noted previously, the lateral horn of the aponeurosis divides the lacrimal gland into an orbital and palpebral lobe. Whitnall's ligament intermingles with the connective tissue of the lacrimal gland and inserts high on the lateral orbital wall. A minor slip of Whitnall's ligament contributes to the lateral retinaculum and inserts near the lateral orbital tubercle. Because of its support for the lacrimal gland, care should be taken to avoid damage to Whitnall's ligament in ptosis surgery (Anderson and Dixon, 1979).

The lacrimal gland, especially the orbital lobe, is enveloped by what appears to be a connective tissue capsule. However, this is not a true fibrous capsule, but is rather an extension of the connective tissue that separates the lobules of the gland. The connective tissue is more prominent on the surface than between the lobules of the gland, which gives the appearance of a true capsule. The concept of a capsule created by connective tissue enveloping or encapsulating the lacrimal gland is an important consideration in lacrimal gland tumor management.

The lacrimal gland is a tubuloracemose structure, that is, it consists of numerous lobules drained by tubules. The acinus is the basic secretory unit of the lacrimal gland. The acinus consists of an irregular circular arrangement of cells around a central lumen (Fig. 5.5). Two types of cells make up the acinus: the inner, low, cylindrical cells, and the outer, flatter, myoepithelial cells. The cylindrical or columnar cells

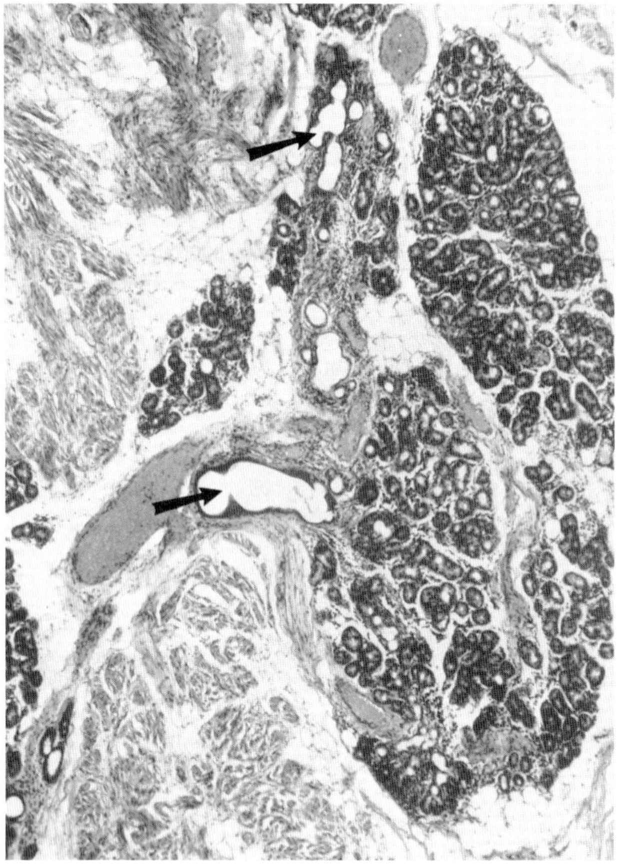

Figure 5.4. Palpebral lobe of the lacrimal gland. Numerous excretory ducts (*arrows*) are adjacent to the palpebral lobe of the lacrimal gland (H&E, ×50).

are the active secretory cells, which demonstrate polarity. These cells have a basally located nucleus with the apical portion of the cell consisting of cytoplasmic organelles—rough-surfaced endoplasmic reticulum, Golgi apparti, and secretory granules—all of which reflect the active secretory capacity of these cells. Histochemical studies of the secretory granules have demonstrated positive staining for both proteins and mucopolysaccharides, which indicate a mixed serous/mucous secretion (Duke-Elder and Wybar, 1961). However, the observation of dense granules on electron micrographic evaluation suggests that the primary secretion of the lacrimal gland is serous (Egeberg and Jensen, 1969; Orzalesi, et al, 1979). The flatter myoepithelial cells form an incomplete layer which surrounds the secretory cells.

The flattened myoepithelial cells line the acini. Contraction of these elements forces glandular secretion into the central canal of the acinus. A connecting network of excretory ducts drain the lacrimal gland. The excretory system of the lacrimal gland can be divided into three portions: the intralobular ducts, the interlobular ducts, and the main excretory ducts. The intralobular collecting ducts are minute, with narrow lumen and epithelial walls of one or two layers. The interlobular ducts have larger lumen and epithelial walls of 2–4 cellular layers. As one would expect, the main excretory duct has the largest lumen and 3–4-layer epithelial walls. Approximately 12 main excretory ducts drain the lacrimal gland, and exit on the conjunctiva.

In addition to the main lacrimal gland, which is mainly for reflex secretion, are accessory lacrimal glands which are said to be mainly responsible for basic tear secretion (Jones, 1966). These consist of the glands of Krause and Wolfring, located in the fornix area and tarsal border, respectively (Fig. 5.2). More recent evidence would indicate that all tear

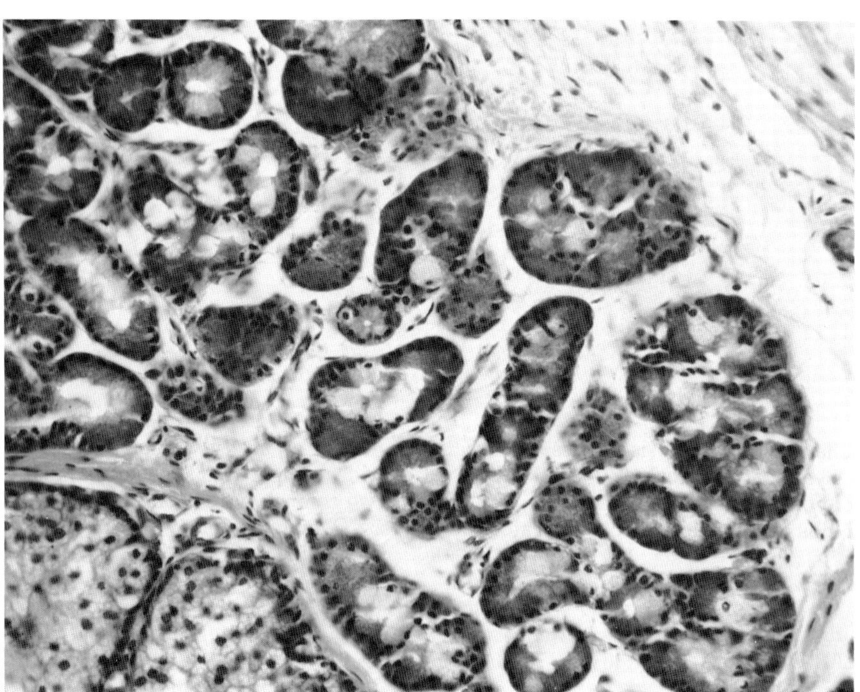

Figure 5.5. Acini of lacrimal gland. The acini are composed of inner, low cylindrical cells and outer, flatter, myoepithelial cells. The superior tip of the tarsus is present (*asterisk*) (H&E, ×250).

secretion is reflex (Scherz and Dohlman, 1975). The accessory lacrimal glands are small lobules located within the substantia propia. These drain directly to the conjunctival surface. The glands of Krause number approximately 20–40 in the upper fornix, and 6–8 in the lower fornix (Jones and Wobig, 1976; Iwamoto and Jakobiec, 1982). The accessory glands of Wolfring are less numerous. Jones and Wobig (1976) noted an average of three glands at the superior border of the upper tarsus, and one at the inferior border of the lower tarsus. Hawes and Dortzbach (1982) failed to identify any accessory glands of Wolfring in the lower eyelid.

CLINICAL NOTE

Tumors of the lacrimal gland present special diagnostic and management problems because of the variety of pathologic conditions which may affect the lacrimal gland. In particular, the potentially aggressive behavior of the benign mixed tumor has necessitated a systematic approach to these tumors. If incompletely excised, the benign mixed tumor will recur with malignant potential (Zimmerman, et al, 1962; Rubin, 1947; Waller, et al, 1973; Forrest, 1971).

Studies of extensive series of lacrimal gland tumors reveal an equal ratio of nonepithelial and epithelial tumors. Nonepithelial tumors include acute dacryoadenitis, idiopathic inflammatory pseudotumor, and lymphoma. Of the epithelial tumors, approximately one-half are malignant (Stewart, et al, 1979; Reese, 1970; Forrest, 1954; Henderson and Neault, 1976; Henderson, 1980).

The benign epithelial tumor is the mixed lacrimal gland tumor. Malignant tumors of the lacrimal gland primarily include adenoid cystic carcinoma and malignant mixed cell tumors. In addition, nonspecific adenocarcinoma, squamous cell carcinoma, and mucoepidermoid carcinomas may arise in the lacrimal gland.

Clinical features can help differentiate inflammatory and epithelial tumors. Radiographs, morphology on CT scan (Fig. 5.6 A–C), and standardized A-scan echography are very useful in classifying and designing an approach to these tumors. An acute inflammatory lesion with a history consistent with dacryoadenitis should be treated with antibiotics. A well-encapsulated lacrimal gland lesion should be excisionally biopsied to avoid possible seeding of a benign mixed tumor (Fig. 5.6 B). Diffuse lesions of the lacrimal gland with or without bony erosion should have a transeptal biopsy before definite therapy is described (Fig. 5.6 C) (Stewart, et al, 1979; Jakobiec, et al, 1982).

The lacrimal nerve provides sensory innervation of the lacrimal gland. The lacrimal nerve is one of three branches of the ophthalmic division of the trigeminal nerve (see Fig. 8.6 A and B). The lacrimal nerve enters the orbit through the lateral portion of the superior orbital fissure, and extends anteriorly in the superior lateral aspect of the orbit. The nerve is associated with the lacrimal artery. The lacrimal nerve divides into two branches prior to its entering the gland. The superior division traverses the lacrimal gland to which it contributes most of its fibers. The superior division then extends superficially to pierce the orbital septum and supply sensation to the conjunctiva and skin in the lateral region of the upper eyelid. The inferior division of the lacrimal nerve anastomoses with the zygomaticotemporal branch of the zygomatic nerve and helps provide sensation to the temple.

Lacrimal gland secretion is produced by stimulation of a reflex arc, as well as central stimulation. The afferent limb of the reflex pathway is supplied by the fifth cranial nerve. The efferent limb is through the facial (seventh) cranial nerve. Interruption of the reflex arc results in a loss of reflex tear production (Ruskin, 1930).

Stimulation of any of the branches of the trigeminal nerve results in reflex tearing. Stimulation of the ophthalmic division of the trigeminal nerve, which supplies the globe, results in reflex tearing, such as when defects of the corneal epithelium or spasm of the ciliary body (iritis) are present. Additional stimuli for tearing include bright lights (stimulating the retina) and stimuli directed to the frontal cortex, basal ganglia, thalamus, hypothalamus, and the cervical sympathetic ganglia (Czermak, 1895).

The efferent pathway to the lacrimal gland follows a long and complicated route, the first part of which is in association with the facial nerve (Fig. 5.7). The nerve fibers originate at the lacrimal nucleus in the pons, which is adjacent to the superior salivatory nucleus (Crosby, et al, 1962). Because these fibers arise in a nucleus distinct from that of the facial nerve, some feel this nerve should be considered a separate cranial nerve. The fibers from the lacrimal nucleus travel in the pons as the nervus intermedius. The nervus intermedius exits from the ventrolateral portion of the brainstem at the cerebellopontine angle. The fine nervus intermedius lies between the motor root of the facial nerve and the acoustic nerve. These nerves enter the internal auditory canal and proceed to the geniculate ganglion. At the geniculate ganglion, the greater superficial petrosal nerve is formed, which runs along the petrous bone. Sympathetic fibers, via the deep petrosal nerve from the carotid artery, join the greater superficial petrosal nerve to form the Vidian nerve, or nerve of the pterygoid canal. The Vidian nerve synapses in the sphenopalatine ganglion. These postganglionic fibers join the maxillary nerve and travel via the zygomatic then zygomaticotemporal branch to join the lacrimal nerve posterior to the lacrimal gland (Fig. 5.8).

TEAR DISTRIBUTION SYSTEM

The precorneal tear film is responsible for maintaining a clear, healthy cornea. The tear film prevents drying of the underlying corneal surface while providing a smooth optical surface. The maintenance of a proper precorneal tear film depends upon the secretion, mixing, distribution, and drainage of the tear fluid. In addition, certain enzymes in the tear film, notably lysozyme and beta lysin, are known to possess bactericidal properties (Ford, et al, 1976).

The tear film consists of three layers: lipid, aqueous, and mucin (Fig. 5.9). The outermost, or lipid, layer is composed of low-polarity lipids, such as waxy and cholesterol esters (Andrew, 1970). The lipid layer is produced primarily by the Meibomian glands, and to a lesser degree by the glands of Zeis and the apocrine glands of Moll. The average thickness of the lipid layer is estimated at 0.1 μm; however, this varies

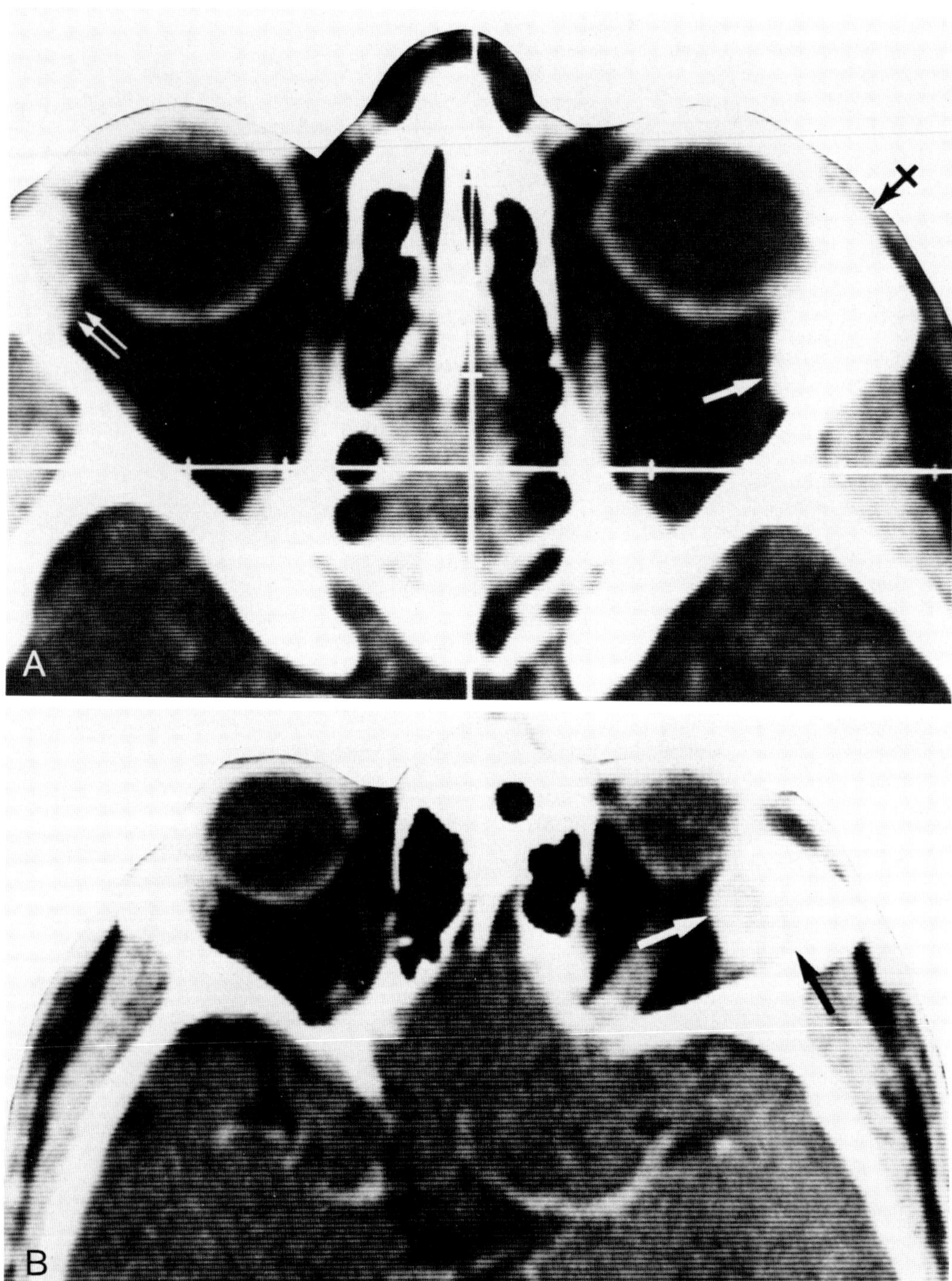

Figure 5.6.   Computerized axial tomography of lacrimal gland tumors: (*A*), Acute dacryoadenitis. Sudden onset of bilateral tender lacrimal gland enlargement in a 23-year-old female which resolved with systemic prednisone. (*B*), Mixed tumor of lacrimal gland. Gradual proptosis of 2 years in a 36-year-old male. CT scan reveals rounded configuration of the tumor. (*C*), Adenoid cystic carcinoma. Six month history of painful proptosis with erosion of the lateral wall of the orbit to enter the temporal fossa. (Reproduced from Jacobiec FA, et al: *American Journal of Ophthalmology* 94:785–807, 1982.)

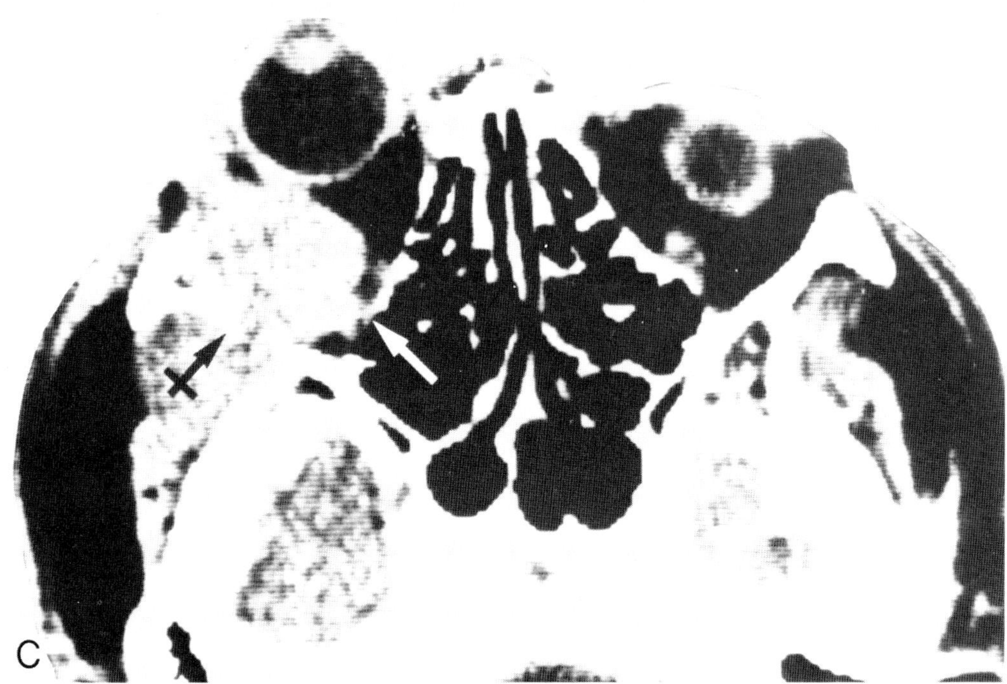

Figure 5.6   C.

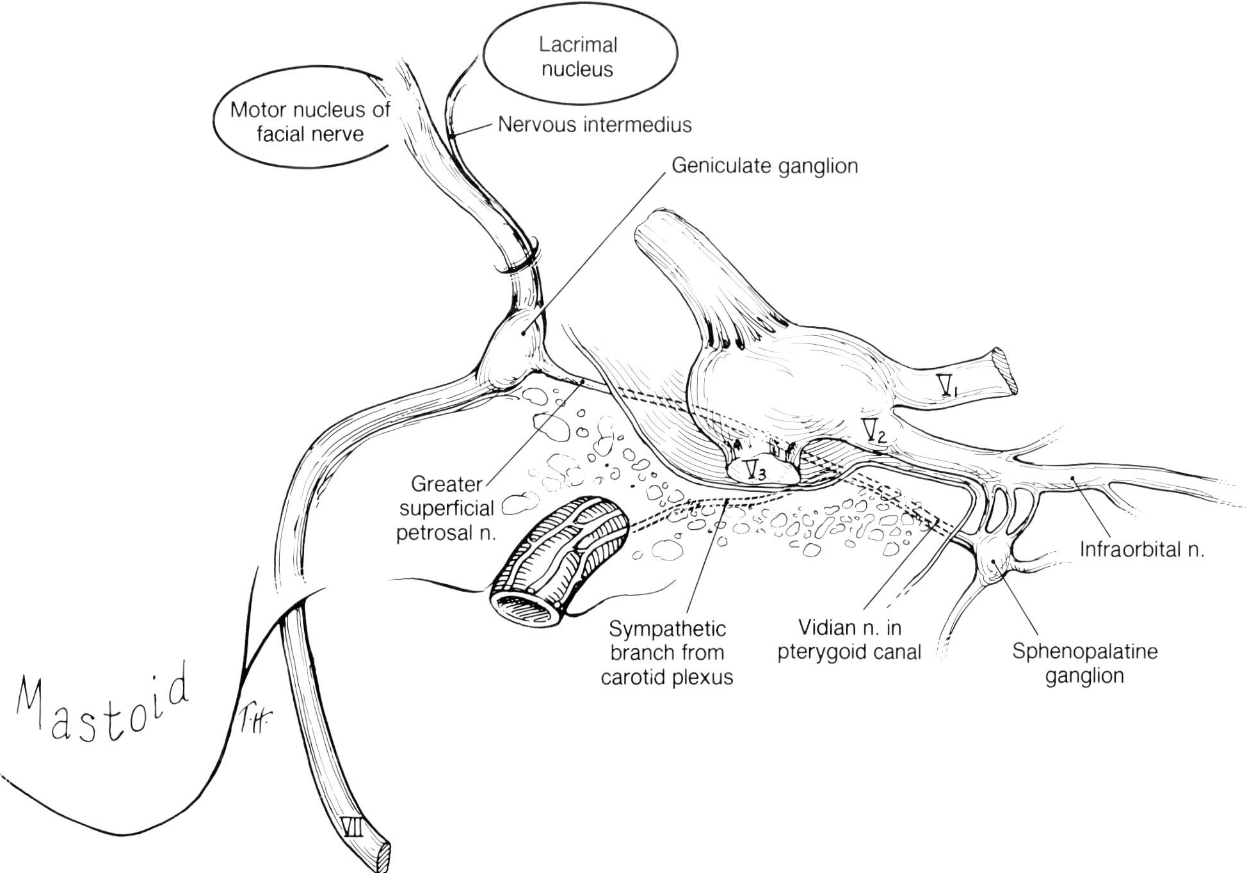

Figure 5.7. Pathway of parasympathetic secretomotor innervation to the lacrimal gland from its nuclear origin to the infraorbital nerve.

thelium to a wettable surface. The epithelial cells are capable of secreting a small amount of glycoprotein themselves; however, the bulk of this is produced by the conjunctival goblet cells.

The tear film distribution system consists of the eyelids and the tear meniscus along the eyelid margins of the open eye. Blinking spreads the tear film, resurfaces areas of dryness, adds fresh components to tears, and removes debris. The tear film is so thin that its surface forces are far greater than gravitational forces, so a bulk flow of tears does not occur during the blinking cycle. Blinking produces a vertical distribution of tears, while horizontal tear flow occurs in the tear meniscus at the eyelid margin. So efficient is the tear distributional system that a drop of concentrated fluorescein solution, carefully placed at the lateral canthus, is well-mixed after only a single blink (Doane, 1980).

Opening of the eyelids is followed by a rapid distribution of the lipid layer. With time, the superficial lipid layer migrates to the surface epithelium producing a dry spot (Fig. 5.10). This produces areas of high interfacial tension and more dry spots develop. The time between a blink and the appearance of the first dry spot is the tear breakup time, and is generally greater than the interval between two blinks (Lemp, 1973; Lemp and Hamil, 1973). Norm (1969) reported the average tear breakup time to be 20–30 sec. The production of dry spots is accelerated by abnormalities in the superficial lipid layer or by gross irregularities in the epithelial surface. The tear film becomes unstable in these areas and produces dry spots. Lipid contamination of the mucin-lined epithelial surface will further exacerbate the condition, producing a dry-eye state.

CLINICAL NOTE

Appreciation of blinking dynamics and the three layers of precorneal tear film provide for the efficient diagnosis and management of tear film abnormalities. Blinking abnormalities may be as apparent as orbicularis paralysis, secondary to facial nerve palsy, or as subtle as nocturnal lagophthalmus. Since abnormalities may affect any one or all three layers of the precorneal tear film, a systematic approach to these problems is essential.

Blinking is necessary to resurface the tear film. Interference with the normal blinking mechanism results in areas of corneal drying. Localized disturbances of the tear film adjacent to an elevation will produce dellen, or localized desiccated areas. In instances of

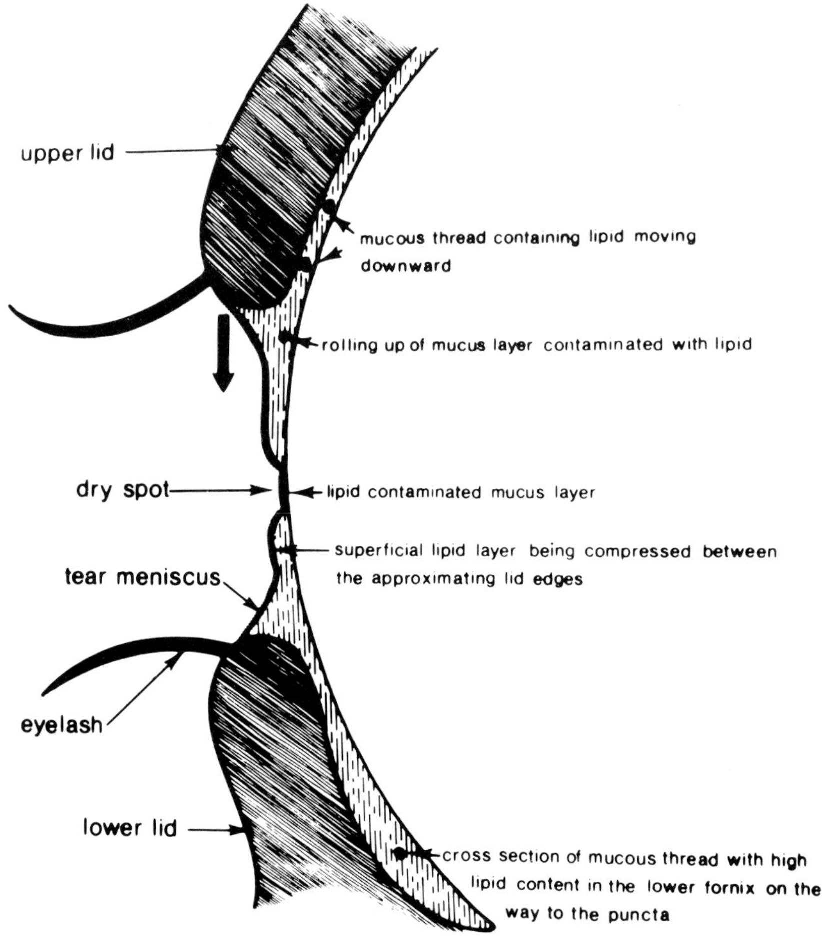

Figure 5.10. Creation of a dry spot and the mechanism for removal of a lipid-contaminated mucus layer. (Reproduced from Holly FJ, Lemp MA: *Survey of Ophthalmology* 22:69, 1977.)

facial nerve paresis, the reflex blinking stops and the resulting exposure keratopathy may progress from punctate keratopathy to corneal ulceration and perforation.

Although less severe, incomplete blinking and incomplete eyelid closure can be the primary cause of corneal disease or may aggravate existing conditions (Katz and Kaufman, 1977; Sturrock, 1976). Doane (1980) has demonstrated with high speed photography that most blinks are, in fact, incomplete, and that during these incomplete blinks the inferior cornea is not resurfaced. However, complete blinks occur frequently enough to prevent desiccation of the inferior cornea in most individuals. Individuals with reduced blink intervals or those with infrequent complete blinks may exhibit inferior corneal desiccation, which may become symptomatic. Nocturnal lagophthalmos may manifest as scratchy, irritated eyes on awakening in the morning, a condition which often persists into the day. With chronicity, this condition may produce substantial corneal damage (Katz and Kaufman, 1977).

Abnormalities which produce a scratchy, irritated, tearing eye, may develop in all three layers of the tear film. It is helpful to recognize a primary abnormality of the corneal tear film so that therapy may be effectively rendered. While there is no specific abnormality of the outermost lipid layer of the tear film, one commonly encounters alterations in lipid composition which produce an unstable tear film. Common bacteria secrete lipase, which is capable of hydrolyzing lipids to release fatty acids (Anderson, et al, 1972). These changes occur secondary to lid margin disease, such as marginal blepharitis, acne rosacea, and Meibomianitis. Elimination of the predisposing condition will stabilize the lipid layer and reduce the symptoms of dry-eye syndrome.

Abnormalities of the aqueous layer are the most common affliction of the corneal tear film. Aqueous deficiency occurs in all age groups, although there is an increase in frequency with advancing age, particularly in middle-aged females (Holly and Lemp, 1977).

In particular, those patients with rheumatoid arthritis manifest an aqueous deficiency, and may develop the triad of Sjögrens syndrome: rheumatoid arthritis, dry mouth, and dry eyes. Additional causes of aqueous deficiency include congenital alacrima (Fleming, 1922) and the Riley-Day syndrome (Riley, et al, 1949).

Schirmer's testing evaluates the aqueous component of tears. A Schirmer's I test is performed without a topical anesthetic and measures both reflex and basic secretion. A basic secretion test requires instillation of a topical anesthetic which reduces reflex secretion, and allows estimates of the basic secretion to be ascertained (Jones, 1966). Although the basic secretion test has been criticized for its lack of precision and variable accuracy, it remains the most commonly accepted test for the diagnosis of dry eye (Jordan and Baum, 1980).

Taylor and Louis (1980) evaluated a variety of tear function tests (rose bengal staining, marginal tear strip, tear break-up time, Schirmer's testing, and tear lysozyme), and suggested that the diagnosis of dry-eye syndrome be based on a variety of findings rather than upon a single test.

Any process which interferes with the production of mucin by conjunctival goblet cells, disturbs the proper wetting of the corneal surface. Conditions associated with reduced goblet cell populations include: avitaminosis A, ocular pemphigoid, Steven-Johnson syndrome, trachoma, chemical burns, and drug-induced disease (Lemp, 1973).

Lemp and co-workers (1970, 1971) have presented evidence indicating that a reduced tear breakup time reflects abnormalities of the mucin-deficient dry eye. Unfortunately, specific therapy that is effective in managing mucin-deficient states has yet to be developed.

LACRIMAL EXCRETORY SYSTEM

The lacrimal excretory system consists of the canaliculi, common canaliculus, lacrimal sac, and the nasolacrimal duct (Fig. 5.11). Tears pool in the medial angle of the eye at the

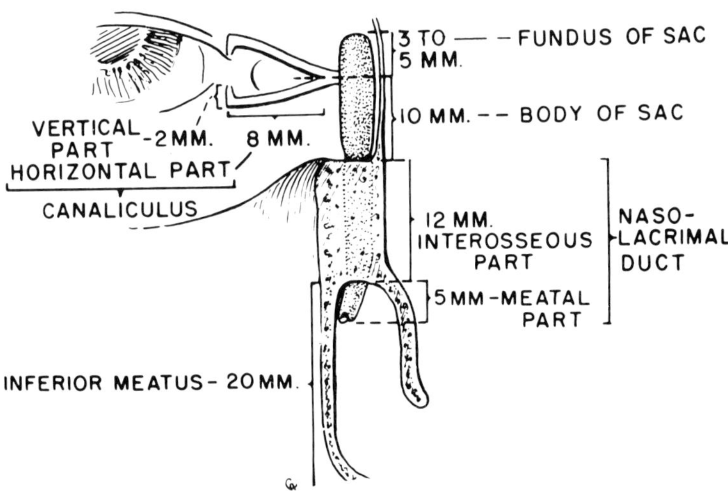

Figure 5.11. The lacrimal excretory system. (Reproduced from Jones LT, Wobig JL: *Surgery of the Eyelids and Lacrimal System.* © 1976, Birmingham, Aesculapius Publishing Company.)

lacus lacrimalis. Tears then drain into the puncta, or openings of the canaliculi, at the eyelid margins. The canaliculi of the upper and lower eyelid join prior to extending into the lacrimal sac to form the common canaliculus. The lacrimal sac collects the tears prior to directing them inferiorly, via the nasolacrimal duct, into the nose.

The lacrimal papillae are located at the mucocutaneous junction of the eyelid. The lacrimal papillae are elevated, pale mounds at the center of which are the puncta, or openings into the canalicular system. The lids must be slightly everted to allow visualization of the puncta. If the puncta are visible on the external examination, a degree of ectropion exists. The lower punctum is actually oriented slightly posteriorly and inferiorly, between the caruncle and the plica semilunaris. This position enables a more efficient siphoning of tears.

The lower canaliculus is located slightly more temporally than the upper, but each is approximately 10 mm in length. The canaliculus has an initial vertical portion that is approximately 2 mm long. At the base of this 2-mm vertical portion, the canaliculus dilates and forms a receptacle for the collection of tears (the ampulla). The canaliculus then turns horizontally along the eyelid margin for 8 mm. Although the punctum is only 0.3 mm in diameter, the lumen of the canaliculus is 1–2 mm in diameter. The horizontal portions of the canaliculi unite prior to entering the lacrimal sac, and form the common canaliculus, or internal common punctum. The common canaliculus, or at least a common internal punctum, enters the lateral wall of the lacrimal sac, slightly posterior and superior to the center of the lateral wall. The internal common punctum lies directly beneath the central portion of the medial canthal tendon.

The opening of the common canaliculus into the sac is preceded by a small dilation called the sinus of Maier. The canaliculus opens into the sac at an angle which produces the valve of Rosenmuller. The valve of Rosenmuller prevents reflux from the lacrimal sac into the canaliculus. Probing the lacrimal system into the lacrimal sac may violate the valve of Rosenmuller, which may permit regurgitation of material from the lacrimal sac, as well as permitting reflux of air on nose blowing.

The puncta and canaliculi are lined with a nonkeratinizing, stratified squamous epithelium (Fig. 5.12). The canaliculi are encased in a dense connective tissue which helps maintain the resiliency of the system.

The lacrimal sac and the nasolacrimal duct are a continuous membranous tube lined with modified, nonciliated respiratory epithelium. A simple conceptualization is that of a tube having orbital, maxillary, and meatal portions. The lacrimal sac lies within the lacrimal fossa, created by the frontal process of the maxillary bone and the lacrimal bone. The lacrimal sac extends 3–5 mm above the superior margin of the medial canthal tendon, forming the fundus of the sac. The body of the lacrimal sac extends from the level of the

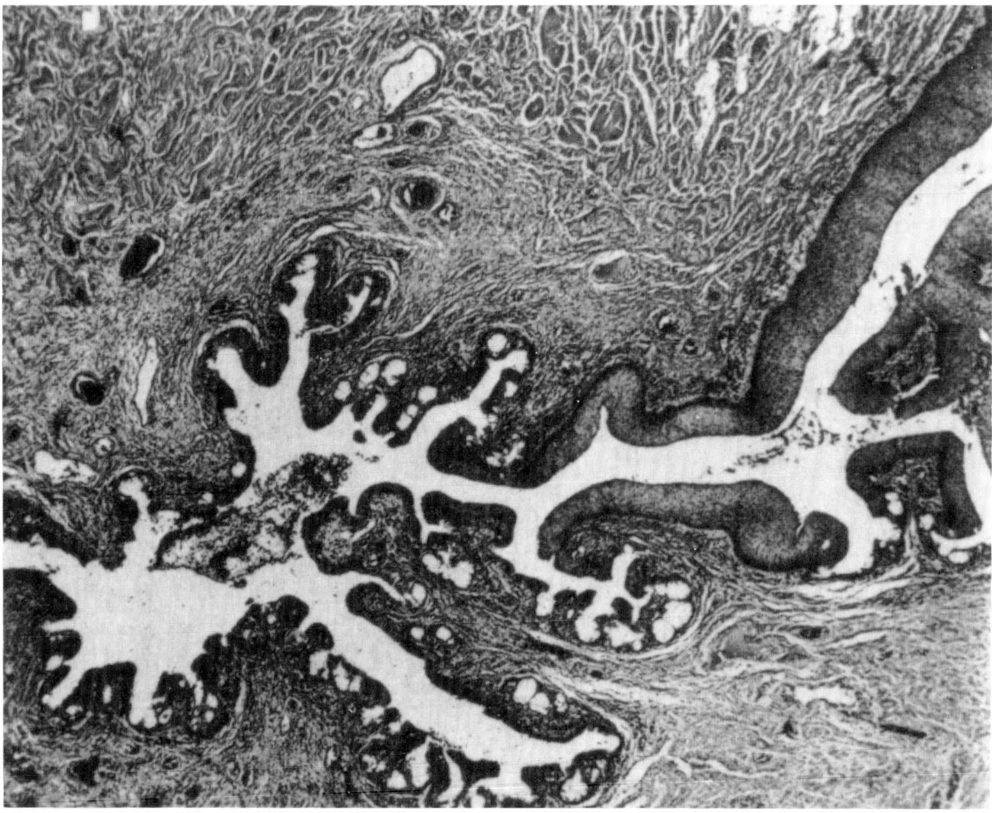

Figure 5.12. Epithelium of the canaliculus and lacrimal sac. The puncta and canaliculi are lined with nonkeratinizing, stratified squamous epithelium. The lacrimal sac is lined with a stratified columnar epithelium containing goblet cells. (Reproduced from Ryan SJ, Font RL: *American Journal of Ophthalmology* 76:73, 1973.)

medial canthal angle to the osseous nasolacrimal canal. Before entering the nasolacrimal canal, the lacrimal sac narrows, forming an isthmus.

The nasolacrimal duct begins at the point where it enters the osseous nasolacrimal canal of the maxillary bone. The nasolacrimal duct extends into the nose, beneath the inferior turbinate (Fig. 5.13). The nasolacrimal duct proceeds in the inferior meatus prior to penetrating the nasal mucosa to produce a substantial meatal portion of the duct. The inferior opening of the nasolacrimal duct is at the ostium lacrimale. The valve of Hasner is a fold of nasal mucosa at the termination of the nasolacrimal duct. The direction of the nasolacrimal canal is vertical. However, it is oriented slightly laterally, which is important to recognize when probing or intubating the nasolacrimal system.

Jones and Wobig (1976) postulated a lacrimal pump which actively pumps tears into the nose (Fig. 5.14). This lacrimal pump has three major components: the superficial and deep heads of the pretarsal orbicularis (Horner's) muscle, the deep heads of the preseptal (Jones') muscle, and the lacrimal diaphragm. Contraction of the pretarsal orbicularis compresses the canaliculus and closes its vertical portion. Contraction of the preseptal orbicularis creates a negative pressure in the lacrimal sac. The lacrimal diaphragm is an extension of the periorbita and forms the lateral wall of the lacrimal sac. Pulling the lacrimal diaphragm laterally creates a negative pressure within the lacrimal sac. Closing the eyelids compresses the lacrimal ampullae, and shortens the canaliculi via the contraction of the deep heads of the pretarsal muscles (Horner's muscle), while expanding the lacrimal sac via the contraction of the deep heads of the preseptal muscle (Jones muscle). Opening the eyelids expands the canaliculus and ampulla to siphon tears into the canaliculus. Closing the eyelids again pushes and sucks

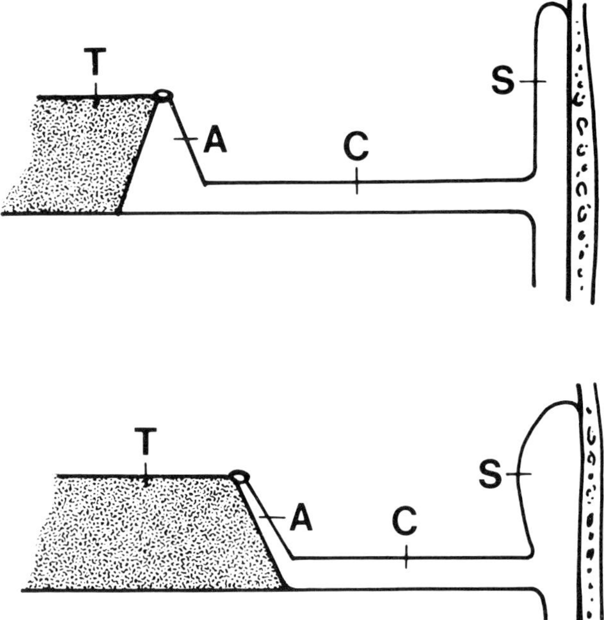

Figure 5.14. The lacrimal pump: The upper diagram represents the position of the excretory system with the eyelids open. The ampulla and canalicular system are expanded and the lacrimal sac is compressed. Closing the eyelids (*bottom diagram*) compresses the ampulla and canaliculus and expands the lacrimal sac. Expansion of the lacrimal sac creates a negative pressure which siphons tears into the sac. (Reproduced from Jones LT, Wobig JL: *Surgery of the Eyelids and Lacrimal System.* © 1976, Birmingham, Aesculapius Publishing Company.)

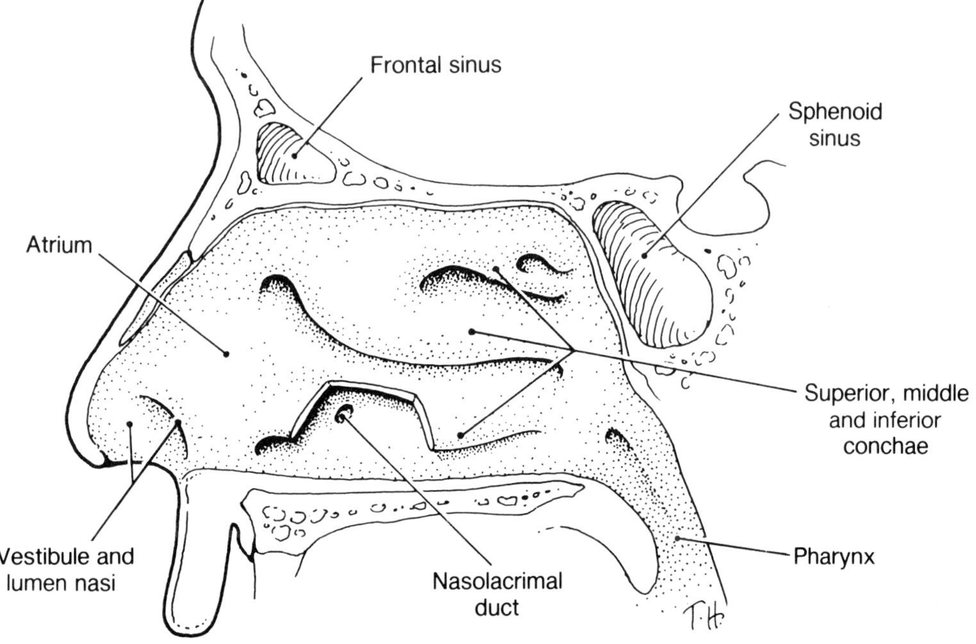

Figure 5.13. The lateral wall of the nose. The nasolacrimal duct enters the nose beneath the inferior turbinate.

the accumulated tears into the lacrimal sac. Reflux from the lacrimal sac into the canalicular system is prevented by the valve of Rosenmuller. Passage of tears down the nasolacrimal duct is probably mainly a result of gravity, although as the lacrimal sac collapses with eyelid opening some tears may be pushed down the nasolacrimal duct.

Using high speed photography, Doane (1981) questioned the Jones-Wobig lacrimal pump theory (Fig. 5.15). Early in the blink cycle, he noted the nasal portion of the eyelids approximate to effectively occlude the puncta, which is followed by compression of the canaliculi and lacrimal sac. At the time of maximum eyelid closure, the volume of fluid within the lacrimal system is at its minimum. Doane's theory of the lacrimal pump differs from that of Jones and Wobig, in that Doane (1981) postulated the complete collapse of the lacrimal excretory system on eyelid closure, while Jones and Wobig (1976) postulated that eyelid closure compressed the ampulla and canaliculi, but expanded the lacrimal sac.

The elastic expansion of the lacrimal channels during the opening phase of the blink cycle creates a partial vacuum within the system, which acts to siphon tears in the first few seconds following the blink. Actions of the canaliculi account for the functional integrity of the lacrimal system following a dacryocystorhinostomy, in which the lacrimal sac has been violated.

CLINICAL NOTE

Obstruction of the lacrimal excretory system is clinically manifested by tearing or dacryocystitis, based upon the level of obstruction. Dacryocystitis results from obstruction of the nasolacrimal duct, while tearing or epiphora occurs from obstruction at any level of the lacrimal excretory system. The frequency with which patients complain of tearing necessitates a standard approach to the evaluation of the lacrimal excretory system. The evaluation of tearing should rule out reflex tearing secondary to a corneal irritation (pseudoepiphora).

Examination of the lacrimal excretory system consists of five steps which allow a thorough evaluation of the system. The first step is history-taking, which indicates the nature of the problem. The second step consists of an external examination including a slit lamp examination, palpation of the lacrimal sac, examination of the nose, and should include studies of tear function to determine secondary causes of tearing. The third and fourth steps are the Jones I and II tests, which determine the functional and anatomical integrity of the lacrimal excretory system. Gentle palpation of the canaliculi with a fine probe is the fifth step of the examination, but should be performed only if required to determine the level of canalicular obstruction. Utilization of this systematic approach in lacrimal excretory system evaluation allows a topographical diagnosis of the site of obstruction.

The nose must always be examined in all cases of nasolacrimal obstruction to rule out nasal pathology as a cause of obstruction. In addition, it should be determined that adequate space is available for nasolacrimal surgery should such be indicated. It is necessary to vasoconstrict the nose to adequately evaluate

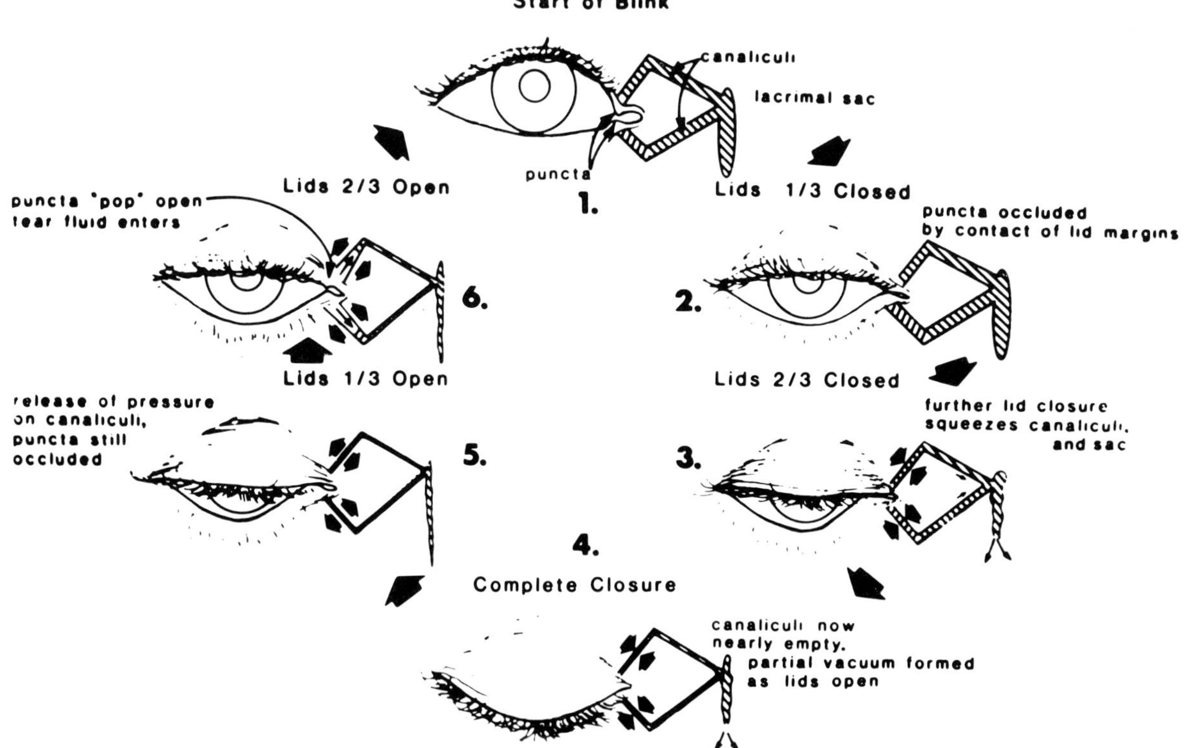

Figure 5.15. Doane's theory of the lacrimal pump. With the eyelids open the lacrimal excretory system is fully expanded, (*1*). Eyelid closure produces a compression of the entire lacrimal system, (*4*). (Reproduced from Doane MG: *Ophthalmology* 88:844, 1981.)

it (preferably with 5% cocaine spray which also provides anesthesia). Tumors, polyps, or septal deviations may be noted and be of great importance in evaluation. A headlight (indirect ophthalmoscope will suffice) and a nasal speculum or a hand-held nasoscope are adequate for most nasal evaluations. If nasal pathology is noted, an otolaryngology consultation is suggested.

In cases of acute dacryocystitis, the blockage is at the level of the nasolacrimal duct and further testing of nasolacrimal function is not productive or warranted. In chronic dacryocystitis, or if upon digital pressure applied over the lacrimal sac mucus or mucopurulent material is expressed from the canaliculus, then the diagnosis of an obstructed nasolacrimal duct with a patent upper system is obvious. If no obvious explanation for tearing is derived from history and external examination, further evaluation of the system is performed.

The Jones I test is a physiologic evaluation of the lacrimal system (Fig. 5.16). Fluorescein dye (2%) is placed in each eye on two occasions, approximately 30 sec apart. After instillation of the dye, the patient should sit forward, tilting his head about 30–45° from its vertical axis. Dye which enters the nose will, therefore, extend anteriorly in the nasal vestibule rather than posteriorly into the nasopharynx. After 5 min, each eye is observed to determine whether the dye has disappeared (dye disappearance test). If the dye has disappeared on one side, and is retained in the cul-de-sac on the other side, a blockage is presumed. The patient is then asked to occlude one nostril and to blow his nose through his other nostril into a facial tissue. If dye is not then recovered on nose-blowing, a nasal speculum is placed in the nose, and a small nasopharyngeal swab is placed under the inferior turbinate, in an attempt to recover the dye. If 5% cocaine spray has been previously used to vasoconstrict and anesthetize the nose, this will cause no discomfort.

The results of this procedure should be recorded as dye recovered or not recovered from the nose, to avoid confusion in recording a positive or negative test. If dye is recovered, the evaluation of the lacrimal excretory system demonstrates functional integrity. No physiologic obstruction is present. Failure to recover dye from the nose signifies a physiologic obstruction of the lacrimal system, and the evaluation continues with the Jones II test.

The Jones II, or secondary dye test, utilizes irrigation through the canaliculus to evaluate the anatomic patency of the lacrimal system (Fig. 5.17 A and B). Balanced salt solution is used for irrigation; no fluorescein dye is used in the irrigation. Several drops of an ophthalmic anesthetic agent are first placed in the cul-de-sac, 30 sec apart, to help anesthetize the puncta and canaliculi. The patient should be positioned in a head-forward position so that any fluid which passes through the nasolacrimal system will run out of the nose and can be caught in a receptacle, such as a white cup.

Attempts to irrigate the canalicular system provide specific information regarding the site of anatomical blockage. If fluid irrigated into the lower canaliculus is immediately regurgitated from the lower canaliculus, obstruction of the lower canaliculus is present. Fluid irrigated into the lower canaliculus and regurgitated through the upper canaliculus, without mucus debris, indicates obstruction in the common canaliculus or internal common punctum (Fig. 5.18). Jones calls this a positive canalicular test. If fluid is irrigated through the lower canaliculus and the lacrimal sac distends, either visibly or palpably, and mucus debris is present in the fluid regurgitated through the upper canaliculus, the nasolacrimal duct is obstructed. If there is a reflux of mucopurulent material during irrigation, chronic dacryocystitis is present. A patient with acute dacryocystitis should not be irrigated since

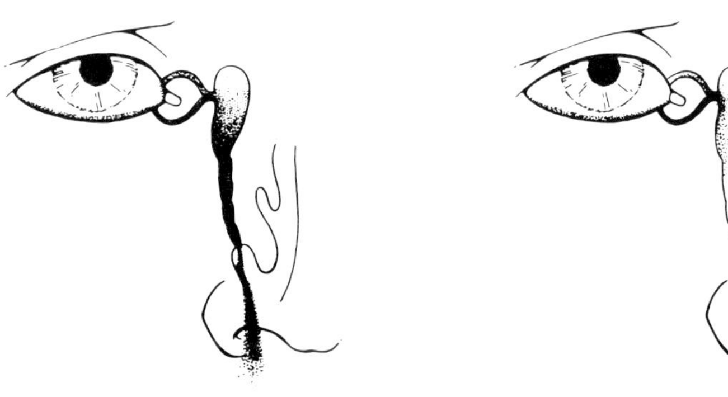

Figure 5.16. Jones I Test: Physiologic evaluation of the lacrimal excretory system.

JONES II TEST
ANATOMIC ANALYSIS

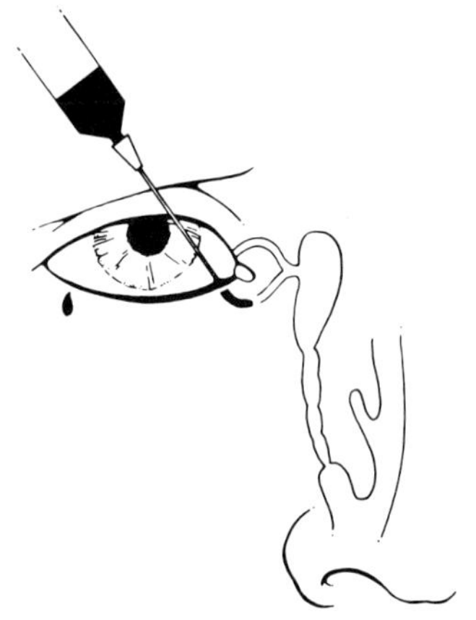
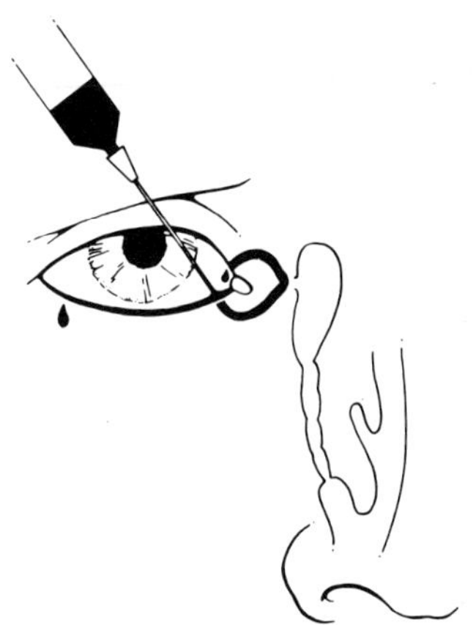

BLOCKAGE OF
LOWER CANALICULUS

BLOCKAGE AT ICP
POSITIVE CANALICULAR TEST

BLOCKAGE OF NLD
SAC DISTENDS

SHOULD CONFIRM LEVEL OF CANALICULAR OBSTRUCTION BY PALPATION WITH A
PROBE

JONES II TEST
ANATOMIC ANALYSIS

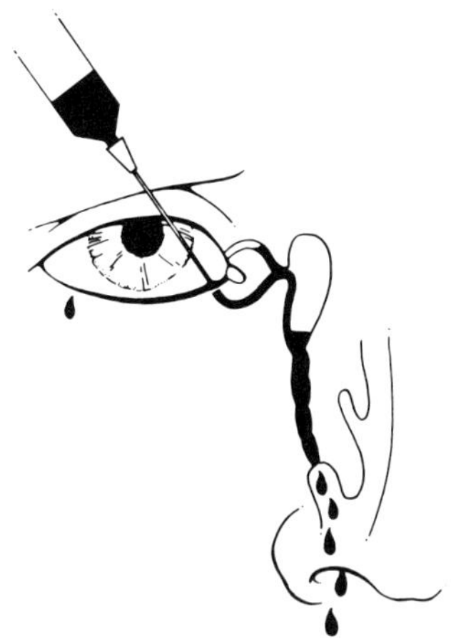

DYE-STAINED FLUID TO NOSE
FUNCTIONAL BLOCK OF
NASOLACRIMAL DUCT

CLEAR FLUID TO NOSE
PUMP FAILURE
FLACCID CANALICULAR SYSTEM
OR PUNCTAL STENOSIS
(DYE NEVER REACHED SAC)

Figure 5.17.   Jones II Test: Anatomic evaluation of the lacrimal excretory system (see text for details).

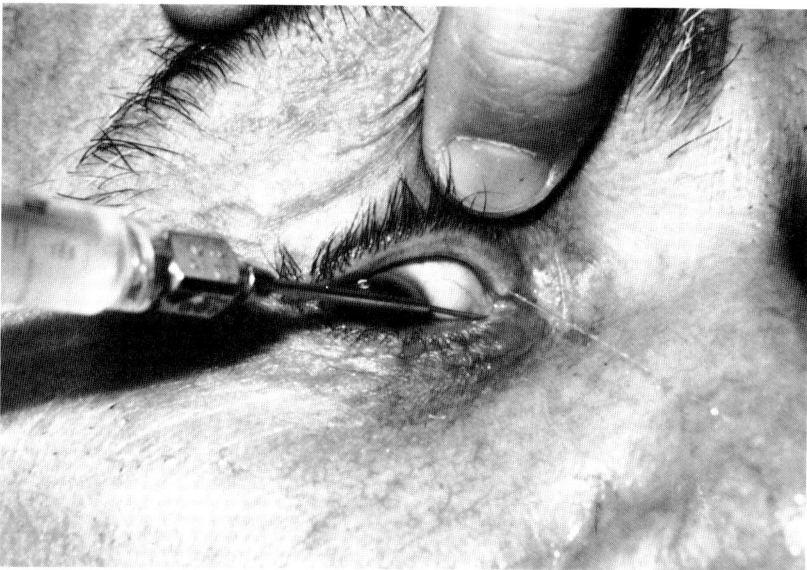

Figure 5.18. Positive canalicular test. Clear fluid injected into the lower canalicular system, indicating obstruction at the internal common punctum.

the obstruction must be present at the nasolacrimal duct. Irrigation provides no additional information and may spread the infection.

If fluid can be irrigated to the nose, a complete blockage of the lacrimal excretory system cannot be present. Fluid irrigated to the nose should be carefully examined to see if the fluid is dye-stained. Dye-stained fluid is an important finding which implies that the fluorescein used for the Jones I test was reservoired in the lacrimal sac. The canaliculus and the internal common punctum are physiologically patent in this circumstance. A relative obstruction, partial obstruction, or functional block must be present at the level of the nasolacrimal duct. Clear fluid irrigated into the nose indicates a failure of the dye to have reached the lacrimal sac. This may be due to eyelid, punctal, or canalicular problems, or a failure of the lacrimal pump to function, as in a facial nerve palsy.

In the elderly, loss of canalicular system resiliency may prevent adequate tear flow. In this situation, the patient will have an anatomically patent, but non-functioning, canalicular system. Jones calls this the flaccid, or floppy, canalicular syndrome. Tightening the eyelids, if lax, may improve this condition (Hill, 1979). Bypass of the lacrimal excretory system by a conjunctivodacryocystorhinostomy should be a last resort for this condition.

The tests outlined have already determined the general level of obstruction. Gentle probe palpation of the lacrimal excretory system is the final step in evaluation, if the block has been determined to be in the canaliculus and the exact site is to be determined. If the site of obstruction is thought to be the canaliculus, common canaliculus, or internal common punctum, gentle palpation with a probe is confirmatory. After instillation of anesthetic drops, the punctum should be dilated to permit easy passage of a 0–0000 size probe. As the canaliculus is probed, one notes the presence of a (soft) stop or resistance to the probe. This resistance indicates the site of obstruction. It can

also be noted that the canaliculus and eyelid are moved medially with gentle force applied to the probe when a canalicular obstruction exists. The distance probed from the punctum to the area of obstruction is noted. A very hard stop to probe passage is bone obstruction, most likely the nasal wall of the lacrimal fossa. In cases of acute dacryocystitis the valve of Rosenmuller is so inflamed that it is seldom beneficial to attempt a canalicular probing to relieve distention of the sac, and such probing may result in scarring or cellulitis. Nasolacrimal duct probings in adults to attempt to relieve nasolacrimal duct obstruction are to be condemned as they may result in scarring of the canaliculus or sac and seldom result in any permanent benefit.

We have found more elaborate routine testing of the lacrimal excretory system, such as dacryocystograms, fluoroscopy, and technetium scanning, to be overused, expensive, and generally provide no additional information. These tests are certainly not a substitute for the ability to perform a clinical evaluation of the lacrimal system. In cases where a lacrimal tumor is suspected, radiographs of this region should be obtained to look for bony erosion. In cases where intermittent epiphora is present, despite a normal Jones I test and the absence of reflex tearing, a dacryocystogram may be indicated to look for dacryoliths of the lacrimal sac or other causes of intermittent obstruction.

In infants, the terminal portion of the nasolacrimal duct extending into the nose may not be patent. This condition is best described as a failure to canalize, rather than an occlusive phenomenon, of the nasolacrimal duct at birth. The incidence of congenital nasolacrimal obstruction may vary from 6–73% (Guerry and Kendig, 1948; Cassidy, 1948). The most common etiology of congenital nasolacrimal duct obstruction is the failure of the nasolacrimal canal to perforate the nasal mucosa (Busse, et al, 1980). There may also be an incomplete penetration of the maxilla by the nasolacrimal canal, or the canal may terminate blindly in the inferior turbinate or the medial wall of the maxillary sinus (Jones and Wobig, 1976). Congenital anomalies of the lacrimal excretory system are reviewed in Chapter 1, "Embryology of the Eyelids, Lacrimal System and Orbit."

Recognition of the precise anatomic location of the lacrimal excretory system obstruction will direct the proper surgical approach for correcting the problem. Four basic procedures are now utilized to treat abnormalities of the lacrimal excretory system: nasolacrimal duct probing (should be performed only in infants and children as it causes scarring in adults); dacryocystorhinostomy (DCR); silicone intubation of the canalicular system; and conjunctivodacryocystorhinostomy (C-DCR). The standard DCR bypasses the nasolacrimal duct and establishes direct continuity of the lacrimal sac and the nasal cavity. However, obstruction of the canaliculi and internal common punctum are not benefited by a DCR. McLachlan and co-workers (1980) reviewed 291 DCR's and found blockage of the internal common punctum to be the most common cause of failure. Localized obstruction of the canaliculi or internal common punctum requires silicone intubation to stent the system. Silicone tubing is left in place for a minimum of 6 months, which permits scar maturation and softening to decrease the chance of obstruction recurrence.

Diffuse canalicular obstruction requires a C-DCR with the placement of a Jones tube to bypass the lacrimal excretory system.

References

Anderson RL, Bozeman MA, Voss JG, et al: Individual and site variation in composition of facial surface lipids. *J Invest Dermatol* 58:369, 1972.

Anderson RL, Dixon RS: The role of Whitnall's ligament in ptosis surgery. *Arch Ophthalmol* 97:705, 1979.

Andrew JS: Human tear film lipids. I. Composition of the principal nonpolar component. *Exp Eye Res* 10:223, 1970.

Busse H, Müller KM, Kroll P: Radiological and histological findings of the lacrimal passages of the newborn. *Arch Ophthalmol* 98:528, 1981.

Cassidy JV: Dacryocystitis in infancy. *Am J Ophthalmol* 31:773, 1948.

Crosby DG, Humphrey T, Lauer EW: *Correlative Anatomy of the Nervous System*. New York, Macmillan, 1962.

Czermak W: Die topographischen Beziehungen der Augenehohle, Augn. Anzliche Unterrichtstafeln (Magnus) 1895, IX, Breslau.

Doane MG: Interactions of eyelids and tears in corneal wetting and the dynamics of the normal human eyeblink. *Am J Ophthalmol* 89:507–516, 1980.

Doane MG: Blinking and the mechanics of the lacrimal drainage system. *Ophthalmology* 88:844, 1981.

Duke-Elder S, Wybar KC: The anatomy of the visual system. In Duke-Elder S (ed): *System of Ophthalmology*, Vol 2. St. Louis, CV Mosby, 1961, p 562.

Egeberg J, Jensen OA: The ultrastructure of the acini of the human lacrimal gland. *Acta Ophthalmol* 47:400, 1969.

Fleming A: On remarkable bacteriolytic element found in tissue and secretions. *Proc Roy Soc London* (b) 93:306, 1922.

Ford LC, DeLange RJ, Petty RW: Identification of a non-lysozymal bactericidal factor (beta lysin) in human tears and aqueous humor. *Am J Ophthalmol* 81:30, 1976.

Forrest AW: Epithelial lacrimal gland tumors: Pathology as a guide to prognosis. *Trans Am Acad Ophthalmol Otolaryng* 58:848, 1954.

Forrest AW: Pathologic criteria for effective management of epithelial lacrimal gland tumors. *Am J Ophthalmol* 71:178, 1971.

Goz A, cited by Whitnall SE: *The Anatomy of the Human Orbit and Accessory Organs of Vision*, ed 2. London, Oxford University Press, 1932, p 209.

Guerry D, Kendig EL: Congenital impatency of the lacrimal duct. *Arch Ophthalmol* 39:193, 1948.

Hawes MJ, Dortzbach RK: The microscopic anatomy of the lower eyelid retractors. *Arch Ophthalmol* 100:1313, 1982.

Henderson JW: *Orbital Tumors*. New York, Brian C Decker, 1980, p 394.

Henderson JW, Neault RW: En bloc removal of intrinsic neoplasms of the lacrimal gland. *Am J Ophthalmol* 82:905, 1976.

Hill JC: Treatment of epiphora owing to flaccid eyelids. *Arch Ophthalmol* 97:323, 1979.

Holly FJ: Formation and rupture of the tear film. *Exp Eye Res* 15:515, 1973.

Holly FJ, Lemp MA: Tear physiology and dry eyes. *Surv Ophthalmol* 22:69, 1977.

Iwamoto T, Jakobiec FA: Lacrimal glands. In Duane TD, Jaeger EA (eds): *Biomedical Foundations of Ophthalmology*, Vol 1, Chapter 30. Philadelphia, Harper & Row, 1982, p 20.

Iwata S, Lemp MA, Holly FJ, Dohlman CH: Evaporation rate of water from the precorneal tear film and cornea in the rabbit. *Invest Ophthalmol* 8:613, 1969.

Jakobiec FA, Yeo JH, Trokel SL, Abbott GF, et al: Combined clinical and computed tomographic diagnosis of primary lacrimal fossa lesions. *Am J Ophthalmol* 94:785, 1982.

Jones LT: The lacrimal secretory system and its treatment. *Am J Ophthalmol* 62:47, 1966.

Jones LT, Wobig JL: *Surgery of the Eyelids and Lacrimal System*. Birmingham, Aesculapius, 1976, p 58.

Katz J, Kaufman HE: Corneal exposure during sleep (nocturnal lagophthalmos). *Arch Ophthalmol* 95:449, 1977.

Lamberts DW, Foster CS, Perry HD: Schirmer test after topical anesthesia and the tear meniscus height in normal eyes. *Arch Ophthalmol* 97:1082, 1979.

Lemp MA: Breakup of the tear film. In Holly FJ, Lemp MA (eds): *The Preocular Tear Film and Dry Eye Syndromes*, Internat Ophthalmol Clin 13 (1). Boston, Little, Brown, 1973, p 97.

Lemp MA: The mucin deficient dry eye. In Holly FJ, Lemp MA (eds): *The Preocular Tear Film and Dry Eye Syndromes*, Internat Ophthalmol Clin 13 (1). Boston, Little, Brown, 1973, p 185.

Lemp MA, Dohlman CH, Holly FJ: Corneal desiccation despite normal tear volume. *Ann Ophthalmol* 2:258, 1970.

Lemp MA, Dohlman CH, Kuwabara T, Holly FJ, Carroll JM: Dry eye secondary to mucin deficiency. *Trans Am Acad Ophthalmol Otolaryng* 75:1223, 1971.

Lemp MA, Hamil JR: Factors affecting tear film breakup in normal eyes. *Arch Ophthalmol* 89:103, 1973.

McDonald JE: Surface phenomena of tear films. *Trans Am Ophthalmol Soc* 66:905, 1968.

McLachlan DL, Shannon GM, Flanagan JC: Results of dacryocystorhinostomy. Analysis of the operations. *Ophthalmol Surg* 11:427, 1980.

Mishima S: Some physiological aspects of the precorneal tear film. *Arch Ophthalmol* 73:233, 1965.

Norn MS: Dessication of the precorneal tear film. I. Corneal wetting time. *Acta Ophthalmol* 47:865, 1969.

Orzalesi N, Riva A, Testa F: Fine structure of the human lacrimal gland. I. The normal gland. *J Submicro Cytol* 3:283, 1971.

Reese AB: Treatment of expanding lesions of the orbit: With particular regard to those arising in the lacrimal gland. *Am J Ophthalmol* 70:767, 1970.

Riley CM, Day RL, Greeley DM, Langford WS: Central autonomic dysfunction with defective lacrimation. Report of five cases. *Pediatrics* 3:468, 1949.

Rubin IE: Adenocarcinoma of the lacrimal gland following a mixed tumor of 25 years' duration. *Arch Ophthalmol* 36:686, 1947.

Ruskin SL: Control of tearing by blocking the nasal ganglion. *Arch Ophthalmol* 4:208, 1930.

Scherz W, Dohlman CH: Is the lacrimal gland dispensable? Keratoconjunctivitis sicca after lacrimal gland removal. *Arch Ophthalmol* 93:281, 1975.

Smith B, Petrelli R: Surgical repair of the prolapsed lacrimal gland. *Arch Ophthalmol* 96:113, 1978.

Stewart WB, Krohel GB, Wright JE: Lacrimal gland and fossa lesions: An approach to diagnoses and management. *Ophthalmology* 86:886, 1979.

Sturrock GD: Nocturnal lagophthalmos and recurrent erosion. *Br J Ophthalmol* 60:97, 1976.

Taylor HR, Louis WJ: Significance of tear function test abnormalities. *Ann Ophthalmol* 12:531, 1980.

van Haeringen NJ: Clinical biochemistry of tears. *Surv Ophthalmol* 26:84–96, 1981.

Waller RR, Riley FC, Henderson JW: Malignant mixed tumor of the lacrimal gland: Occult source of metastatic carcinoma. *Arch Ophthalmol* 90:297, 1973.

Whitnall SE: *The Anatomy of the Human Orbit and Accessory Organs of Vision*, ed 2. London, Oxford University Press, 1932, p 219.

Zimmerman LE, Sanders TE, Ackerman LV: Epithelial tumors of the lacrimal gland: Prognostic and therapeutic significance of histologic types. *Internat Ophthalmol Clin* 2:337, 1962.

CHAPTER 6

Connective Tissue Planes of the Orbit

A diffuse connective tissue framework supports orbital structures, maximizes their function, and maintains anatomic relationships between these structures. Historically, four major connective tissue components in the orbit have been described: the bulbar fascia (Tenon's capsule); fascial sheaths of the extraocular muscles; the common muscle sheath connecting the four rectus muscles; and the medial and lateral check ligaments (Fig. 6.1 *A* and *B*) (Tenon, 1806; Motais, 1905; Hesser, 1913; Whitnall, 1932; Scobee, 1948; Hosokawa, 1956; Neiger, 1960; Fink, 1962).

Early anatomic studies of the human orbital connective tissue were performed on exenteration specimens or on gross orbital dissections. Unfortunately, structural and functional intricacies of orbital connective tissue were not appreciated in these microscopic or gross studies. The collapse of orbital specimens further obscured anatomic relationships and complicated conventional study of orbital connective tissue.

Koornneef (1977, 1977, 1979, 1982) recognized the problems inherent in the study of orbital connective tissue. He first appreciated mechanical and anatomic features of orbital structures, then sought a more efficient means of relating these findings to clinical phenomena. In particular, he desired to explain the frequent association of restricted extraocular motility to an orbital blow-out fracture, despite the failure to discern, on surgical exploration, inferior rectus or oblique muscle entrapment in the fracture.

Koornneef advanced the study of orbital connective tissue by a unique approach. To study the anatomic relationships of structures in the anterior orbit, he dissected an intact orbit. Instead of the standard gross anatomic dissection, he used a cul-de-sac surgical approach and an operating microscope to delineate structural features of the anterior orbit. In particular, he studied attachments of the periorbita and Tenon's capsule. However, this technique was not feasible for examining the posterior orbit, unless orbital walls were removed, which in turn disrupted the very continuities which were to be studied. Therefore, a different method of study was needed.

Koornneef utilized a thick-tissue technique for studying anatomic relationships of posterior orbit structures. Orbital blocks, consisting of the intact orbital contents and walls, were decalcified, and then sectioned into 60–140 μ thick sections (as opposed to the conventional 7–10 μ thick sections). Serial sections through the orbit enabled him to reconstruct orbital tissues and permitted a better appreciation of the anatomic foundation of orbital connective tissue. He identified a spatial and specific architecture of the human orbit with interindividual uniformity and bilateral symmetry.

Koornneef described three essentially distinct connective tissue systems within the orbit. The first, Tenon's capsule or the bulbar fascia, surrounds the globe and the extraocular muscles in the anterior orbit. The second connective tissue system connects Tenon's capsule to the periorbita near the anterior orbital margin. This dense, fibrous tissue septal system helps support the globe and maintains its range of extraocular motility. The third connective tissue system of the orbit is derived from the fascial sheaths of the extraocular muscles. These sheaths produce a diffuse connective tissue framework in the posterior orbit. Anteriorly, the fascial sheaths of the extraocular muscles consolidate to form the medial and lateral check ligaments and the intermuscular fibrous septa.

TENON'S CAPSULE

Tenon's capsule is a dense fibrous layer of elastic connective tissue which surrounds the eye and the extraocular muscles in the anterior orbit (Fig. 6.2). The capsule originates at the limbus and extends posteriorly to the optic nerve. Since Tenon's capsule envelops the globe, the extraocular muscles must penetrate the capsule prior to inserting onto the globe. Tenon's capsule is a barrier between the globe and remaining structures of the orbit which enhances the functional and anatomical autonomy of the globe.

As noted, all six extraocular muscles must penetrate Tenon's capsule prior to inserting onto the globe. The four rectus muscles penetrate Tenon's capsule posterior to the equator of the globe, while the two oblique muscles penetrate anterior to the equator of the globe. The fascial sheath of the extraocular muscles and the superior oblique tendon are attached to the inner surface of Tenon's capsule by a sleeve of tissue. As such, the muscles and tendons have unrestricted mobility, forward and backward, through Tenon's capsule.

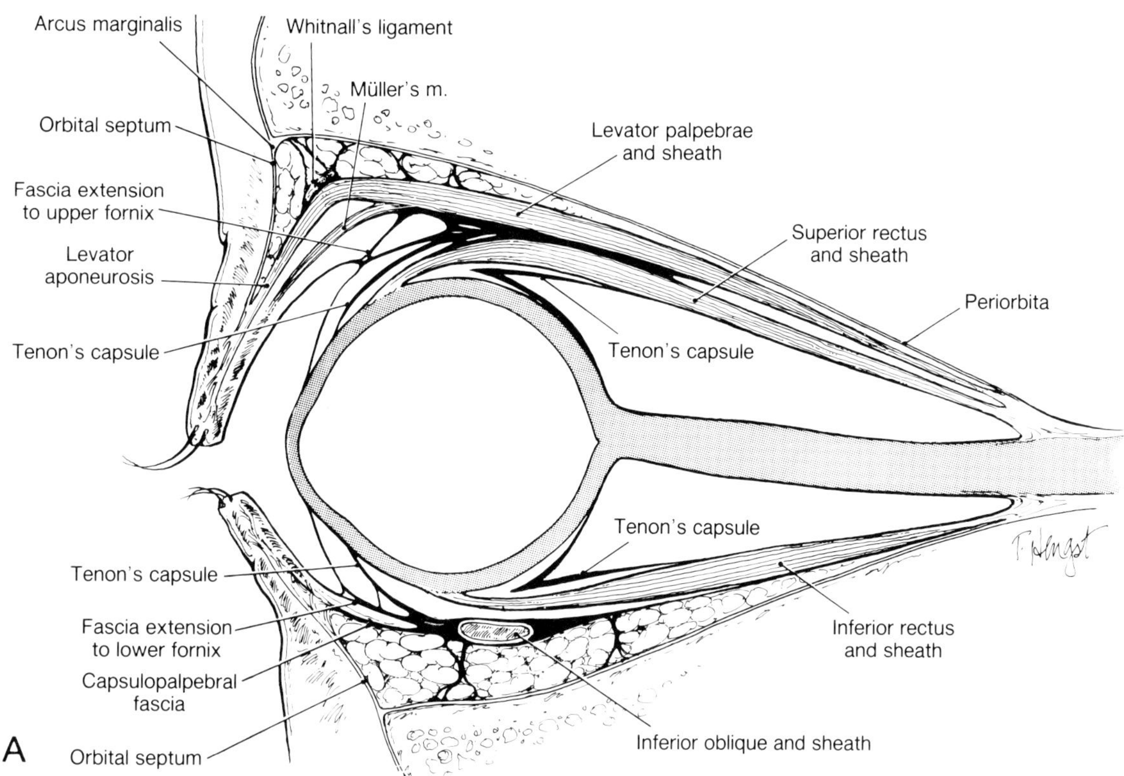

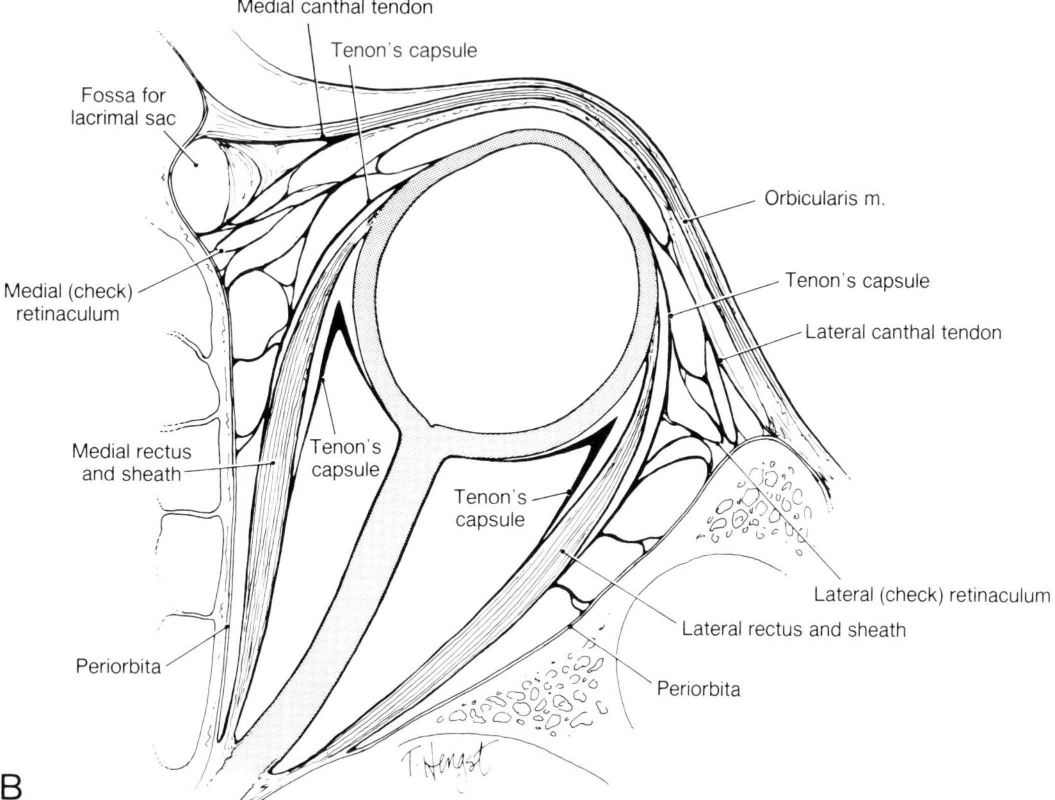

Figure 6.1. Connective tissue system of the orbit. (*A*), Sagittal section of the orbit demonstrating the diffuse attachments of the extraocular muscles to the periorbita and Tenon's capsule. (*B*), Section of the orbit illustrating the contribution of the medial and lateral recti to the corresponding check ligaments and canthal tendons.

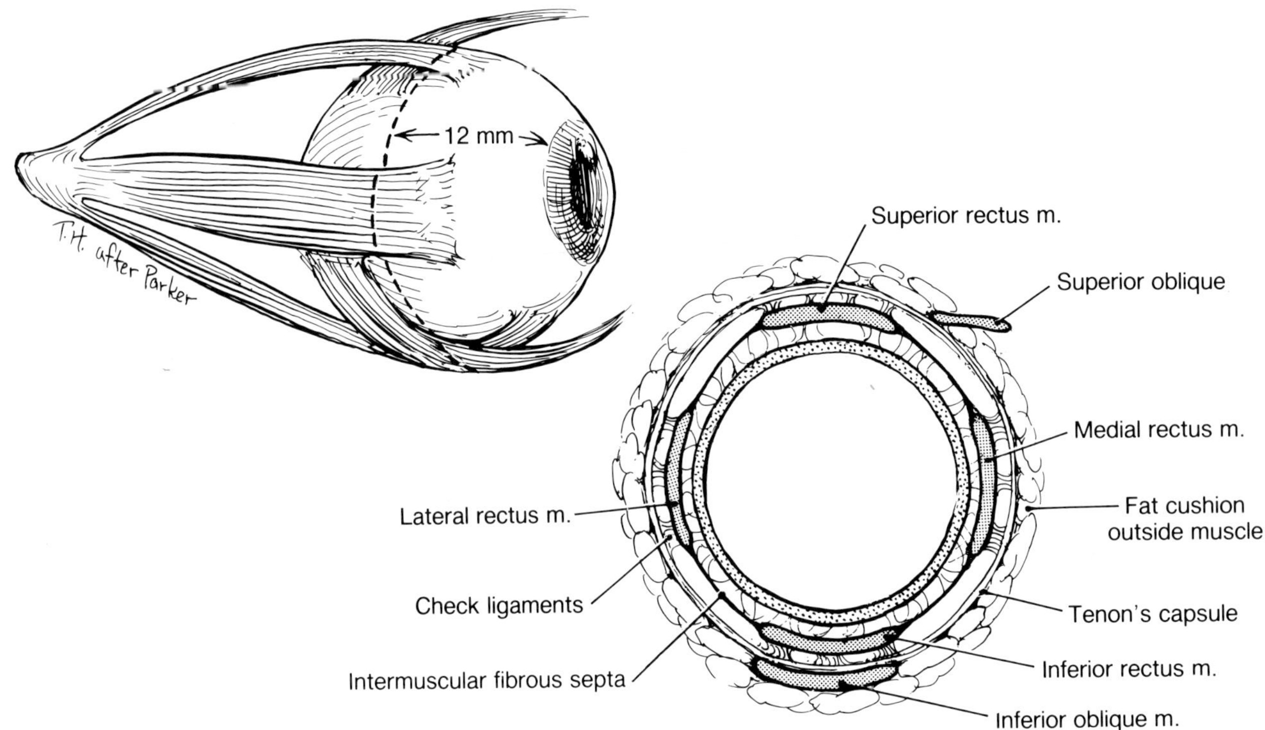

Figure 6.2.   Tenon's capsule. Equatorial section 12 mm from the limbus. (Modified from Parks MM: Ocular motility and strabismus. In Duane TD, Jaeger EA (eds): *Clinical Ophthalmology*, Vol. 1, Chapter 1. © 1982, Philadelphia, Harper & Row.)

The penetration of the rectus muscles divides Tenon's capsule into an anterior and a posterior portion. Anterior Tenon's capsule, which is approximately one-third of the capsule, is adherent to the episcleral tissue extending to the limbus. Extension of the fascial sheaths of the extraocular muscles produces the intermuscular fibrous septa. The thin anterior extension of this structure fuses with the anterior Tenon's capsule to form a single fascial plane. Posterior Tenon's capsule is that portion of the bulbar fascia posterior to the penetration of the rectus muscles, and it extends to the optic nerve. The inner lining of Tenon's capsule is smooth and is an ideal surface to oppose the sclera since it offers minimal resistance with ocular motility.

Tenon's capsule separates the globe and muscles from intraconal and extraconal fat. The intraconal fat is within the anatomically defined potential space within the extraocular muscles. In the anterior orbit, the fascial sheaths of the rectus muscles produce confluent intermuscular fibrous septa. However, posterior to the globe, the intermuscular fibrous septa are not continuous, and therefore a true muscle cone is not present (Whitnall, 1932; Koornneef, 1977). The extraconal fat proceeds into the anterior orbit, approximately 10 mm from the limbus. The extraconal fat may prolapse from the inferior or superior conjunctival fornix (Fig. 6.3).

CLINICAL NOTE

Tenon's capsule produces an anatomic barrier between the globe and the retrobulbar fat. Extraocular mobility is thereby enhanced due to the lack of restrictive forces. Retrobulbar fat introduced into sub-Tenon's space may produce an intense lipogranulomatous inflammation. A secondary proliferative inflammatory response may result in adhesions and cicatrization, with marked reduction in extraocular motility (Parks, 1982). Conditions which allow this extrusion of fat into sub-Tenon's space are trauma such as orbital fractures, and incisions in Tenon's capsule more than 10 mm from the limbus. In addition, inadvertent penetration of Tenon's capsule during strabismus surgery may result in an inflammation due to the introduction of fat into sub-Tenon's space (Parks, 1982).

ANTERIOR ORBITAL CONNECTIVE TISSUE SYSTEM

Koornneef's unique approach to orbital dissections demonstrated various connective tissue systems in the orbit. Via dissection of the orbit through the conjunctival fornix, he noted attachments between the globe and periorbita. Careful dissection in this area revealed connective tissue septa with enclosed adipose tissue (Fig. 6.4). By exposure of the cul-de-sac, he noted connective tissue septa in a 360° encapsulation of the globe (Fig. 6.5).

The connective tissue septa encircled and supported the globe within the bony orbit. In addition, septa were also recognized between the superior and inferior oblique muscles and Tenon's capsule, and the corresponding rectus muscles. A connective tissue mass was identified between the inferior oblique and inferior rectus muscle which contributes to Lockwood's suspensory ligament (Lockwood, 1886). The diffuse anterior connective tissue septal system supports and stabilizes the globe within the central aspect of the orbit.

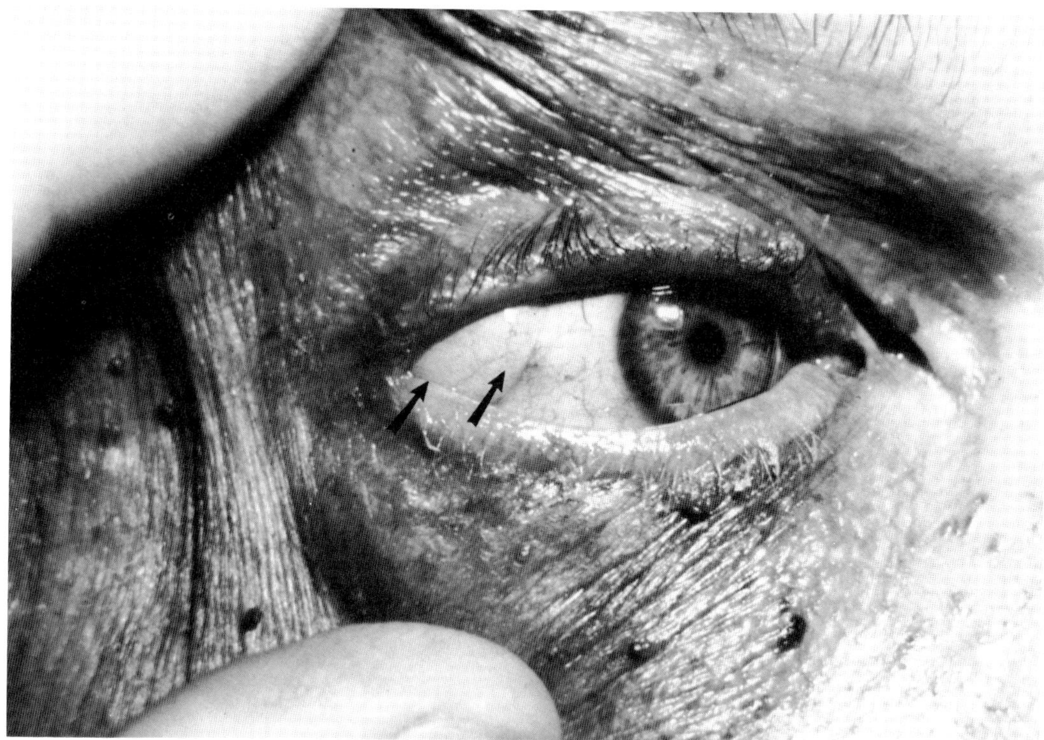

Figure 6.3. Prolapsed retrobulbar orbital fat (*arrows*).

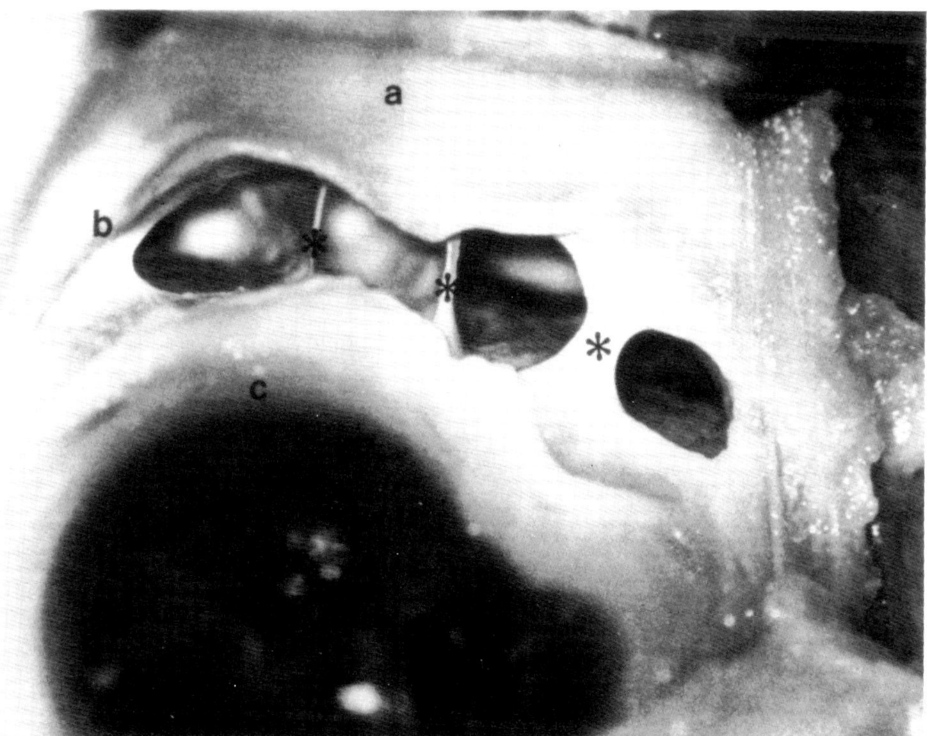

Figure 6.4. Connective tissue septa of the anterior orbit. Exposure of the superior conjunctival fornix reveals connective tissue septa between Tenon's capsule and the periorbita. Abbreviations: (*a*) reflected upper eyelid; (*b*), periorbita; (*c*), bulbar conjunctiva. (Reproduced from Koornneef L: Orbital connective tissue. In Duane TD, Jaeger EA (eds): *Biomedical Foundations of Ophthalmology*, Vol. 1, Chapter 32. © 1982, Philadelphia, Harper & Row.)

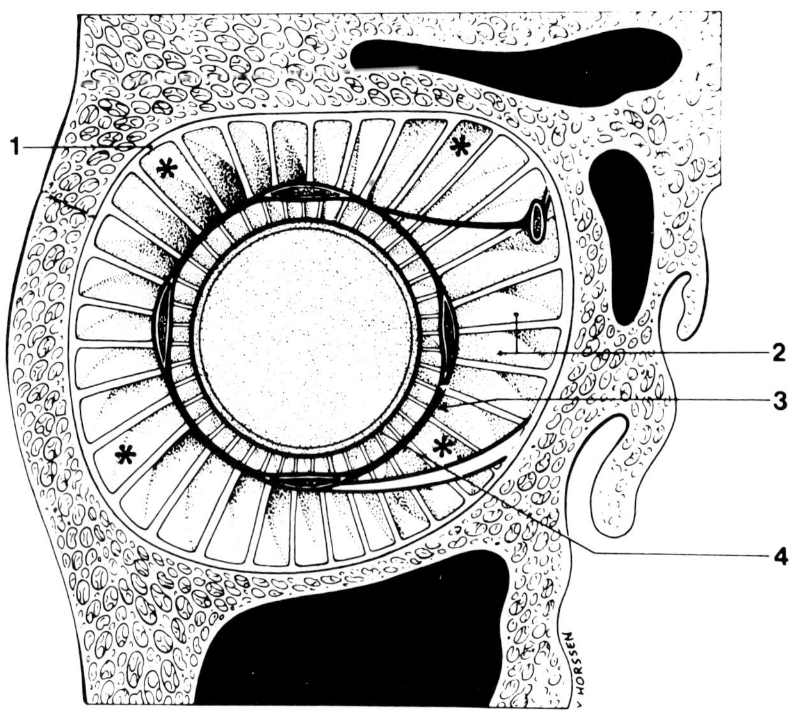

Figure 6.5. Schematic representation of the anterior orbital connective tissue septa. *1*, Periorbita; *2*, Common muscle sheath around the eye; *3*, Fibrous septa. (Reproduced from Koornneef L: *Archives of Ophthalmology* 95:1269, 1977, American Medical Association.)

Koornneef states that traction on the globe best illustrates the function of the anterior connective tissue system. The firm connective tissue strands between Tenon's capsule and the periorbita change direction with traction on the globe (Fig. 6.6). Thus, one of the limiting factors in globe displacement is the connective tissue septa. Koornneef believed the connective tissue septa in the superior region played a significant role in preventing a downward displacement of the eye.

EXTRAOCULAR MUSCLE CONNECTIVE TISSUE SYSTEM

A connective tissue system derived from the fascial sheaths of the extraocular muscles was described by Koornneef (1977). In the posterior orbit, the extraocular muscle fascial sheaths serve as the origin of a highly characteristic connective tissue framework. The connective tissue becomes better defined and organized in the immediate retrobulbar region. In the anterior orbit, the fascial sheaths of the extraocular muscles contribute to additional connective tissue systems. The fascial sheaths of the medial and lateral rectus muscles contribute to the corresponding check ligaments which anchor these muscles to the periorbita. Connective tissue extensions of the fascial sheaths produce the intermuscular fibrous septum, which becomes a confluent structure at the level of the globe.

The posterior orbital connective tissue is poorly developed at the orbital apex. Serial section reconstruction of the orbital connective tissue system reflects progressive complexity of the system from the orbital apex to the anterior orbit. Koornneef depicted the connective tissue system at various levels of the orbit in four schematic drawings (Fig. 6.7).

At the level of the orbital apex (Fig. 6.7*A*) the connective tissue attachments of the extraocular muscles are already developed. The superior rectus and levator complex is attached to the orbital roof by connective septa extending from its superolateral border. The superior oblique muscle is supported by an arch-like band of connective tissue in the superior medial angle of the orbit. The medial rectus connects primarily to the orbital floor, and has a few connections to the medial orbital wall. The inferior rectus muscle attaches to the periorbita overlying the smooth muscle, which covers the inferior orbital fissure (Müller's muscle). The lateral rectus attaches both to the lateral orbital wall and to the periorbita overlying Müller's muscle in the inferior orbital fissure.

As the muscles proceed anteriorly, attachments become more diffuse, enhancing the supporting structure and confluent nature of the orbital connective tissue system. The levator/superior rectus complex maintains its attachment to the periorbita of the frontal bone until it reaches the level of the equator of the globe. Connective tissue septa from this complex form a hammock for the superior ophthalmic vein (Fig. 6.7*B*). The medial and lateral rectus muscles also contribute to the supporting structure for this vein. Prior to its attachment to the periorbita of the frontal bone, the connective tissue extension of the fascial sheath of this complex fans first to the lateral, then the medial rectus muscles. At the level of the globe, the fascial sheath of the superior rectus/levator complex contributes to the intermuscular fibrous septa which envelopes the globe (Fig. 6.7 *C* and *D*).

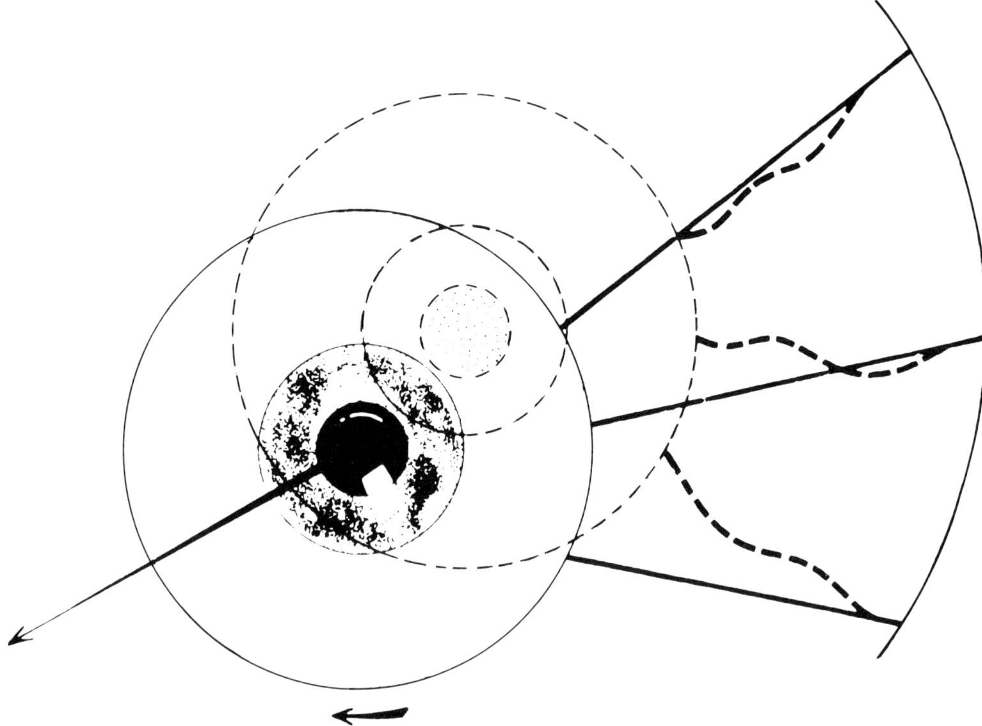

Figure 6.6. Anterior orbital connective tissue system. Alteration of fiber course with varying traction on the globe. (Reproduced from Koornneef L: *Archives of Ophthalmology* 95:1269, 1977, American Medical Association.)

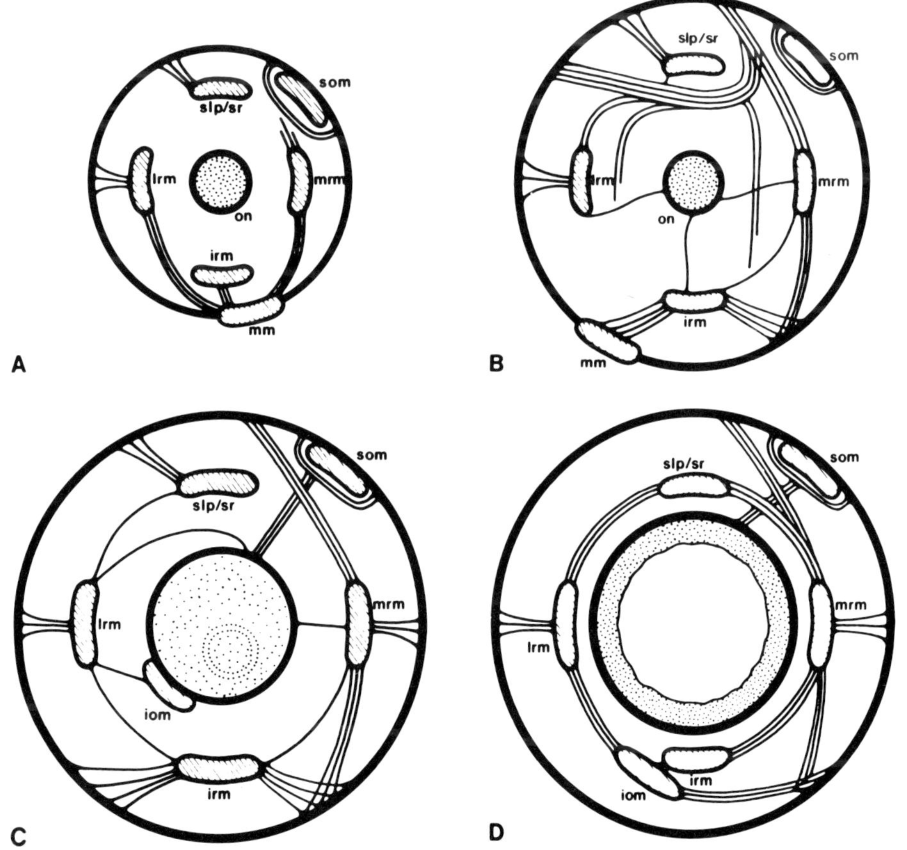

Figure 6.7. Extraocular muscle connective tissue system. Highly schematic representation of the connective tissue system of the extraocular muscles. (*A*), Coronal section near the orbital apex. (*B*), Coronal section near the posterior portion of the globe. (*C*), Coronal section lying just anterior to the posterior portion of the globe. (*D*), Coronal section near the equator of the globe. Abbreviations: *slp/sr*, levator palpebrae superioris, superior rectus complex; *lrm*, lateral rectus muscle; *iom*, inferior oblique muscle; *irm*, inferior rectus muscle; *mm*, Müller's muscle; *mrm*, medial rectus muscle; *som*, superior oblique muscle; *on*, optic nerve. (Reproduced from Koornneef L: *Ophthalmology* 86:76, 1979.)

The medial rectus has the most diffuse attachments of all the extraocular muscles. At the orbital apex (Fig. 6.7A), the medial rectus muscle exhibits cephalocaudal connections to the orbital floor. No connective tissue to the medial wall or the roof of the orbit is identified posteriorly. In addition, loose connective tissues connecting the medial rectus to the inferior rectus muscle are also identified. In coronal sections from near the posterior pole of the globe (Fig. 6.7B), the diffuse connective tissue system of the medial rectus is evident. In addition to the previously recognized attachments to the orbital roof and floor, septal attachments connect the medial rectus muscle to the medial periorbita, the medial aspect of the globe and inferior rectus muscle. At the level of the equator of the globe (Fig. 6.7D), the medial rectus muscle has lateral connections with the levator-superior rectus complex. As the medial rectus proceeds anteriorly in the orbit, a connective tissue replaces the polychamber-like architecture of the fascial extensions of the muscle. The consolidation of connective tissue produces the medial check ligament.

The inferior rectus muscle has connective tissue attachments to Müller's muscle over the inferior orbital fissure at the orbital apex (Fig. 6.7A). The connective tissue system surrounds the inferior rectus and firmly attaches it to the periorbita of the orbital floor. At the level of the posterior portion of the globe (Fig. 6.7C), the inferior rectus has attachments to the medial and lateral rectus muscles. At the level of the equator (Fig. 6.7D), the inferior rectus has firm attachments with the inferior oblique muscle, as well as better defined connections with the medial and lateral rectus muscles.

The fascial extension of the inferior rectus proceeds anteriorly, beyond its insertion on the globe, as the capsulopalpebral fascia, to form the lower eyelid aponeurosis. Near its origin, the capsulopalpebral fascia encapsulates the inferior oblique muscle. The aponeurotic extension of the inferior rectus can be traced medially and laterally toward the corresponding rectus muscles, and beyond, to its attachments to the orbital walls. These fibers contribute to Lockwood's suspensory ligament which provides support for the globe.

At the orbital apex (Fig. 6.7A) the lateral rectus muscle has connective tissue attachments to the periorbita of the lateral orbital wall, and the periorbita covering Müller's muscle in the inferior orbital fissure. Proceeding anteriorly in the orbit (Fig. 6.7B), the lateral rectus muscle looses its attachment to the orbital floor, but maintains the close connections with the lateral orbital wall. In addition, superior extension of the fascial sheath of the lateral rectus muscle contributes to the septal system supporting the superior ophthalmic vein. A fine attachment to the optic nerve can also be identified. At the posterior portion of the globe (Fig. 6.7C), the lateral rectus has firm attachments with the inferior rectus. The superior extension of the lateral rectus fascial sheath joins with the levator/superior rectus complex, and the connective tissue of the superior oblique tendon. Fine attachments with the inferior oblique muscle occur prior to its global insertion. At the level of the equator of the globe, (Fig. 6.7D), the lateral rectus contributes to the intermuscular fibrous septa, the "muscle cone," while maintaining the attachments to the lateral orbital wall.

The superior oblique is autonomous in the superior medial quadrant of the orbit, until it reaches the posterior portion of the globe (Fig. 6.7C). Connective tissue septa intermingle with the superior extension of the medial rectus muscle and attach to the posterior Tenon's capsule. At the level of the equator (Fig. 6.7D), diffuse connective tissue attachments of the superior oblique are better defined. Progressing anteriorly, successive sections identify the anterior orbital connective tissue system, which produces a greater mechanical resistance to traction.

The inferior oblique muscle is unique in that it originates in the inferior nasal portion of the anterior orbit. The muscle proceeds from its origin in a posterior-lateral direction, to insert on the posterior portion of the globe. The inferior rectus lies above the inferior oblique, to which it displays firm attachments. In the posterior extension of the inferior oblique, toward its insertion, connective tissue attachments are solely to the lateral rectus muscle. As noted previously, the inferior oblique muscle forms part of the anterior connective system (Fig. 6.5). The fascial sheath of the inferior oblique muscle also contributes to Lockwood's suspensory ligament (Lockwood, 1886).

CLINICAL NOTE

Koornneef has helped define a structural organization of the human orbital connective tissue. Each extraocular muscle has its own separate, distinct connective tissue system, each with large areas of attachment and support. The diffuse connective tissue systems of the extraocular muscles probably have a role in coordinating extraocular motility.

The connective tissue system, particularly that of the inferior rectus muscle, accounts for the restriction of vertical motility following an orbital "blow-out" fracture. Clinical observations rarely identify the inferior rectus or oblique muscles in the orbital floor fracture. However, the tissue herniated into the maxillary sinus consists of the connective tissue septa and fat cushions of the inferior rectus muscles. Restriction of extraocular motility is therefore produced by the incarcerated fascial sheaths of the extraocular muscles, particularly that of the inferior rectus muscle (see Fig. 2.10A–D).

Although ingrained in our anatomic concepts, a distinct "muscle cone" does not exist in the orbit. The sheaths of the extraocular muscles extend anteriorly toward the globe to anatomically and functionally demarcate the orbit by forming the intermuscular fibrous septa. However, in the posterior orbit the intermuscular fibrous septa are poorly developed. Whitnall (1932) recognized the intermuscular septa to be nearly imperceptible in the posterior orbit and Koornneef (1977) was not able to identify a common muscle sheath posterior to the globe.

The muscle sheaths also contribute fibrous septa to the periorbita. The fibrous septa serve as check ligaments for the extraocular muscles. The check ligaments are the most pronounced for the lateral and medial rectus (Fig. 6.1A). The fibrous septa also extend from the superior and inferior rectus to insert at the conjunctival fornix, helping to form the suspensory ligaments of the upper and lower fornix.

The fascial extensions of the rectus muscles maximize the efficient action of these muscles. The periorbital attachments of the rectus muscles in the anterior orbit creates an efficient direction of action for these muscles. This is analogous to the role of Whitnall's ligament in creating an efficient direction of action for the levator muscle (Anderson and Dixon, 1979).

In summary, few areas of orbital anatomy have been so influenced by an individual as is represented by Koornneef's exhaustive study of the orbital connective tissue system. He has been a pioneer in this area and further studies may help clarify some still unexplained and undiscovered facets of orbital anatomy. His studies represent a unique approach to orbital dissection, and they identify and reconstruct the orbital connective tissue system. Koornneef delineated three basic systems of connective tissue: Tenon's capsule, the anterior connective tissue system, and the connective tissue system of the extraocular muscles.

References

Anderson RL, Dixon RS: The role of Whitnall's ligament in ptosis surgery. *Arch Opthalmol* 97:705, 1979.

Fink WH: *Surgery of the Vertical Muscles of the Eye.* Springfield, IL, Charles C Thomas, 1962.

Hesser C: Der Bindegewebsapparat und die glatte Muskulatur der Orbita beim Menschen in normalen Zustande. *Anat Hefte* 49:1, 1913.

Hosokawa H: A note on the fibrous apparatuses surrounding the human eyeball. *Okajimas Folia Anat Jap* 28:165, 1956.

Koornneef L: *Spatial Aspects of Orbital Musculo-fibrous Tissue in Man.* Amsterdam, Swets & Zeitlinger, 1977.

Koornneef L: New insights in the human orbital connective tissue: Results of a new anatomic approach. *Arch Ophthalmol* 95:1269, 1977.

Koornneef L: Details of the orbital connective tissue system in the adult. *Acta Morphol Neerl Scand* 15:1, 1977.

Koornneef L: Orbital septa: Anatomy and function. *Ophthalmology* 86:876, 1979.

Koornneef L: Orbital connective tissue. In Duane TD, Jaeger EA (eds): *Biomedical Foundations of Ophthalmology,* Vol 1, Chapter 32. Philadelphia, Harper & Row, 1982.

Lockwood CB: The anatomy of the muscles, ligaments, and fasciae of the orbit, etc. *J Anat Physiol* 20:1, 1886.

Motais M: De l'Appareil moteur oculair. In *Encyclopedie Francaise d'Ophthalmologie,* Vol. 2. Paris, Octave Dion, 1905.

Neiger M: Les structures conjunctives de l'orbite et le coussinet graisseux orbitaire. *Acta Anat (Basel)* (suppl.) 39:1, 1960.

Parks MM: Extraocular muscles. In Duane, TD, Jaeger EA (eds): *Clinical Ophthalmology,* Vol 1, Chapter 1. Philadelphia, Harper & Row, 1982, p 3.

Scobee RG: Anatomic factors in the etiology of heterotrofia. *Amer J Ophthalmol* 31:781, 1948.

Tenon JR: *Memoires sur l'anatomie, la pathologie et la chirurgie et surl'organe de la vue.* Paris, 1806, p 193. Cited by Koornneef L: New insights in the human orbital connective tissue: Result of a new anatomical approach. *Arch Opthalmol* 95:1269, 1977.

Whitnall SE: *Anatomy of the Human Orbit and Accessory Organs of Vision,* ed 2. London, Oxford University Press, 1932.

CHAPTER 7

Extraocular Muscles

Ocular muscles can be divided into two basic groups: intrinsic and extrinsic. The intrinsic ocular muscles are those associated with the movement of the lens and iris (ciliary body, iris dilator, and iris sphincter). The extrinsic muscles of the eye are the extraocular muscles. These include four rectus (superior, medial, inferior, lateral) and two oblique (superior, inferior) muscles. The levator palpebrae superioris may be considered as a specialized extraocular muscle. However it is considered in detail in Chapter 4, "Eyebrows, Eyelids, and Anterior Orbit." The extraocular muscles provide the eye with an active motility system that is capable of producing coordinated ocular movement.

Initially, the gross anatomic features of the extraocular muscles will be reviewed. This will include the origin, insertion, and course of each muscle. The microscopic anatomy of the extraocular muscles will be discussed to provide an understanding of their special contractile properties. Finally, functional anatomy of the extraocular muscles will be reviewed to provide insight into the nature of extraocular motility.

GROSS ANATOMY

The rectus muscles originate from or around a fibrous ring at the orbital apex, the annulus of Zinn. The annulus of Zinn encircles the optic foramen and the medial aspect of the superior orbital fissure (Fig. 7.1). The annulus is formed by a lower portion (tendon of Zinn) and an upper portion (tendon of Lockwood). The tendon of Zinn is attached to the inferior root of the lesser wing of the sphenoid, below the optic foramen. The attachment site may develop into a small tubercle, the infraoptic tubercle. At its base, the annulus of Zinn is contiguous with the dura mater lining the middle cranial fossa, which enters the orbit through its apical apertures. The tendon of Zinn serves as the origin of the inferior rectus and the inferior portions of the medial and lateral rectus muscles.

The upper portion of the annulus of Zinn (tendon of Lockwood) is not as well developed as the lower portion. The tendon of Lockwood attaches to the greater wing of the sphenoid and bridges over the superior orbital fissure. A bony spur, the spina recti lateralis, is observed at the posterior tip of the lateral orbital wall (see Fig. 2.14). The superior rectus and the superior portions of the medial and lateral rectus muscles arise from the superior portion of the annulus of Zinn.

The lateral rectus muscle arises from the spina recti lateralis and some authors feel a second head arises from a portion of the annulus of Zinn which bridges the superior orbital fissure. Merkel (1901) described a dual origin of the lateral rectus: one portion from the spina recti lateralis, and the other from the annulus, adjacent to the optic nerve sheath. Whitnall (1932) did not recognize a discontinuity in the annulus at this area and dispelled the concept of the lateral rectus having two heads.

The levator muscle originates above the superior rectus at the orbital apex. It is not considered to arise from the annulus of Zinn, although it proceeds anteriorly in the orbit as a functional unit with the superior rectus muscle.

As noted previously, the annulus of Zinn envelops the optic foramen and the medial aspect of the superior orbital fissure (Fig. 7.1). The portion of the orbital apex enclosed by the annulus is called the oculomotor foramen. Arteries, veins, and nerves transmitted through the optic foramen and the medial portion of the superior orbital fissure enter the orbit through the oculomotor foramen to supply structures within the "muscle cone." The optic nerve, ophthalmic artery, and the ocular sympathetics enter the orbit through the optic canal and foramen. Through the medial end of the superior orbital fissure, various venous and nervous structures enter; these include the inferior and superior divisions of the oculomotor nerve, the abducens nerve, the nasociliary branch of the ophthalmic division of the trigeminal nerve, the sensory root of the ciliary ganglion, and the superior ophthalmic vein.

CLINICAL NOTE

The superior and medial rectus muscles are adherent to the optic nerve sheath at the annulus of Zinn (Fig. 7.1). Because of this anatomical relationship at the orbital apex, painful ocular motility occurs with retrobulbar neuritis (Whitnall, 1932). Painful ocular motility does not occur with inflammation of other areas of the optic nerve.

The oblique muscles originate separately. The superior oblique originates superomedial to the annulus of Zinn near the frontoethmoidal suture. The muscle courses forward at the junction of the medial wall and roof of the orbit. The inferior oblique muscle is unique in that it originates at the anterior nasal orbital floor. The muscle arises in a shallow depression lateral to the origin of the nasolacrimal duct.

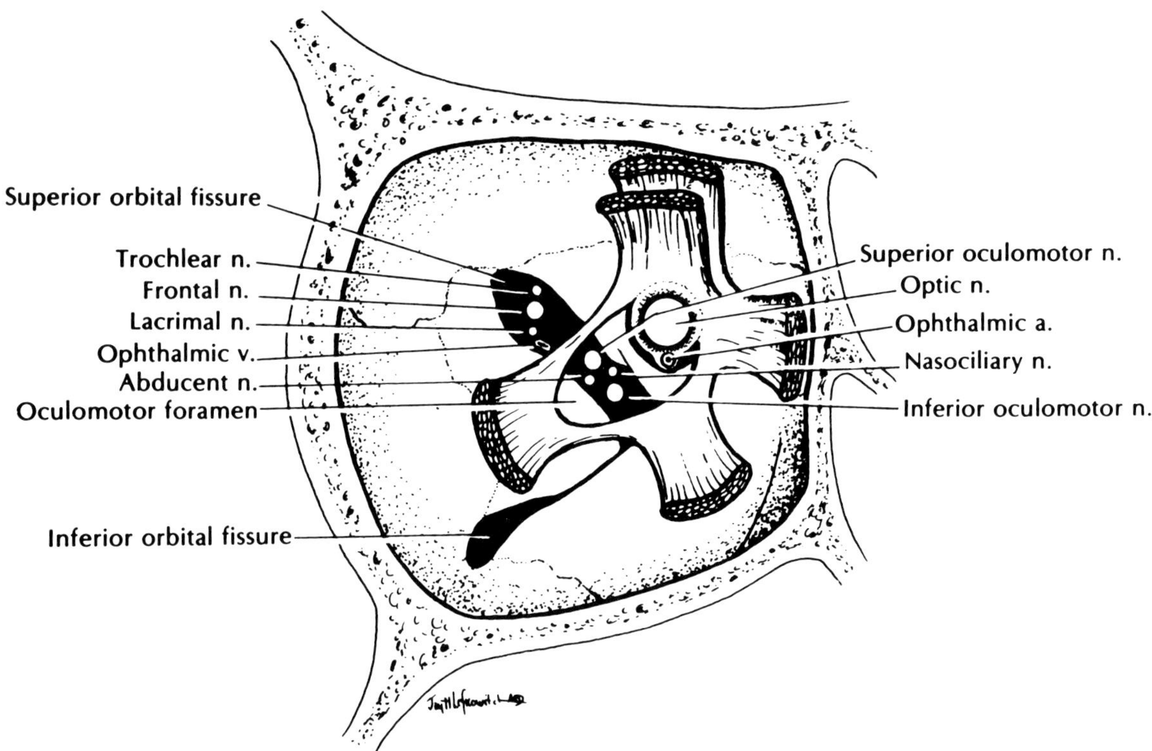

Figure 7.1. The orbital apex. The rectus muscles arise from the annulus of Zinn. (Reproduced from Jakobiec FA, Iwamoto T: Ocular adnexa. Introduction to lids, conjunctiva and orbit. In Duane TD, Jaeger EA: *Biomedical Foundations of Ophthalmology*, Vol. 1, Chapter 28. Philadelphia, Harper & Row, 1982.)

The rectus muscles have similar points of origin, length, and insertion (Fig. 7.2, *A* and *B*). As noted, the rectus muscles originate from or near the annulus of Zinn at the orbital apex. The length of each muscle, lengths of muscular tendons, widths of insertions, and distances of insertion from the limbus are summarized in Table 7.1. Each muscle is approximately 40 mm in length, and has a tendonous insertion onto the globe (Fig. 7.3 *A* and *B*). The muscles interdigitate and firmly attach to the sclera anterior to the equator of the globe overlying the ora serrata. The distance from the limbus to the extraocular muscle insertion progressively increases from the medial rectus (5.5 mm) around the globe to the superior rectus (7.7 mm) and forms the spiral of Tillaux (Fig. 7.4). The concept of the spiral of Tillaux is limited by the variability in the distance of insertion and oblique nature of insertion of the extraocular muscles. Tendon insertion width is the least for the lateral rectus (9.2 mm), and progressively increases to that of the superior rectus muscle (10.6 mm) (Fig. 7.4). The sclera is thinnest just posterior to the insertion of the extraocular muscles (0.3 mm).

The extraocular muscles pierce Tenon's capsule, and are covered by sleeve-like extensions of the capsule. This intracapsular portion of the rectus muscles is 7–10 mm in length and has fascial connections to overlying Tenon's capsule. In addition, there are fine attachments between the undersurface of the muscle and the sclera. This arrangement allows the muscles to move forward and backward within Tenon's capsule, thereby preventing restricted ocular motility.

Each extraocular muscle has its own connective tissue system, which has been discussed in detail in Chapter 6, "Connective Tissue Planes of the Orbit."

The motor nerve to each rectus muscle enters the internal surface of the muscle approximately at the junction of the posterior one-third and the anterior two-thirds of the muscle (Fig. 7.5). The lateral rectus is supplied by the abducens nerve (CN VI). The inferior division of the oculomotor nerve supplies the inferior rectus, while the superior division supplies the medial and superior rectus muscles. The superior oblique muscle is supplied by the trochlear nerve (CN IV), which enters the posterior one-third of the muscle on its orbital surface. The inferior oblique muscle is supplied by the inferior division of the oculomotor nerve, which enters the muscle at the middle third of its posterior border.

The muscular branches of the ophthalmic artery supply the extraocular muscles. Basically, these muscles are supplied by two main branches of the ophthalmic artery, but muscular contributions may also arise from the lacrimal and infraorbital arteries. The two muscular trunks arising from the ophthalmic artery have been called the superior (upper, lateral) and the inferior (lower, medial) muscular branches (Thane, 1892; Meyer, 1887; Bedrossian, 1958; Duke-Elder, 1961). The superior muscular branch supplies the lateral and superior rectus, superior oblique, and levator muscles. The inferior muscular branch supplies the medial and inferior rectus and inferior oblique muscles. The infraorbital artery has contributions extending to the inferior oblique muscle. The lacrimal artery proceeds in the orbit between the superior and lateral recti, and sends branches to each. The muscular branches are divided into the anterior ciliary

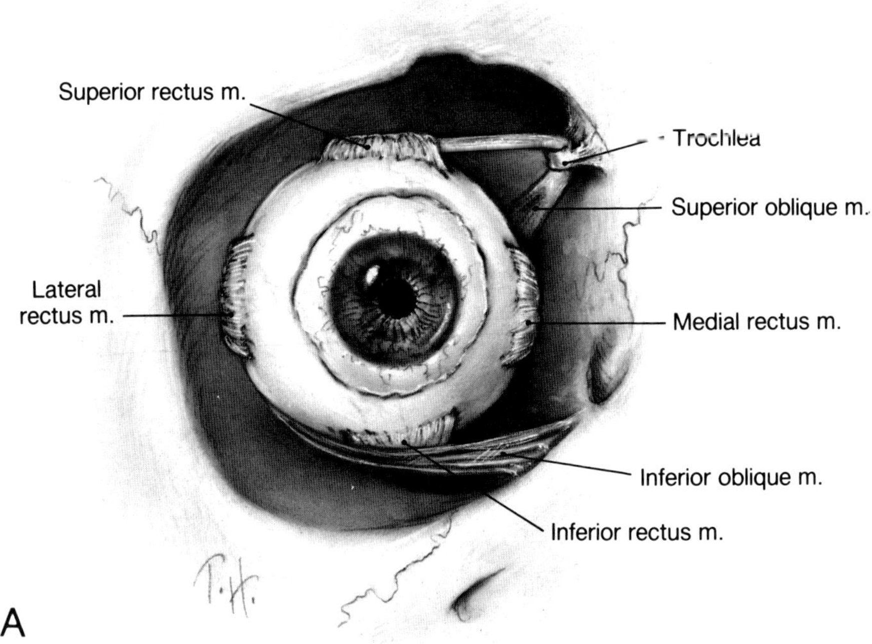

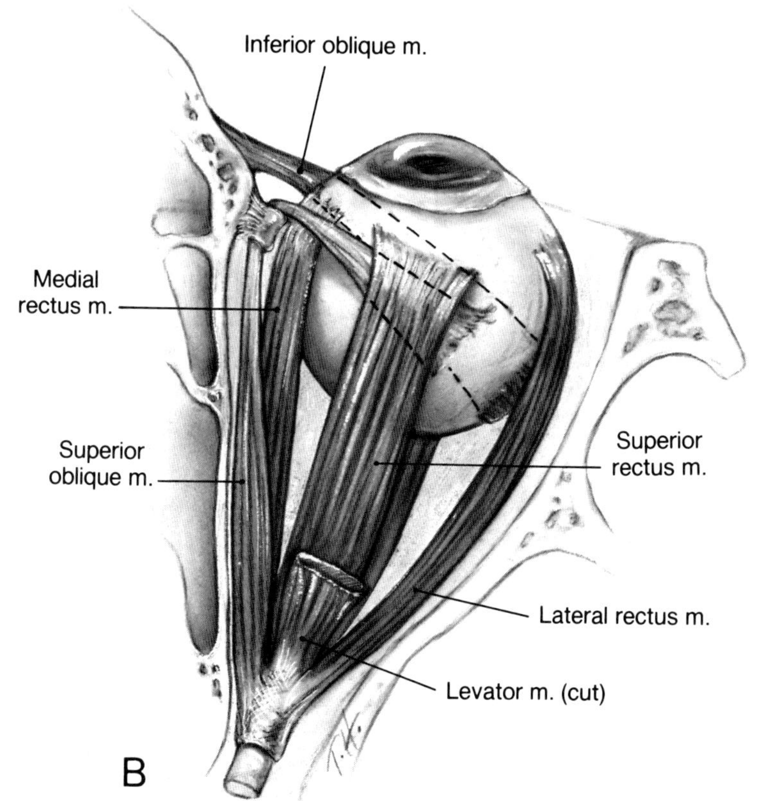

Figure 7.2.   The extraocular muscles, orientation in the orbit. (*A*), frontal view; (*B*), view from above.

Table 7.1.

	Distance from limbus	Width of insertion	Tendon of EOM	Length of muscle (mm)
Medial (Inf. 3)	[a]5.5 (5.3)[b]	10.3 (11.3 ± 0.8)	4	40.8
Inferior (Inf. 3)	6.5 (6.8)	9.8 (10.5 ± 0.8)	10	40.0
Lateral (6)	6.9 (6.9)	9.2 (10.1 ± 0.8)	9	40.6
Superior (Sup. 3)	7.7 (7.9)	10.6 (11.5 ± 0.8)	6	41.8

[a] Fuchs (1884).
[b] Apt (1980).

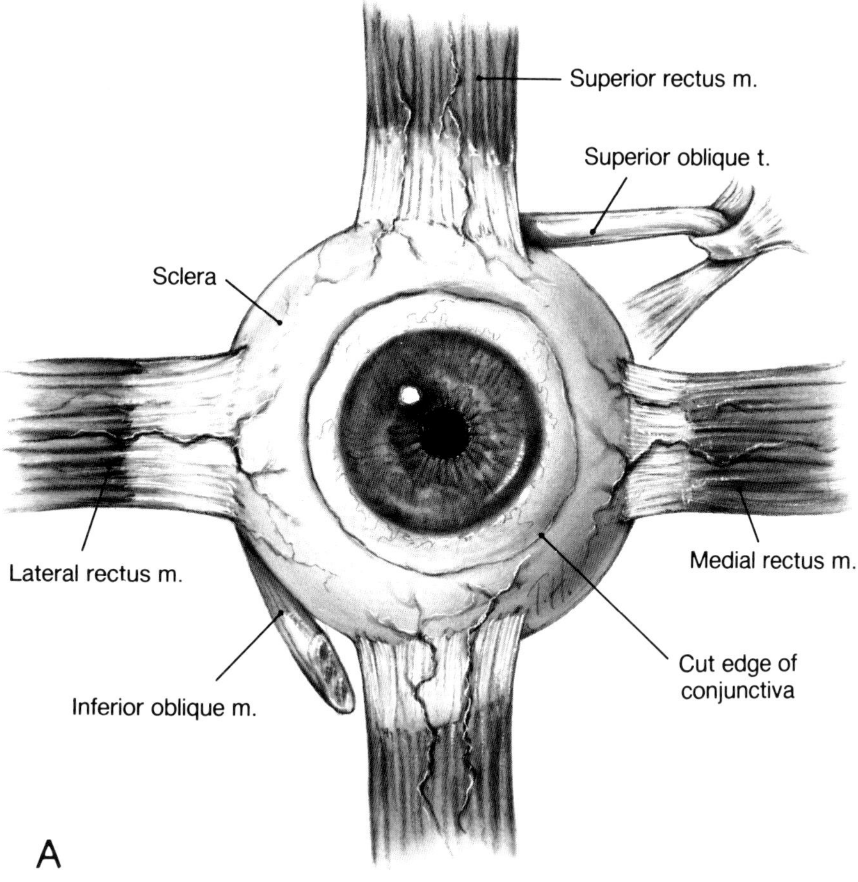

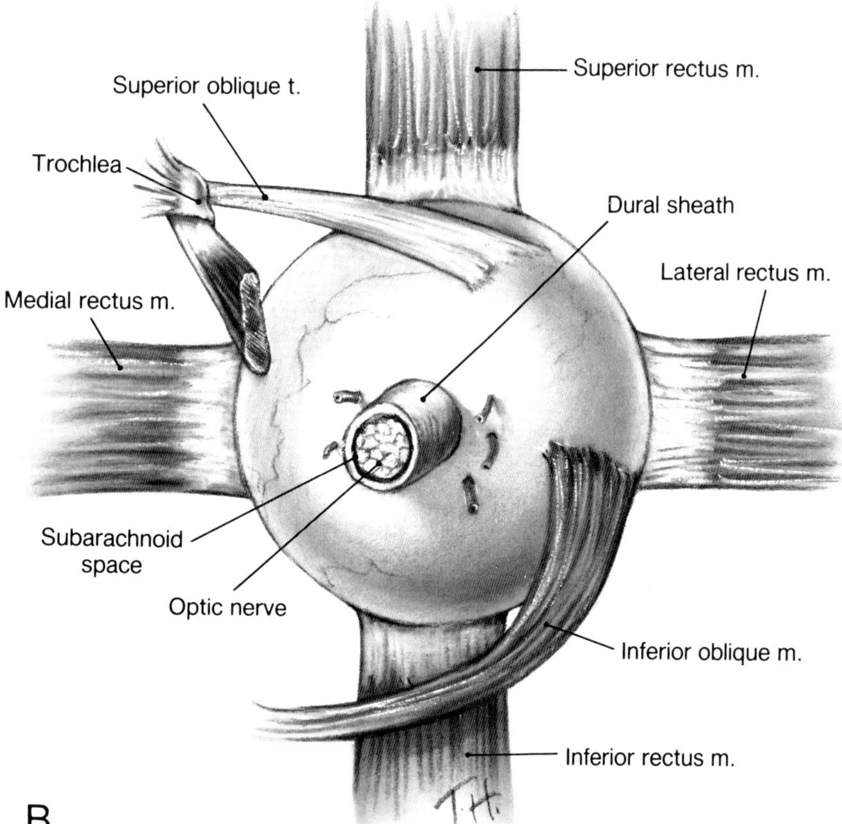

Figure 7.3. Insertion of the extraocular muscles on the globe.

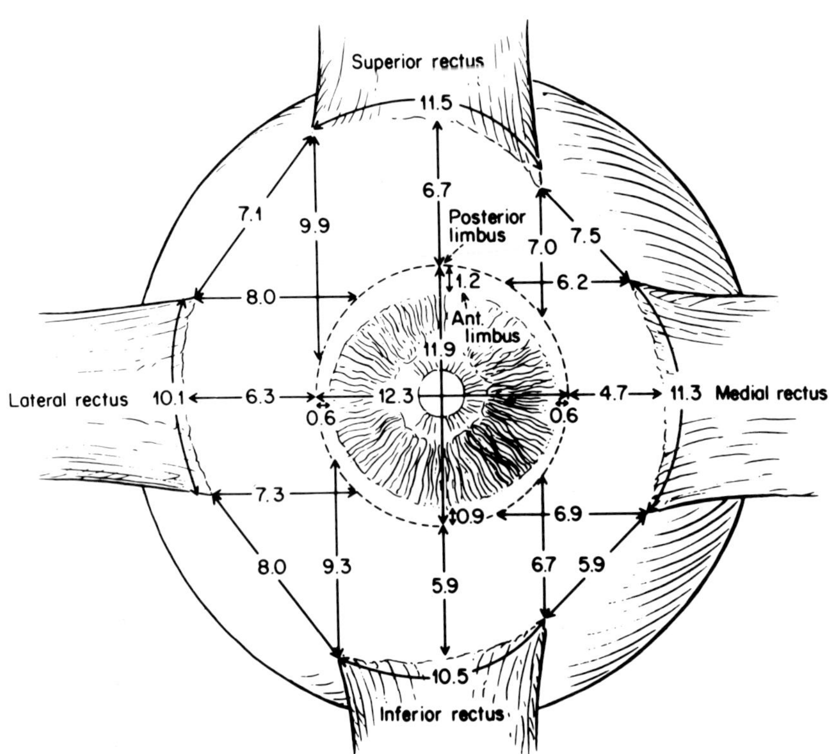

Figure 7.4.　Spiral of Tillaux. A spiral is formed by the distance between the limbus and the rectus muscles. This illustration also demonstrates the widths of the rectus muscle insertions on the globe. (Reproduced from Apt L: *Transactions of the American Ophthalmological Society* 78:365, 1980.)

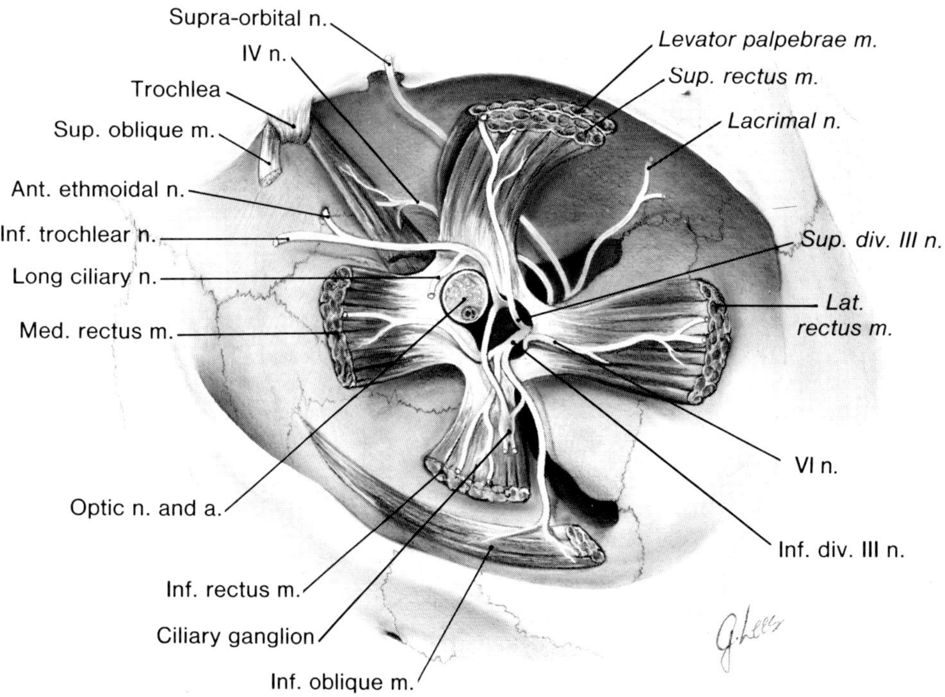

Figure 7.5.　Innervation of the extraocular muscles. The extraocular muscles are innervated from their conal surface except the superior oblique muscle, which is innervated on the orbital surface. (Reproduced from Miller NR: *Walsh and Hoyt's Clinical Neuro-Ophthalmology* (ed 4), Vol. 1. Baltimore, Williams & Wilkins, 1982.)

arteries. Each rectus muscle has two anterior ciliary arteries, except the lateral rectus, which has only one. The anterior ciliary arteries penetrate the sclera at the insertion of the extraocular muscles to supply the anterior segment of the eye.

Clinical Note

Anterior segment ischemia, due to an interruption of the vascular supply to the globe, is an occasional complication of strabismus surgery. Ocular ischemia generally occurs following removal of all four rectus muscles (Forbes, 1959; Girard and Beltranena, 1960; Uribe, 1968; Helveston, 1971), but may also occur following surgery on only three rectus muscles (Girard, Beltranena, 1960; Hiatt, 1978; Easty and Chignell, 1973). Symptoms and signs of anterior segment ischemia are corneal edema, corneal ulceration, uveitis, iris atrophy, ectopic pupil, posterior synechiae, cataract hypotony, and phthisis bulbi.

The patients reported upon by Saunders and Sandall (1982) developed anterior segment ischemia following full tendon transposition 9–20 years after horizontal strabismus surgery. Helveston and co-workers (1980) and Saunders and Sandall (1982) therefore question the indications for removal of all four rectus muscles from the globe. Girard and Beltranena (1960) and Harley (1971) recommend a cautious approach to strabismus surgery in which no more than two rectus muscles are approached during a single procedure. Since the anterior ciliary vessels enter the anterior segment of the globe through the rectus muscles scleral insertion, interruption of the muscle insertion also ablates the vascular supply.

The superior and inferior oblique muscles provide torsional movements for the globe. Due to the unique nature of these muscles, each will be discussed separately.

The superior oblique muscle lies in the angle formed by the roof and the medial wall of the orbit (Fig. 7.2 *B*). The muscle arises from a short tendon, superior to the annulus of Zinn, at the frontoethmoidal suture. As the superior oblique proceeds anteriorly, approximately 10–15 mm from the orbital margin, it becomes a tendon. Prior to reaching the anterior orbital margin, the tendon of the superior oblique passes through a cartilaginous ring, the trochlea. The basic function of the trochlea is to redirect the oblique tendon so that it pulls in an anterior-medial direction approximately 55° nasal to the saggital plane.

The reflected superior oblique tendon is intimately involved with the connective tissue system of the anterior orbit. As noted in Chapter 6, "Connective Tissue Planes of the Orbit," this firm connective tissue system helps support the globe within the orbit. The superior oblique tendon pierces Tenon's capsule approximately 3–4 mm nasal to the medial border of the superior rectus muscle, approximately 3–4 mm posterior to insertion. However, when the globe is rotated downward, the anterior border of the superior oblique tendon is approximately 8 mm from the superior rectus insertion (Parks, 1982). Under the superior rectus, the superior oblique becomes thin and nearly transparent. The posterior-most insertion of the superior oblique tendon

is approximately 5 mm from the optic nerve, near the superior temporal vortex vein exit from the sclera.

Helveston and co-workers (1982) studied the anatomy and physiology of the trochlea. The authors described four major components of the trochlea: a cartilage saddle, an intratrochlear portion of the superior oblique tendon, a fibrillovascular sheath surrounding the tendon, and a dense, fibrous condensation which secures the trochlea to the periosteum of the medial orbital wall (Fig. 7.6). These components of the trochlea are arranged in a way that maximizes the movement of the superior oblique tendon through the trochlea.

The cartilage saddle of the trochlea is a biconcave structure measuring 4×6 mm. It is attached to the periorbita by a dense connective tissue. The tendon of the superior oblique passes through the cartilaginous saddle of the trochlea and a bursa-like space created by a loose, fibrillovascular connective tissue. The bursa-like space maximizes superior oblique tendon mobility through the trochlea by preventing adhesion of the tendon to surrounding structures. The outermost layer envelops the trochlea and attaches it firmly to the periosteum at the fovea trochlearis. The trochlea is the only cartilaginous structure in the orbit.

The sliding motion of the superior oblique tendon through the trochlea is accompanied by stretching and elongation of the paratendinous tissue in the direction of pull as the tendon emerges from the trochlea. Similarly, tissue at the entrance of the trochlea is telescoped and becomes redundant. Clinically, the range of motion of the superior oblique is approximately 16 mm, or 8 mm on either side of the primary position. Manual manipulation of an excised trochlea, however, failed to demonstrate this range of mobility.

Helveston and co-workers, therefore, postulated a sliding theory of superior oblique function. The sliding action of the superior oblique tendon is analogous to a laminar flow pattern, with the central fibers exhibiting the greatest excursion (Fig. 7.7). With the use of electron microscopy, they confirmed the sliding theory of the superior oblique tendon. These authors found the fibers of the tendon to be autonomous with few lateral attachments. Their relative autonomy allowed the fibers to slide in relation to one another, which enhances their motility.

CLINICAL NOTE

Brown (1950, 1957, 1962) first described a motility defect manifested by the inability to raise the adducted eye above the midhorizontal plane. The restricted upgaze is not as marked in the midline and lateral fields of gaze. Brown originally postulated that the restricted upgaze simulated an inferior oblique palsy, with a secondary contracture anomaly of the superior oblique tendon. Electromyography, however, demonstrated normal innervation to the inferior oblique muscle on attempted upgaze.

Brown (1973), therefore, redefined the superior tendon sheath syndrome. A true sheath syndrome is congenital, permanent, and invariably associated with a positive traction test (inability to manually elevate an adducted eye). Most patients will not have vertical

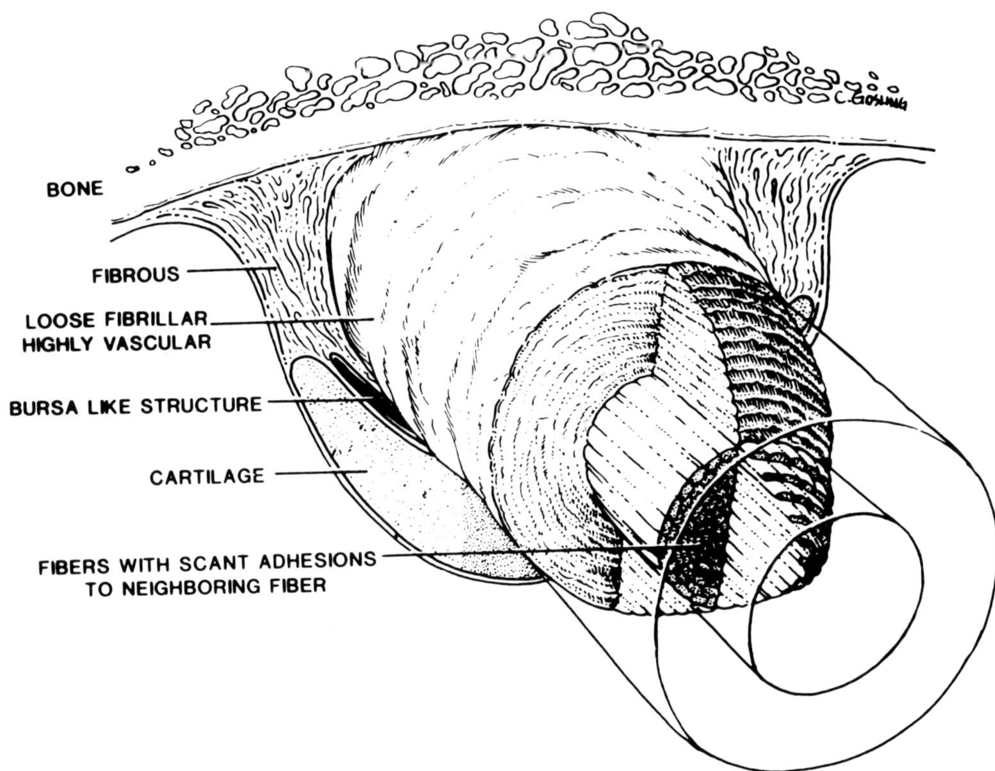

Figure 7.6. The trochlea. Four major components comprise the trochlea: cartilage saddle, an intratrochlear portion of the superior oblique tendon; a fibrillovascular sheath; and a dense fibrous condensation which secures the trochlea to the periosteum. (Reproduced from Helveston EM, et al: The trochlea: A study of the anatomy and physiology. *Ophthalmology* 89:124, 1982.)

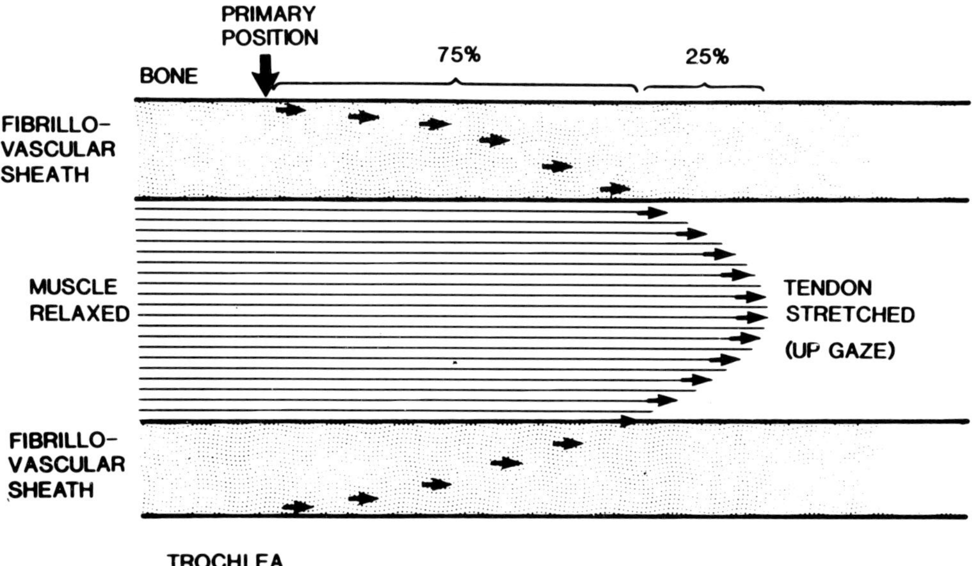

Figure 7.7. Sliding theory of superior oblique function. Individual tendon fiber mobility permits a greater excursion of the superior oblique tendon through the trochlea. (Reproduced from Helveston EM, et al: The trochlea: A study of the anatomy and physiology. *Ophthalmology* 89:124, 1982.)

misalignment in primary gaze. However, some patients may exhibit a hypotropia which causes a compensatory chin-up position to minimize diplopia. Another group of patients are recognized who manifest restriction to elevation of an adducted eye, but may elevate the involved eye with an audible snap. Brown described these patients as having a simulated sheath syndrome, which may arise secondary to inflammation from the contiguous ethmoidal sinus.

Helveston and co-workers (1982) felt the superior

tendon sheath syndrome could be produced by the accumulation of fluid or solid material within the fibrillovascular sheath of the superior oblique tendon in the area of the trochlear saddle. Distortion or non-conforming of this space could produce restricted movement of the superior oblique tendon in the trochlea.

The inferior oblique muscle originates from the inferior-anterior nasal orbit and courses posteriorly and laterally at a 51° angle with the medial orbital wall (Fig. 7.2 *B*). The inferior oblique muscle penetrates Tenon's capsule lateral to the inferior rectus muscle, then arcs laterally around the globe. The inferior oblique is firmly attached to the inferior rectus by a fusion of their fascial sheaths, and this combined structure contributes to Lockwood's suspensory ligament. The inferior oblique muscle has a muscular insertion to the sclera inferior to the macula (Fig. 7.3 *B*). The total length of the inferior oblique muscle is approximately 37 mm (Whitnall, 1932).

CLINICAL NOTE

Enlargement of extraocular muscles has been identified with greater ease since the advent of sonography and computed tomography. Trokel and Hilal (1979) reported 70 patients with enlarged extraocular muscles as determined on computed tomography. The most common cause of enlarged extraocular muscles was Graves' disease (65.7%), followed by carotid-cavernous sinus fistula (17.1%), pseudotumor (10%), and orbital tumor (7.1%). Associated clinical and radiographic features will facilitate the differentiation of these conditions.

The principal pathologic abnormality in Graves' disease is a chronic inflammatory infiltration of the extraocular muscles, with glycoprotein and mucopolysaccharide deposition (Werner, 1972). The extraocular muscles may reach up to eight times their normal size and the orbital volume may increase by 400% (Werner, 1979) (Fig. 7.8). As a consequence of the dramatic increase in orbital volume there is proptosis and herniation of orbital fat anterior to the orbital septum. Extraocular muscle enlargement at the orbital apex may compromise the function of the optic nerve. Restricted extraocular motility may result from the fibrotic contracture of the involved rectus muscles. Three major patterns of extraocular muscle involvement have been recognized with Graves' disease (Jakobiec and Jones, 1979). The inferior rectus is the most commonly involved extraocular muscle, and fibrosis may simulate a double elevator palsy. Medial rectus fibrosis will mimic a sixth nerve palsy. Superior rectus fibrosis, limiting downgaze, is the least common abnormality. Generally, a combination of the above patterns exist, producing noncommittent ocular deviations.

Extraocular enlargement associated with cavernous sinus fistula and malformation results in increased venous pressure in the orbital tissues. In the Trokel and Hilal (1979) series, of six of twelve patients with vascular malformations, the diagnosis was not suspected prior to computed tomography. Uniform enlargement of the medial and lateral rectus muscles were observed; however, the pathognomonic finding of a carotid-cavernous sinus fistula was an enlarged superior ophthalmic vein (see Fig. 9.13).

Idiopathic orbital myositis (pseudotumor) presents as an acute orbital syndrome. Characteristic features are rapid onset periorbital pain, eyelid swelling, proptosis and diplopia associated with ductional restric-

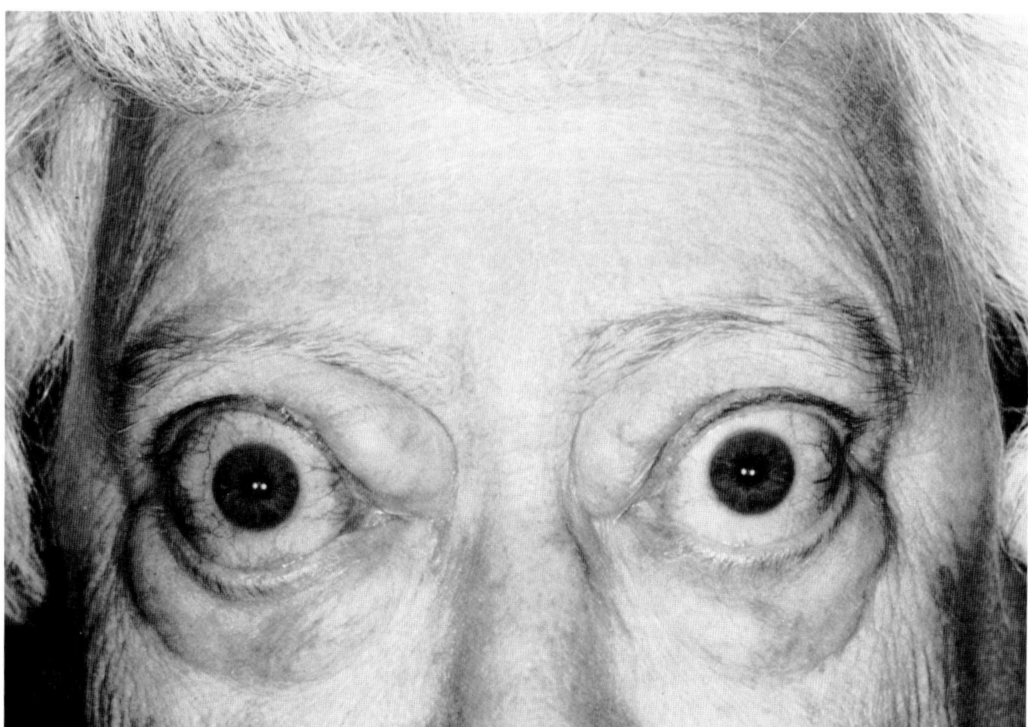

Figure 7.8. Enlargement of the extraocular muscles in thyroid disease producing marked eyelid retraction and proptosis.

tions. This syndrome responds rather dramatically to systemic steroids. Computed tomography or ultrasonography will demonstrate enlargement of one or more of the extraocular muscles, with sparing of the other orbital soft tissues (Fig. 7.9) (Slavin and Glaser, 1982).

Tumors may also produce enlargement of the extraocular muscles. The neoplasms involving extraocular muscles in Trokel and Hilal's series were lymphosarcoma, plasmacytoma, metastatic breast carcinoma, metastatic neuroblastoma, and angioma. The enlargement in extraocular muscles can be secondary to direct infiltration or compression of the muscle obstructing venous outflow.

Rhabdomyosarcoma is the most common primary orbital malignancy in children. Rhabdomyosarcomas of the orbit usually do not arise in extraocular muscles; rather, they originate from primitive mesenchyme in soft tissues adjacent to muscles (Stout, 1946). Knowles

and co-workers (1978) reviewed and summarized previous reports (Frayer and Enterline, 1959; Porterfield and Zimmerman, 1962; Ashton and Morgan, 1965; Jones, et al, 1966) and confirmed the tendency of these tumors to occur in children and adolescents. The diagnosis of rhabdomyosarcoma should be suspected in any child with a history of a rapidly developing lid swelling or proptosis. Radiographic evaluation may demonstrate subtle erosion or destruction of the orbital walls (Henderson, 1980).

MICROSCOPIC ANATOMY

The extraocular muscles have unique functional properties—the maintenance of constancy of activity with the capability of fine rapid movements. The histologic evaluation of the extraocular muscles reflects these functional demands. Normal skeletal muscle fibers exhibit consistency

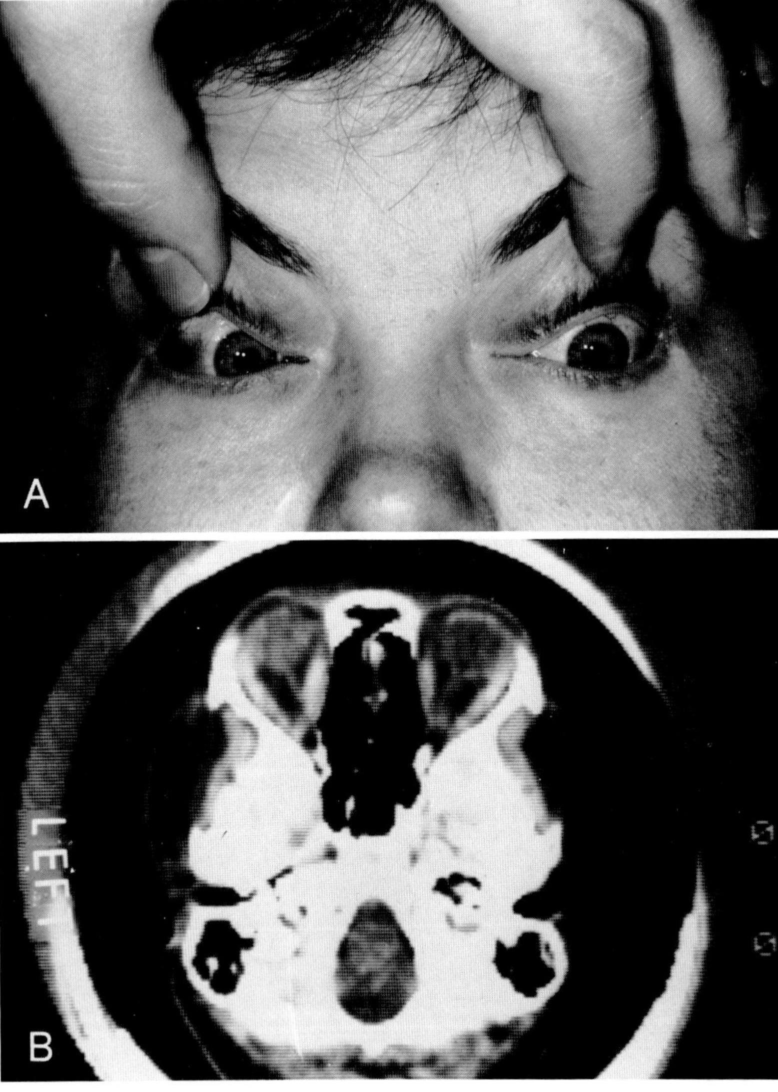

Figure 7.9. Orbital myositis. (*A*), Acute onset of orbital pain with restricted lateral gaze. Note the injection over the lateral rectus muscle. (*B*), CT scan of the orbit demonstrates the enlargement of the medial and lateral rectus muscles near their insertion onto the globe.

of size and shape. Extraocular muscles, however, display striking variations in fiber size (Ringel, et al, 1978). In addition, extraocular muscles have a lower innervation ratio, greater vascularity (Wooten and Reis, 1972), and more abundant connective tissue between fibers than is found in skeletal muscle.

The extraocular muscles are groups of fibers forming fasciculi, bound by a connective tissue system, called the epimysium. Each individual fiber is bound by a sheath called the sarcolemma. Sarcoplasm, the ground substance surrounding an individual fiber, contains the fiber's organelles. Extraocular muscle fibers are longitudinally oriented and extend through most of the length of the muscle, although some fibers terminate or arise along the length of the muscle. (Lockhart and Brandt, 1938; Hines, 1931, Cooper and Daniel, 1949; Alvarado and van Horn, 1975). A cross section of extraocular muscle shows a lamellar organization. The orbital surface of the muscle contains smaller diameter fibers (5–15 μm), while muscle adjacent to the globe contains larger diameter fibers (10–40 μm) (Ringel, et al, 1978).

Light and electron microscopic evaluation of the myofibril reveals alternating light and dark staining zones. The light zone is called the isotropic (I) band, and the dark zone, the anistropic (A) band, based upon their respective birefrin-gence to polarized light (Fig. 7.10). The I band is bisected by a dark line, the Z line. The A band is bisected by a light zone, the H zone, with a vertically placed dark line, the M line. The sarcomere is the basic contractile unit of muscle, and is measured from one Z line to another. Each sarcomere has a tubular arrangement called the T system which acts as a conduit for electrical stimulation.

The unique features of the extraocular muscles are a reflection of their two fiber system: the fibrillenstruckur and the felderstrukur (Sommerkamp, 1928; Brandt and Leeson, 1966; Dietert 1965). Fibrillenstrukur fiber resembles regular skeletal muscle fiber. These fibers are characterized by small, organized myofibrils surrounded by abundant sarcoplasm, large concentrations of mitochondria, and peripheral nuclei. Each sarcomere has an orderly T system. The unique fiber found in extraocular muscle is the felderstrukur fiber, which contains large myofibrils, reduced amounts of sarcoplasm, and few mitochondria. These fibers have a rudimentary T system.

The variations in morphologic appearance of the two basic muscle fibers is a reflection of the functional demands of the extraocular muscles. The fibrillenstrukur fibers are responsible for rapid (twitch) contractions, with a short contraction-relaxation cycle. These fibers permit rapid changes

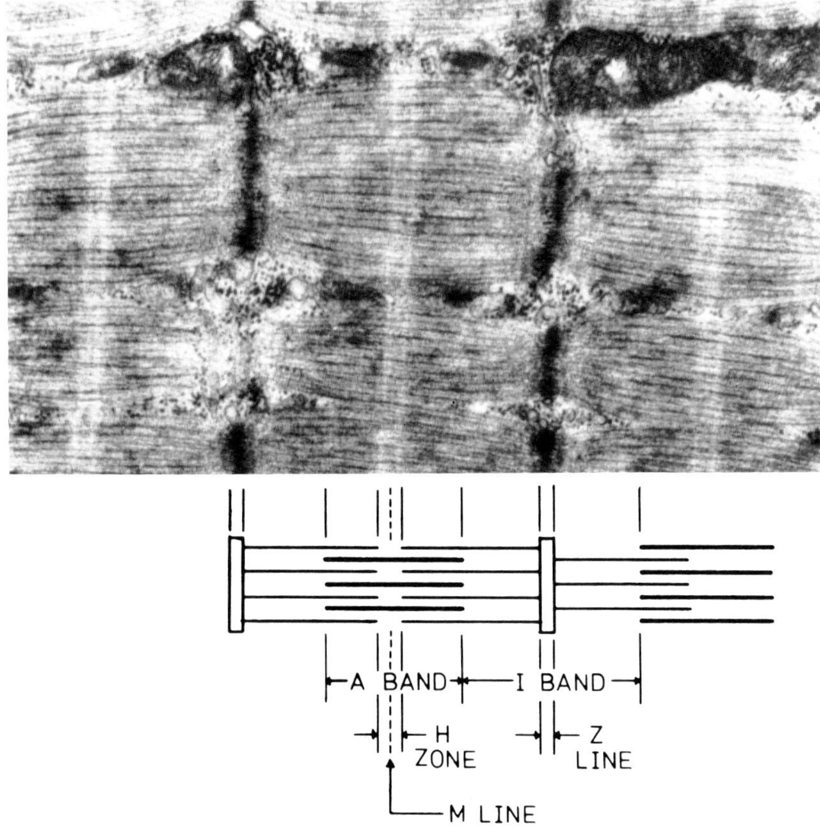

Figure 7.10. Microscopic features of the extraocular muscles. Electron micrograph of a twitch extraocular muscle fiber. The mitochondria occur in rows between the myofibrils. The schematic diagram below the electron micrograph demonstrates how the filaments produce the banding pattern. (Reproduced from Eggers HM: Functional anatomy of the extraocular muscles. In Duane TD, Jaeger EA (eds): *Biomedical Foundations of Ophthalmology*, Vol. 1, Chapter 31. © 1982. Philadelphia, Harper & Row Publishers.)

in eye position that are necessary for saccadic and pursuit eye movements. The fibrillenstruckur fiber contracts briskly to a single impulse. The T system, which transmits the nervous impulse, is therefore well-developed in this fiber. In addition, the abundant mitochondria reflect the high metabolic demands of the twitch fibers. The muscles are innervated by thicker nerves, which terminate on the muscles at typical motor end plates, with junctional folds at the synapse.

Two types of fibrillenstrukur fibers are distinguished by histologic and functional attributes. The larger fibrillenstrukur fiber is responsible for fast twitch movements. Histologically, the fiber is a whitish color, has fewer mitochondria, and has high glycolytic activity. The smaller fibrillenstrukur fiber is responsible for slow twitch movements. Histologically, this fiber is reddish, has greater numbers of mitochondria, and has high oxidative activity.

The felderstrukur fibers are unique to the extraocular muscles. These fibers contract slowly and smoothly, to maintain tonicity of the muscles which coordinate and maintain the position of the eyes. The grade of contraction is proportional to the repetitive stimulation the fiber receives. The neuromuscular junction in the felderstrukur fiber resembles a bunch of grapes and is called the en grappe (Tschiriew, 1979), as opposed to the typical en plaque neuromuscular junction. Since an individual nerve impulse does not stimulate the felderstrukur fiber, the T system, designed to rapidly transmit the nerve impulse is not well developed. Reduced metabolic demands of the felderstrukur fiber are reflected in less sarcoplasm and fewer mitochondria than in fibrillenstrukur fiber.

FUNCTIONAL ANATOMY

The synergistic action of the extraocular muscles produces coordinated ocular motility. Ocular motility is produced by rotation of the globe around the axes of Fick. Basic movements of the eye are: abduction, adduction, supraduction, infraduction, and cycloductions. The actions of the extraocular muscles are summarized in Table 7.2.

Ocular movement in the horizontal plane is primarily controlled by the medial and lateral rectus muscles. Contraction of the medial rectus muscle adducts the globe, while contraction of the lateral rectus abducts the globe. A small amount of adduction is also produced by the vertical rectus muscles in their direction of gaze. Adduction is a result of the orientation of the vertical rectus muscles which extend anteriorly from the orbital apex in an angle 23° to the medial orbital wall. A small amount of abduction is provided by the contraction of the oblique muscles. Therefore, abduction is produced by the inferior oblique muscle in upgaze, and by the superior oblique in downgaze. The abduction produced by the oblique muscle results from their direction of insertion.

Supraduction, or elevation of the eye, is produced by the superior rectus and inferior oblique muscles. As the eye moves into an abducted position, the superior rectus becomes the primary elevator of the eye, while the inferior oblique becomes the primary elevator of the adducted eye.

Table 7.2.
Actions of the Extraocular Muscles[a]

Muscle	Primary action	Secondary action	Synergists[b]
MR	Adduction	None	SR and IR
LR	Abduction	None	SO and IO
SR	Elevation		IO
		Adduction	MR and IR
		Intorsion	SO
IR	Depression		SO
		Adduction	MR and SR
		Extorsion	IO
SO	Intorsion		SR
		Depression	IR
		Abduction	LR and IO
IO	Extorsion		IR
		Elevation	SR
		Abduction	LR and SO

[a] From: Harley RD (ed): *Pediatric Ophthalmology.* Saunders, 1975, p 138.

[b] Abbreviations: MR, medial rectus; LR, lateral rectus; SR, superior rectus; IR, inferior rectus; SO, superior oblique; IO, inferior oblique.

Since the superior rectus proceeds from the orbital apex at a 23° angle from the medial orbital wall, the motility induced by its contraction will vary depending upon the horizontal position of the eye. For example, if the eye is abducted 23° corresponding to the axis of the superior rectus muscle, contraction produces vertical movement. In primary gaze, contraction produces elevation, intorsion, and adduction. With extreme adduction, contraction of the superior rectus produces intorsion. The inferior oblique function will be discussed in the section on excycloduction.

Infraduction or depression of the eye is produced by the combined action of the inferior rectus and superior oblique. The inferior rectus is the primary depressor of the abducted eye, and the superior oblique is the primary depressor of the adducted eye. The axis of the inferior rectus muscle, as previously noted, forms a 23° angle with the medial orbital wall, and the resultant forces produced by its contraction will vary with horizontal position of the globe. In abduction, the axis of the globe corresponds to that of the inferior rectus, so contraction produces depression only. In primary position, the movement resulting from contraction is primarily depression; however, a slight amount of adduction and excycloduction is produced. In adduction, the only movement produced is excycloduction. The superior oblique muscle will be discussed in the section on incycloduction.

Incycloduction is a torsional movement of the globe that rotates the superior pole of the eye nasally. This motion is produced by the combined contraction of the superior oblique and superior rectus muscles. As the eye moves horizontally, the intorting action of the muscles vary. In the abducted position, the superior oblique becomes the primary intorter, while the superior rectus is the primary intorter of the adducted eye.

The superior oblique is redirected at the trochlea, with an

axis directed at a 51° angle from the axis of the globe. The oblique angle of insertion is reflected in the various motility patterns produced in horizontal gaze. At abduction of 39°, the only movement of the globe is incycloduction. With the globe in primary position, action of the superior oblique produces depression and intorsion, plus slight abduction. In the adducted position, the only movement of the globe is depression.

Excycloduction is a torsional movement of the globe that rotates the superior pole of the eye temporally. This motion is produced by combined action of the inferior oblique and inferior rectus muscles. In the abducted position, the inferior oblique muscle is the prime extorter, while the inferior rectus is the prime extorter of the adducted eye.

Contraction of the inferior oblique produces ocular motility based upon the initial horizontal position of the eye. The inferior oblique originates at the inferior nasal portion of the anterior orbit, and inserts onto the globe. A 51° angle is produced between the axis of the globe and the inferior oblique muscle. In abduction, approximately 39°, the only movement produced by the inferior oblique is extortion. In primary position, inferior oblique contraction produces elevation, extortion, and abduction. The adducted globe is elevated by contraction of the inferior oblique.

Jampel (1966, 1970) reported a different concept in vertical and torsional ocular movements. After mathematical evaluation of ocular motility, work with animal models, and clinical observations, Jampel attributed vertical and torsional ocular movements to the rectus and oblique muscles, respectively. In addition, Jampel reported that the abducting component of the oblique muscles increased, rather than decreased, on adduction.

CLINICAL NOTE

Overaction of the inferior and superior oblique muscles is not uncommon. Overaction of the inferior oblique muscle results in hyperdeviation of an adducted eye. The hyperdeviation can be secondary to weaknesses of the depressor in the ipsilateral eye, or of the elevator in the contralateral eye. Since the involved eye is adducting, this would reflect a weakness in the ipsilateral superior oblique or the contralateral superior rectus. Weakness of these muscles is generally not observed; most often, primary overaction is in the inferior oblique. Patients with an overacting inferior oblique muscle secondary to a paretic muscle (secondary overaction) will display a hyperdeviation in primary gaze, while patients with a primary overaction usually do not manifest a hyperdeviation in primary gaze. Patients with congenital esotropia have a tendency to develop overaction of the inferior oblique muscle. (Parks, 1982).

Overaction of the superior oblique is generally primary in nature since palsies of the inferior rectus or inferior oblique muscles are exceedingly rare. Overaction of the superior oblique produces an "A" pattern exodeviation, since increased abduction is produced in the infraduction.

Palsies of the superior oblique muscles are not uncommon due to the long intracranial course of the trochlear nerve. Superior oblique palsies present with a combination of vertical and torsional tropias which are the result of varied action of the superior oblique in various fields of gaze. Torsional deviation increases with abduction of the involved eye and vertical deviation increases with adduction of the involved eye. The resultant deviation is minimized by a head tilt away from the field of action of the involved superior oblique muscle. Since the superior oblique muscle is an intorter and depressor, its use is diminished by upgaze and tilting of the head toward the shoulder opposite the involved eye. (Parks, 1958).

Two basic laws, Hering's and Sherington's, facilitate one's understanding of ocular motility. Hering's law states that equal and simultaneous innervation flows to synergistic muscles or groups of muscles. In pathologic condition, in which an extraocular muscle is paretic or weak, excessive innervation is required to fixate the eye. Fixation of the nonparetic eye produces a primary deviation, and fixation of the paretic eye produces a secondary deviation. The secondary deviation produces a greater abnormality in ocular alignment than does the primary deviation, since greater innervation is required to fixate the paretic eye. Moreover, Hering's law is important in explaining the inhibition palsy of a contralateral antagonist. The antagonist of a paretic muscle requires less than normal innervation to promote greater range of motion in the eye because the paretic muscle lacks normal tone. The yolk muscle of the antagonist of the paretic muscle will, therefore, receive less than normal innervation with resultant paretic eye fixation. This reveals an apparent underaction of the yolk muscle. This deviation occurs only when the paretic eye is fixated. Hering's law is also important regarding the levator muscle. This is clinically important in ptosis and thyroid lid retraction (Fig. 7.11 *A* and *B*). In unilateral ptosis, maximum innervation is being transmitted not only to the ptotic eyelid but also to the contralateral lid. Thus the contralateral lid may be higher preoperatively than following correction of the ptotic lid. This must be considered preoperatively and is especially important in evaluating asymmetric bilateral ptosis cases where one lid is markedly ptotic and the other appears to be only minimally ptotic preoperatively. The minimally ptotic eyelid may fall considerably after elevating the most ptotic lid. Bilateral surgery is frequently indicated in these cases. In asymmetrical thyroid lid retraction, the reverse is true. Due to pathologic lid retraction minimal innervation is required to the levator muscles. After lowering the markedly retracted eyelid a more normal amount of innervation is required, which may make the retraction of the contralateral eyelid more apparent.

Sherington's law states that contraction of an extraocular muscle is accompanied by a diminution of contractile activity in its antagonist. Co-contraction of antagonistic muscles will retract the globe into the orbit. This is most commonly observed in Duane's retraction syndrome. In addition, rhythmic co-contractions of the medial and lateral rectus muscles produce a retractory nystagmus.

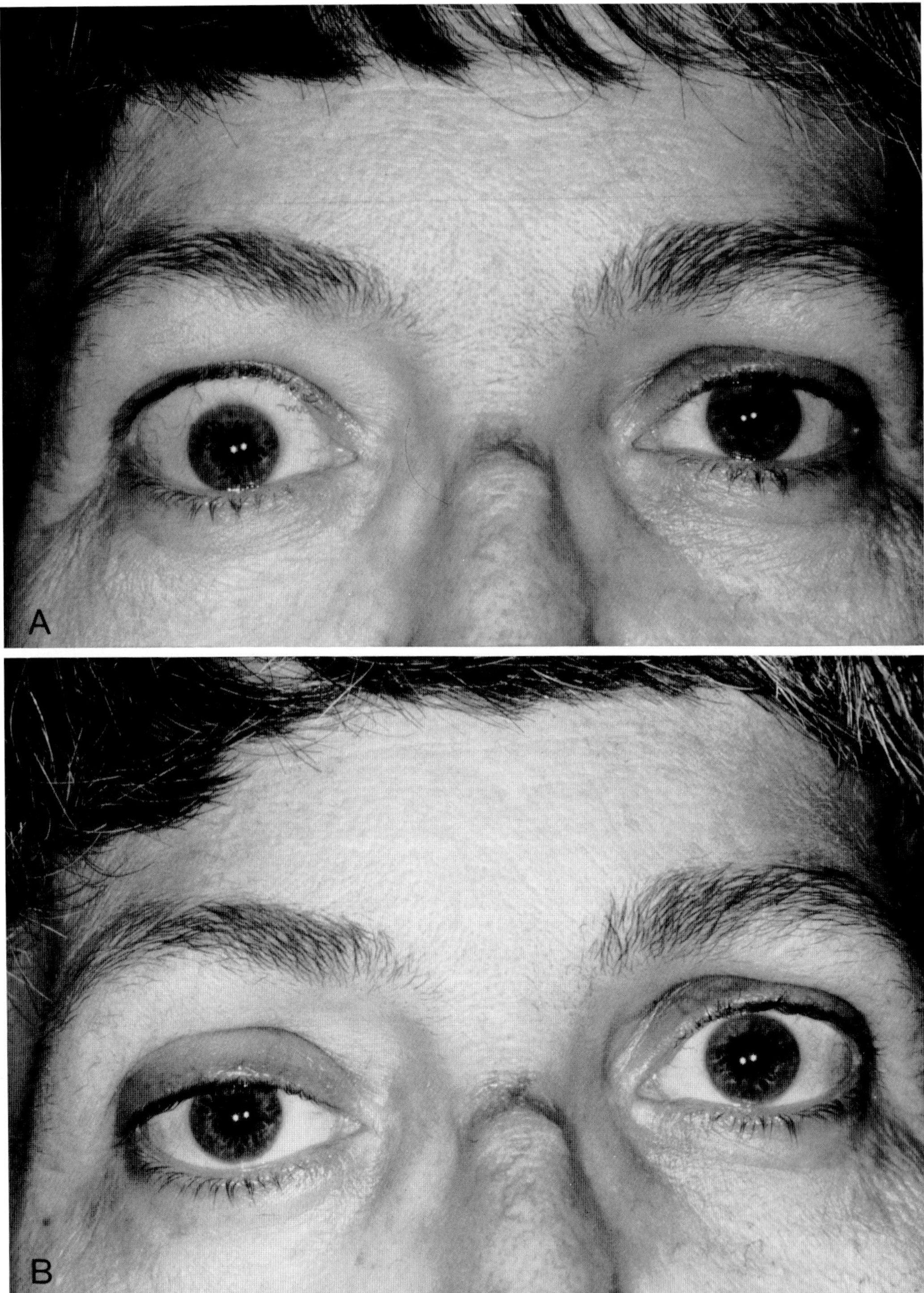

Figure 7.11.   Hering's law of equal innervation. (*A*), Preoperative clinical photograph of patient with right upper eyelid retraction secondary to thyroid disease. (*B*), Postoperative appearance following retractor recession of the right upper eyelid. Note the increased retraction of the left upper eyelid due to the increased bilateral innervation to the levator muscles with the right upper eyelid lowered to its normal position. (Reproduced from Harvey JT, Anderson RL: The aponeurotic approach to eyelid retraction. *Ophthalmology* 88:513, 1981.)

CLINICAL NOTE

Duane's retraction syndrome is a congenital eye movement disorder characterized by a limitation or absence of abduction. Additional clinical features include a variable limitation of adduction, and palpebral fissure narrowing and globe retraction on attempted adduction. Huber (1974) classified Duane's syndrome into three types based upon clinical and electromyographic findings:

Type I: Absence of abduction with restricted adduction; retraction of globe on adduction, and eyelid retraction with attempted abduction; no electromyographic evidence of electrical activity in the lateral rectus on attempted abduction, however, electrical activity is present on adduction.

Type II: Marked reduction of adduction and abduction; frequently exotropia; electromyographic evidence of electrical activity with both abduction and adduction.

Type III: Severe restriction of abduction and adduction; may manifest exotropia, esotropia, or orthophoria; electromyographic evidence of cocontraction of the horizontal rectus muscles on abduction and adduction.

Duane's syndrome results from a developmental neurogenic anomaly in which the abducens nucleus does not develop (Matteuci, 1946; Hotchkiss, et al, 1980; Miller, et al, 1982). As a result, the lateral rectus muscle receives innervation from branches of the oculomotor nerve, which accounts for the electromyographic changes observed in Duane's syndrome.

Möbius' syndrome, a multisystem disorder, produces abnormal ocular motility. The abducens, facial, and glossopharyngeal nerves are aplastic in this syndrome. The horizontal vergences are also defective in this syndrome, which suggests an abnormality of the medial longitudinal fasciculus. Esotropia and faulty abduction are the characteristic ocular motility abnormalities. Since the affected patient has poor horizontal vergences, maintaining fixation on a lateral moving target is accomplished by voluntarily converging. Secondarily increased accommodation and pupillary constriction will result.

Bilateral seventh nerve palsies will result in a loss of facial expression and poor eyelid closure. An additional feature that is consistent with Möbius syndrome is an atrophied, fissured tongue, with secondary difficulty in sucking. The abnormal sucking reflex produces feeding problems during infancy. Mental retardation and deafness are frequently observed. Additional defects include syndactyly, brachydactyly, clubbed feet, rocker-bottom feet, and atrophy of peroneal muscles.

References

Alvarado JA, van Horn C: Muscle cell types of the cat inferior oblique. In Lennerstand G, Bach-y-Rita P (eds): *Basic Mechanisms of Ocular Motility.* Oxford, Pergamon Press, 1975, p 15.

Apt L: An anatomical reevaluation of rectus muscle insertions. *Trans Am Ophthalmol Soc* 78:365, 1980.

Apt L, Call NB: An anatomical reevaluation of the rectus muscle insertion. *Ophthalmol Surg* 13:108, 1982.

Ashton N, Morgan G: Embryonal sarcoma and embryonal rhabdomyosarcoma of the orbit. *J Clin Pathol* 18:699, 1965.

Bedrossian EH: *The Eye.* Springfield, IL, Charles C Thomas, 1958.

Brandt DE, Leeson CR: Structural differences of fast and slow fibers in human extraocular muscle. *Am J Ophthalmol* 62:478, 1966.

Brown HW: Congenital structural muscle anomalies. In Allen JH (ed): *Strabismus Ophthalmic Symposium I.* St. Louis, CV Mosby, 1950, p 205.

Brown HW: Isolated inferior oblique paralysis: An analysis of 97 cases. *Trans Am Ophthalmol Soc* 55:415, 1957.

Brown HW: Strabismus in the adult. In Haik GM (ed): *Strabismus Symposium of the New Orleans Academy of Ophthalmology.* St. Louis, CV Mosby, 1962.

Brown HW: True and simulated superior oblique tendon sheath syndrome. *Doc Ophthalmol* 34:123, 1973.

Cooper S, Daniel PM: Muscle spindles in human extrinsic eye muscles. *Brain* 72:1, 1949.

Dietert SE: The demonstration of different types of muscle fibers in human extraocular muscle by electron microscopy and cholinesterase staining. *Invest Ophthalmol* 4:51, 1965.

Duke-Elder S: Anatomy of the visual system. In Duke-Elder S: *System of Ophthalmology,* Vol 2. London, Klimpton, 1961, p 467.

Easty DL, Chignell AH: Fluorescein angiography in anterior segment ischemia. *Br J Ophthalmol* 57:18, 1973.

Forbes SB: Muscle transplantation for external rectus paralysis. Report of a case with unusual complications. *Am J Ophthalmol* 48:248, 1959.

Frayer WC, Enterline HT: Embryonal rhabdomyosarcoma of the orbit in children and young adults. *Arch Ophthalmol* 62:203, 1959.

Fuchs E: Beitrage zur normulen Anatomie des Augapfels. *Albrecht von Graefes Arch Klin Exp Ophthalmol* 30:1, 1884.

Girard LJ, Beltranena F: Early and late complications of extensive muscle surgery. *Arch Ophthalmol* 64:576, 1960.

Harley RD: Complete tendon transplantation for ocular muscle paralysis. *Ann Ophthalmol* 3:459, 1971.

Helveston EM: Muscle transposition procedures. *Surv Ophthalmol* 16:92, 1971.

Helveston EM, Merriam WW, Ellis FD: Extraocular muscle-tendon transfer with scleral augmentation. *Am J Ophthalmol* 89:819, 1980.

Helveston EM, Merriam WW, Ellis FD, Shellhamer RH, Gosling CG: The trochlea: A study of the anatomy and physiology. *Ophthalmology* 89:124, 1982.

Henderson JW: *Orbital Tumors,* ed 2. (In collaboration with Farrow, G.M.) New York, Thieme-Stratton, 1980, p 497.

Hiatt RL: Production of anterior segment ischemia. *J Pediatr Ophthalmol & Strab* 15:197, 1978.

Hines M: Studies on innervation of skeletal muscle. III. Innervation of extrinsic eye muscles of rabbit. *Am J Anat* 47:1, 1931.

Hotchkiss MG, Miller NR, Clark AW: Bilateral Duane's retraction syndrome. *Arch Ophthalmol* 98:870, 1980.

Huber A: Electrophysiology of the retraction syndromes. *Br J Ophthalmol* 58:293, 1974.

Isenberg S, Urist MJ: Clinical observations in 101 consecutive patients with Duane's retraction syndrome. *Am J Ophthalmol* 84:419, 1977.

Jakobiec FA, Jones IS: Orbital inflammations. In Snow IS, Jakobiec FA (eds): *Diseases of the Orbit.* Hagerstown, MD, Harper & Row, 1979, p 187.

Jampel RS: The action of the superior oblique muscle. *Arch Ophthalmol* 75:535, 1966.

Jampel RS: The functional principle of action of the oblique ocular muscles. *Am J Ophthalmol* 69:623, 1970.

Jones IS, Reese AB, Kraut J: Orbital rhabdomyosarcoma. An analysis of 62 cases. *Am J Ophthalmol* 61:721, 1966.

Knowles DM II, Jakobiec FA, Potter GD: The diagnosis and treatment of rhabdomyosarcoma of the orbit. In Jakobiec FA (ed): *Ocular and Adnexal Tumors,* Chapter 49. Birmingham, Aesculapius, 1978.

Lockhart RD, Brandt W: Length of striated muscle fibers. *J Anat* 72:470, 1938.

Manley DR: Strabismus. In Harley RD (ed): *Pediatric Ophthalmology.* Philadelphia, WB Saunders, 1975, p 138.

Matteuci P: I difetti consenti di abduzione con particolare risuardo alla patogenesi. *Rassegna Ital Ottalmol* 15:345, 1946.

Merkel F, Kallius: *Makroskopische Anatomie des Auges: Graefe Saemisch Handbuch der gesamten Augenheilkunde*, ed 2. 1901, Chapter 1, p 1.

Meyer F: Zur Anatomie der Orbitalarterien. *Morphol Jahrbuch* 12:414, 1887.

Miller NR, Kiel SM, Green WR, Clark AW: Unilateral Duane's retraction syndrome (type 1). *Arch Ophthalmol* 100:1468, 1982.

Parks MM: Isolated cyclovertical muscle palsy. *Arch Ophthalmol* 60:1027, 1958.

Parks MM: Oblique muscle dysfunctions. In Duane TD, Jaeger EA (eds): *Clinical Ophthalmology*, Vol 1, Chapter 17. Philadelphia, Harper & Row Publishers, 1982, p 3.

Porterfield JT, Zimmerman LE: Rhabdomyosarcoma of the orbit: A clinicopathologic study of 55 cases. *Virchows Arch Pathol Anat* 335:329, 1962.

Ringel SP, Wilson WB, Barden MT, Kaiser KK: Histochemistry of human extraocular muscle. *Arch Ophthalmol* 96:1067, 1978.

Saunders RA, Sandall GS: Anterior segment ischemia syndrome following rectus muscle transposition. *Am J Ophthalmol* 93:34, 1982.

Sevel D: A reappraisal of the origin of the human extraocular muscles. *Ophthalmology* 88:1330, 1981.

Slavin ML, Glaser JS: Idiopathic orbital myositis. Report of six cases. *Arch Ophthalmol* 100:1261, 1982.

Sommerkamp H: Das substrat der Dauerverkürzung am Froschmuskel. *Arch Exp Path Pharmakol* 128:99, 1928.

Stout AP: Rhabdomyosarcoma of the skeletal muscles. *Ann Surg* 123:447, 1946.

Thane GD: In Schafer EA, Thane GD (eds): *Quain's Elements of Anatomy*, ed 10. London, Oxford University Press, 1892, pp 408, 524.

Trokel SL, Hilal SK: Recognition and differential diagnosis of enlarged extraocular muscles in computed tomography. *Am J Ophthalmol* 87:503, 1979.

Tschiriew S: Sur les terminaisons nerveuses dans les muscles stries. *Arch de Physiol normal et Path* 6:89, 295, 1879.

Uribe LE: Muscle transplantation in ocular paralysis. *Am J Ophthalmol* 65:600, 1968.

Werner SC: Orbital changes in Grave's disease. In Jones IS, Jakobiec FA (eds): *Diseases of the Orbit*. Hagerstown, MD, Harper & Row, 1979, p 266.

Whitnall SE: *The Anatomy of the Human Orbit and Accessory Organs of Vision* (ed 2). London, Oxford University Press, 1932, p 266.

Wooten GF, Reis DJ: Blood flow in extraocular muscle of the cat. *Arch Neurol* 26:350, 1972.

Wright KW, Silverstein D, Marrone AC, Smith RE: Acquired inflammatory superior oblique tendon sheath syndrome: A clinicopathologic study. *Arch Ophthalmol* 100:1752, 1982.

CHAPTER 8

Nerves of the Orbit

In this chapter the nerves of the orbit will be considered. The orbital nerves will be discussed in groups that are based on common features. The nerves to the extraocular muscles (oculomotor, trochlear, and abducens nerves) constitute one group. The sensory, motor, and autonomic nerves to the orbit constitute other groups. In addition, the optic nerve will be briefly discussed. The discussion of nerves begins with their nuclear origin, proceeds to their intracranial and orbital course, and continues to their termination. The function of the nerves and the pathologic processes which affect them will then be discussed.

NERVES TO THE EXTRAOCULAR MUSCLES

Oculomotor Nerve

The oculomotor nerve (CNIII), innervates the levator palpebrae superioris and the extraocular muscles, with the exception of the superior oblique (trochlear) and lateral rectus muscles (abducens nerve). The oculomotor nerve also provides the pathway for parasympathetic fibers to the globe where they innervate the ciliary muscle and the pupillary sphincter muscles.

The nucleus of the oculomotor nerve rests in the gray substance of the brainstem (mesencephalon), ventral to the aqueduct of Sylvius (Fig. 8.1). The nucleus is 6–10 mm long and extends rostrally to the posterior portion of the floor of the third ventricle, and caudally to the level of the trochlear nucleus. The oculomotor nucleus is intimately associated with the medial longitudinal fasciculus, which lies against its ventrolateral aspect. The medial longitudinal fasciculus coordinates and modulates horizontal and vertical eye movements.

Warwick (1953) delineated the complex organization of the oculomotor nucleus, which consists of a somatic and an autonomic portion. The somatic nucleus is composed of definite groups of neurons which are topographically traced to individual extraocular muscles (Fig. 8.2). Unlike the trochlear and abducens nucleus, the oculomotor nucleus consists of unpaired medial nuclei and paired lateral nuclei. The medial rectus nucleus forms the ventral border of the oculomotor nucleus, while the inferior rectus nucleus forms its dorsal border. The nucleus of the inferior oblique muscle is located between the medial and inferior rectus nuclei. The nucleus of the superior rectus is situated medially in the caudal two-thirds of the somatic nucleus. The nucleus of the

levator palpebrae superioris muscle is represented by a single, centrally placed, dorsal caudal nucleus. Recent studies utilizing retrograde tracing techniques have characterized the oculomotor nucleus to be more complex than postulated by Warwick (Gacek, 1974; Bienfans, 1975; B'auuttner-Ennever and Akert, 1981).

Fibers extending from the oculomotor nucleus will, therefore, be topographically oriented. Fibers proceeding to the inferior rectus, inferior oblique, and medial rectus do not decussate, and supply the ipsilateral eye. Fibers originating from the superior rectus nucleus decussate to proceed to the contralateral eye (Gacek, 1974; Akagi, 1978). The levator palpebrae superioris muscle is supplied by a single, centrally-placed, dorsal-caudal nucleus, which provides bilateral innervation.

The autonomic nucleus of the oculomotor nerve is rostral and dorsal to the somatic nucleus (Warwick, 1953; Jampol and Mindel, 1967). A lateral portion of this segment forms the Edinger-Westphal nucleus. This dorsal cell mass can be further divided into a rostral portion concerned with accommodation, and a caudal portion which produces pupillary constriction. The mid-portion of the dorsal autonomic nucleus is associated both with accommodation and constriction. The fibers extending from the autonomic nucleus extend, uncrossed, to supply the ipsilateral globe. Fibers from the autonomic nucleus travel with the inferior division of the oculomotor nerve and extend to the ciliary ganglion as preganglionic fibers. Postganglionic fibers from the ciliary ganglion proceed to the globe as the short posterior ciliary nerves. The vast majority of the post ganglionic fibers extend to the ciliary body, with approximately 5% of fibers extending to the pupillae sphincter (Warwick, 1953).

Axons extending from the oculomotor nucleus consolidate to form the fascicular portion of the nerve, that portion of the nerve within the brainstem. These fibers originate from the ventral border of the nucleus and extend laterally through the red nucleus and the cerebral peduncle. Multiple fibers emerge from the interpeduncular fossa on the ventral surface of the brainstem, then immediately coalesce to form two large nerve trunks.

Near its brainstem origin, the oculomotor nerve is contiguous with two major branches of the basilar artery, the superior cerebellar artery, and the posterior cerebral artery. It then runs parallel to the posterior communicating artery of the circle of Willis (Fig. 8.3). The intracranial course of the oculomotor nerve is approximately 25 mm prior to

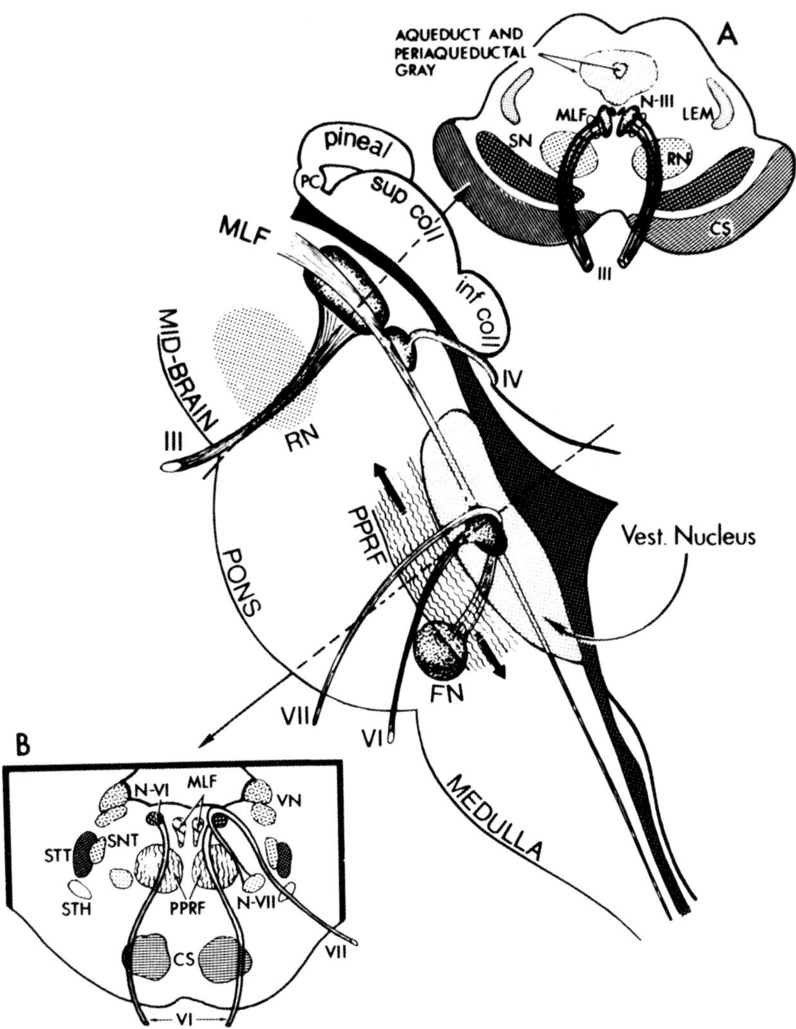

Figure 8.1. Schematic diagram of the brainstem. (*A*), Cross-section at the level of the superior colliculus. Abbreviations: *N-III*, oculomotor nucleus; *MLF*, medial longitudinal fasciculus; *LEM*, medial lemniscus; *RN*, red nucleus; *SN*, substantia nigra; *CS*, corticospinal tract. (*B*), Cross-section at the level of the abducens nucleus. Abbreviations: *N-VI*, abducens nucleus; *N-VII*, facial nucleus; *PPRF*, pontine paramedian reticular formation; *VN*, vestibular nuclear complex; *SNT*, spinal nucleus of trigeminal; *STT*, spinal tract of trigeminal; *STH*, spinothalamic tract. (Reproduced from Glaser JS: Intranuclear disorders of eye movements. In Duane TD, Jaeger EA (eds): *Clinical Ophthalmology*, Vol. 1, Chapter 12. Philadelphia, Harper & Row, 1978.)

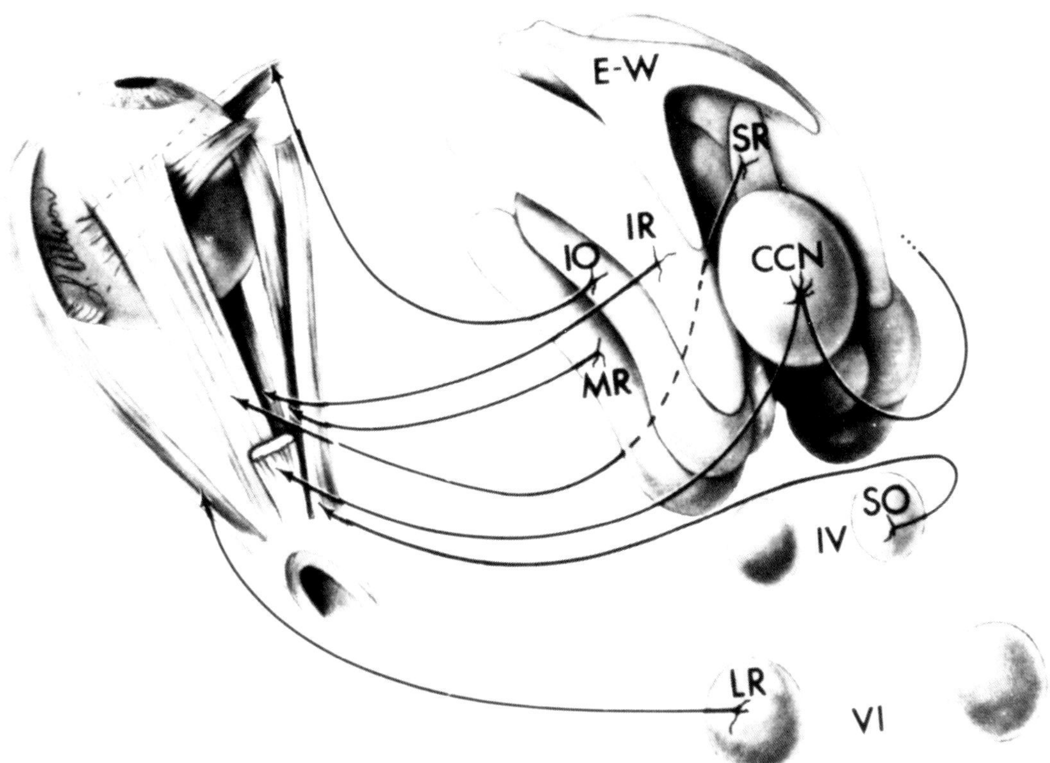

Figure 8.2. Organization of the oculomotor nucleus. Abbreviations: *E-W*, Edinger-Westphal parasympathetic subnucleus; *IR*, inferior rectus; *IO*, inferior oblique; *MR*, medial rectus; *SR*, superior rectus; *SO*, superior oblique; *CCN*, caudal nucleus to both levator palpebral superioris; *LR*, abducens nucleus for lateral rectus. (Reproduced from Glaser JS: Internuclear disorders of eye movements. In Duane TD, Jaeger EA (eds): *Clinical Ophthalmology*, Vol. 2, Chapter 12. Philadelphia, Harper & Row, 1982, illustration adapted from Warwick, 1953.)

piercing the dura at the top of the clivus, just lateral to the posterior clinoid process, to enter the lateral portion of the cavernous sinus. In the posterior portion of the cavernous sinus, the oculomotor nerve is located superiorly, but shifts inferiorly.

As the oculomotor nerve enters the lower portion of the superior orbital fissure, it divides into superior and inferior branches. The superior branch of the oculomotor nerve supplies the superior rectus and the levator palpebrae superioris muscles. It enters the superior rectus in its posterior third on its inner, or conal surface. Some of the fibers pass through the superior rectus or around its lateral border to enter the levator on its inner surface, posteriorly (Sacks, 1933). The inferior branch of the third nerve enters the orbit through the annulus of Zinn. It divides into three branches. One branch passes beneath the optic nerve and enters the medial rectus muscle in its posterior third, on its conal surface. Another branch enters the inferior rectus posteriorly on its conal surface. The third branch runs anteriorly along the lateral border of the inferior rectus muscle and innervates the inferior oblique near its midpoint as it crosses the inferior rectus muscle. In the apex of the orbit, this branch carries the parasympathetic fibers to the ciliary ganglion, which are given off as a small branch directed superiorly to the ciliary ganglion.

CLINICAL NOTE

Oculomotor nerve function may be interrupted at any point from its nuclear origin to final termination.

This may occur at the nuclear level, in the fascicular portion within the midbrain, in the interpeduncular space, in its course forward alongside the posterior communicating artery, in the cavernous sinus, or in the orbit. Associated clinical findings will reflect the anatomic site of pathology to the oculomotor nerve.

Oculomotor palsy of nuclear origin is rare. Based upon Warwick's (1953) anatomic organization of the oculomotor nucleus, Daroff (1971) established ground rules for oculomotor nuclear lesions. He described two clinical presentations which are obligatory oculomotor nuclear lesions. The first is a unilateral third-nerve palsy with contralateral superior rectus and bilateral partial ptosis. The second obligatory nuclear lesion is a bilateral third-nerve palsy (with or without internal ophthalmoplegia) associated with spared levator function. Conditions which may be nuclear are bilateral total third-nerve palsy, bilateral ptosis, bilateral internal ophthalmoplegia, and bilateral medial rectus palsy. Bilateral internuclear ophthalmoplegia associated with exotropia and loss of convergence indicates a midbrain lesion involving medial rectus subnuclei. Lubow (1982) coined the term "WEBINO" (wall-eyed bilateral internuclear ophthalmoplegia) to describe patients with bilateral medial rectus palsy and exotropia.

Fascicular lesions of the oculomotor nerve are associated with additional brainstem signs. Oculomotor nerve palsy associated with contralateral hemiplegia (Weber's syndrome) will indicate corticospinal tract involvement. Contralateral ataxia and intention tremor (Benedickt's syndrome) reflect involvement of the red nucleus. Vascular accidents are the most com-

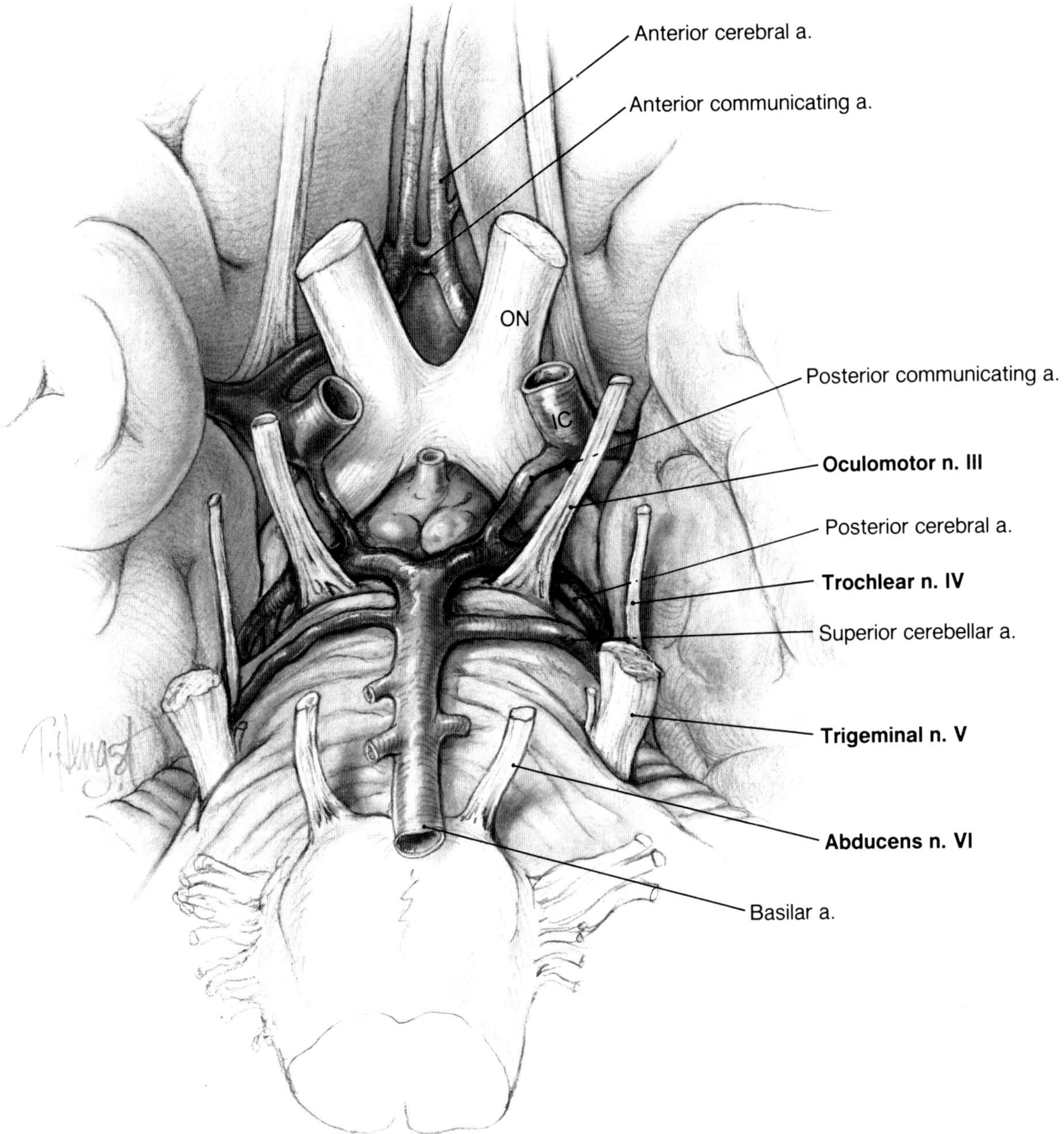

Figure 8.3. Schematic representation of the base of the brain. The close association of arteries and cranial nerves are demonstrated.

mon cause of nuclear and fascicular oculomotor lesions.

Interpeduncular lesions may be secondary to slowly growing masses, but are more likely due to aneurysms of the posterior cerebral artery and the posterior communicating artery. Trobe and co-workers (1978) reported isolated oculomotor palsies associated with basilar artery aneurysms. After exiting the interpeduncular sulcus, the oculomotor nerve is contiguous with the posterior communicating artery (Fig. 8.4 *A–C*). Aneurysm of the posterior communicating artery, with accompanying hemorrhage, will therefore acutely compromise the oculomotor nerve (Hyland and Barnett, 1956). These patients generally present with an intensely painful, complete unilateral third-nerve palsy and a fixed dilated pupil.

Oculomotor palsy may result from trauma, although less frequently than does trochlear palsy. Closed-head injury generally has to be extensive to induce oculomotor palsy. Heinze (1969) dissected cases of fatal head trauma and reported three sites of pathology: avulsion from the proximal portion of the nerve trunk, contusion and necrosis of the proximal portion of the nerve, and intracranial and perineural hemorrhages at the level of the superior orbital fissure.

The cavernous sinus may be involved with inflammatory disease, tumor, or vascular disorders such as thrombosis, arteriovenous fistula, and aneurysm. The

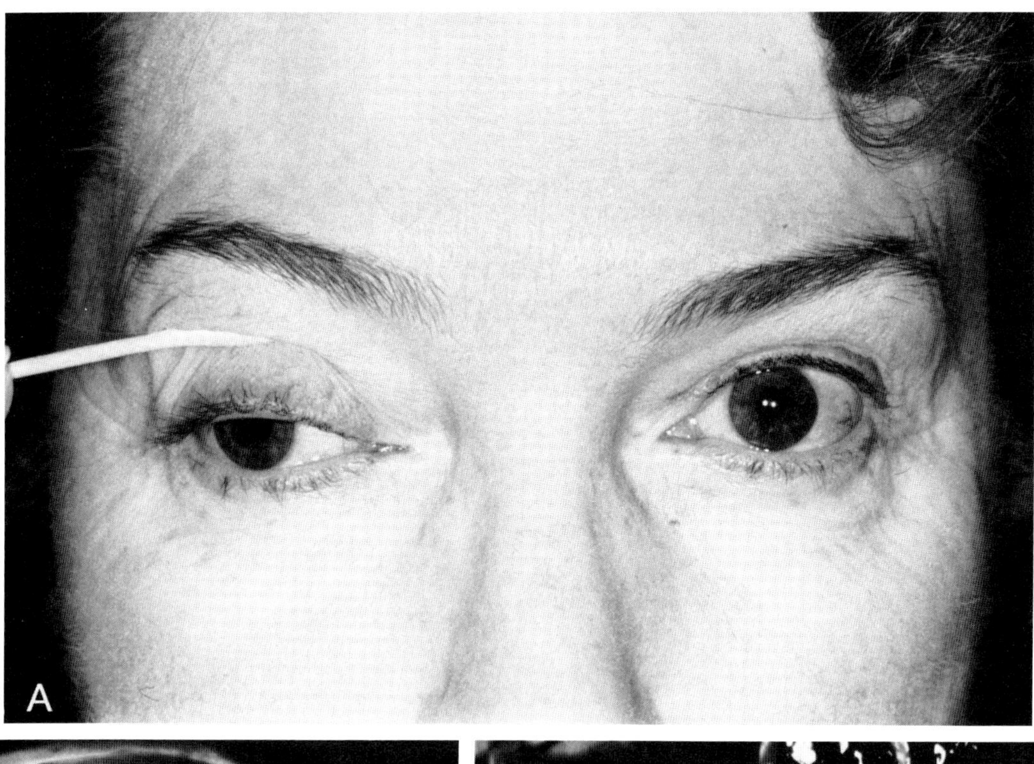

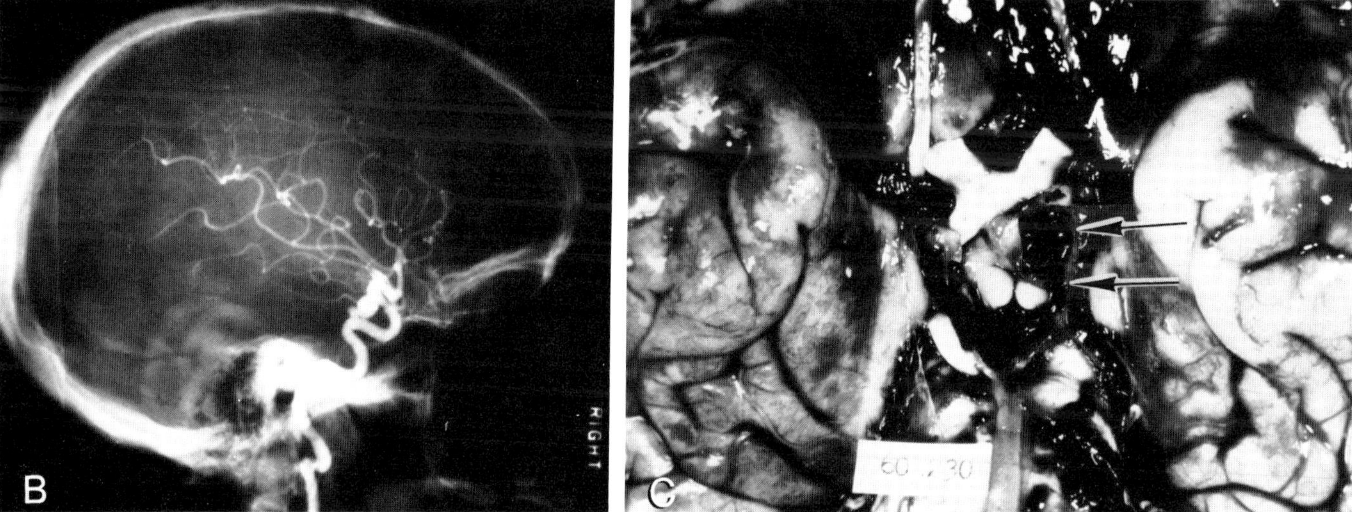

Figure 8.4. Oculomotor nerve palsy secondary to an aneurysm of the posterior communicating artery. (*A*), Clinical presentation with ptosis and a prominent exodeviation. (*B*), Carotid arteriogram demonstrating a posterior communicating artery aneurysm (PCA). (*C*), Postmortem view of a large PCA aneurysm (*arrows*). (Reproduced with permission from R. Aouchiche.)

proximity of the oculomotor, trochlear, abducens, and ophthalmic division of the trigeminal nerves is reflected in painful, multiple ocular motor palsies associated with cavernous sinus lesions. Sympathetic paresis may also accompany cavernous sinus lesions due to its contribution from the carotid plexus within the cavernous sinus. Third-nerve palsies, secondary to cavernous sinus lesions, tend to be partial and the pupil is frequently spared. However, slowly expanding cavernous lesions (meningioma, infraclinoid aneurysm) may produce a progressive paresis of the oculomotor nerve.

Oculomotor palsy may be the presenting manifestation of diabetes. Although recurrences are not un-

common, coexistent ocular motor nerve involvement suggests additional pathology. Generally, patients present with a painful ophthalmoplegia, which spares the pupil (Goldstein and Cogan, 1960; Nadeau and Cogan, 1983). Recovery is generally spontaneous within 3 months of its onset. The underlying pathology is ischemia due to occlusion of intraneural arterioles producing focal demyelinization (Weber, et al, 1970; Asbury, et al, 1970).

Pupillary involvement is an important finding in the evaluation of a painful oculomotor palsy. Involvement of the pupil suggests an aneurysm actively impinging upon the oculomotor nerve, while pupil sparing suggests a diabetic etiology. Pupillary fibers arise from the autonomic nucleus of the oculomotor nerve. In the subarachnoid space the pupillary fibers are on the periphery or superficial aspects of the oculomotor nerve. Within the cavernous sinus the pupillary fibers coalesce to extend with the inferior division of the oculomotor nerve as it enters the orbit. Therefore, expanding lesions which impinge upon the oculomotor nerve (aneurysm, tumor) within the subarachnoid space will involve the pupillary fibers of the nerve. However, localized ischemia or demyelinization of the oculomotor nerve, which occurs with diabetic oculomotor palsy, does not affect pupillary fibers because of their auxiliary vascular supply from the epineurium (Weber, et al, 1970; Asbury, et al, 1970).

Trochlear Nerve

The trochlear nerve contains somatic motor fibers for the superior oblique muscle. The nucleus of the trochlear nerve (CN IV) is formed by a small clump of cells at the lower, or caudal end of the oculomotor nucleus (Fig. 8.1). The nucleus is situated in the central gray matter near the floor of the aqueduct of Sylvius, at the level of the inferior colliculus. The medial border of the trochlear nucleus is closely associated with the medial longitudinal fasciculus.

The fascicular portion of the nerve extends from the trochlear nucleus dorsocaudally around the paraventricular gray matter. The fibers abruptly redirect themselves dorsally prior to reaching the level of the pons. The fibers decussate, then exit the dorsal surface of the brainstem, caudal to the inferior colliculus at the anterior medullary velum. The trochlear nerve is the only cranial nerve which exits on the dorsum of the brainstem. Since the trochlear fibers decussate in the brainstem, it supplies the contralateral superior oblique muscle.

Intracranially, the trochlear nerve is approximately 40 mm long, and is the longest of the nerves to the extraocular muscles. The nerve winds around the cerebral peduncle above the pons to assume a position in the subarachnoid space between the pons and temporal lobe. The nerve then runs along the free border of the tentorium cerebelli for 1–2 cm. The trochlear nerve penetrates the dura to enter the cavernous sinus beneath the posterior clinoid process inferior and lateral to the oculomotor nerve. In the lateral wall of the cavernous sinus the trochlear nerve lies between the oculomotor nerve and the ophthalmic division of the trigeminal nerve. Within the cavernous sinus, the trochlear nerve passes superiorly, crossing the oculomotor nerve, and enters the orbit through the superior orbital fissure outside the annulus of Zinn. Once within the orbit, the nerve extends nasally to innervate the orbital surface of the superior oblique muscle. Thus the trochlear nerve is the only extraocular nerve which does not penetrate the conal surface of its muscle.

CLINICAL NOTE

The trochlear nerve is the only oculomotor nerve to exit the brainstem on its dorsal surface. Because of the dorsal location of the trochlear nucleus and exit of the nerve from the dorsum of the brainstem, the fascicular portion of the nerve is extremely short. In addition, the trochlear nucleus is not adjacent to significant neural structures. As such, it is virtually impossible to recognize an isolated trochlear palsy of a nuclear origin.

The trochlear nerve has a long intracranial course prior to entering the cavernous sinus. This long course predisposes the trochlear nerve to trauma. Countercoup injuries to the nerve occur at its exit at the brainstem, or at some point along its course around the midbrain (Lindenberg, 1966; Burger, et al, 1970; Miller, et al, 1970; Younge and Sutla, 1977). The tectum of the midbrain is subject to contusion at the tentorial notch when the forehead or skull vortex strikes a stationary object. However, trochlear nerve palsies may also occur with seemingly insignificant trauma; even a fall on the buttocks, due to the cerebellum being thrust against the tentorium (Lindenberg, 1966).

Isolated trochlear palsies, other than those traumatically induced, are rare. Localized orbital pathology should be excluded, as well as myasthenia gravis and diabetes. In most cases, the etiology of these nontraumatic trochlear palsies cannot be determined, and most are thought to be of vascular origin. The symptoms and signs of a superior oblique palsy are reviewed in Chapter 7, "Extraocular Muscles."

Abducens Nerve

The abducens nerve (CN VI) carries motor fibers to the lateral rectus muscle. The nucleus is situated beneath the floor of the fourth ventricle near the midline, and in the upper portion of the rhomboid fossa (Fig. 8.1). The abducens nucleus is caudal to the pontine paramedian reticular formation and is surrounded by the arch of the facial nerve. The medial longitudinal fasciculus is located at the medial border of the trochlear nerve. Fibers from the abducens nucleus pass ventrally, caudally, and laterally through the pons on the lateral side of the pyramidal tract. The fascicular portion of the abducens nerve is in close proximity to the motor nucleus of the facial nerve, the motor nucleus of the trigeminal nerve, the spinal tract of the trigeminal nerve, the superior olivary nucleus, and the central tegmental tract. The nerve emerges from the groove between the pons and medulla-oblongata (medullary-pontine sulcus).

The abducens nerve initially is flattened by the pons following its exit from the brainstem at the medullary pontine sulcus (Fig. 8.3). The abducens nerve passes between the pons and the anterior inferior cerebellar artery. The abducens nerve ascends through the subarachnoid space along the clivus to pierce the dura of the clivus about 1 cm below the crest of the sphenoid bone. The abducens nerve passes through the inferior petrosal sinus, which is formed by the junction of the petrous portion of the temporal bone and the basilar part of the occipital bone. The nerve then makes a sharp bend over the petrous ridge and passes beneath the petrosphenoidal (petroclinoid) ligament of Gruber (Fig. 8.5). This passageway into the cavernous sinus, known as Dorello's canal, lies on the lateral side of the internal carotid artery, and medial to the trigeminal ganglion. Initially, the abducens is closely associated with the internal carotid artery within the cavernous sinus. The nerve

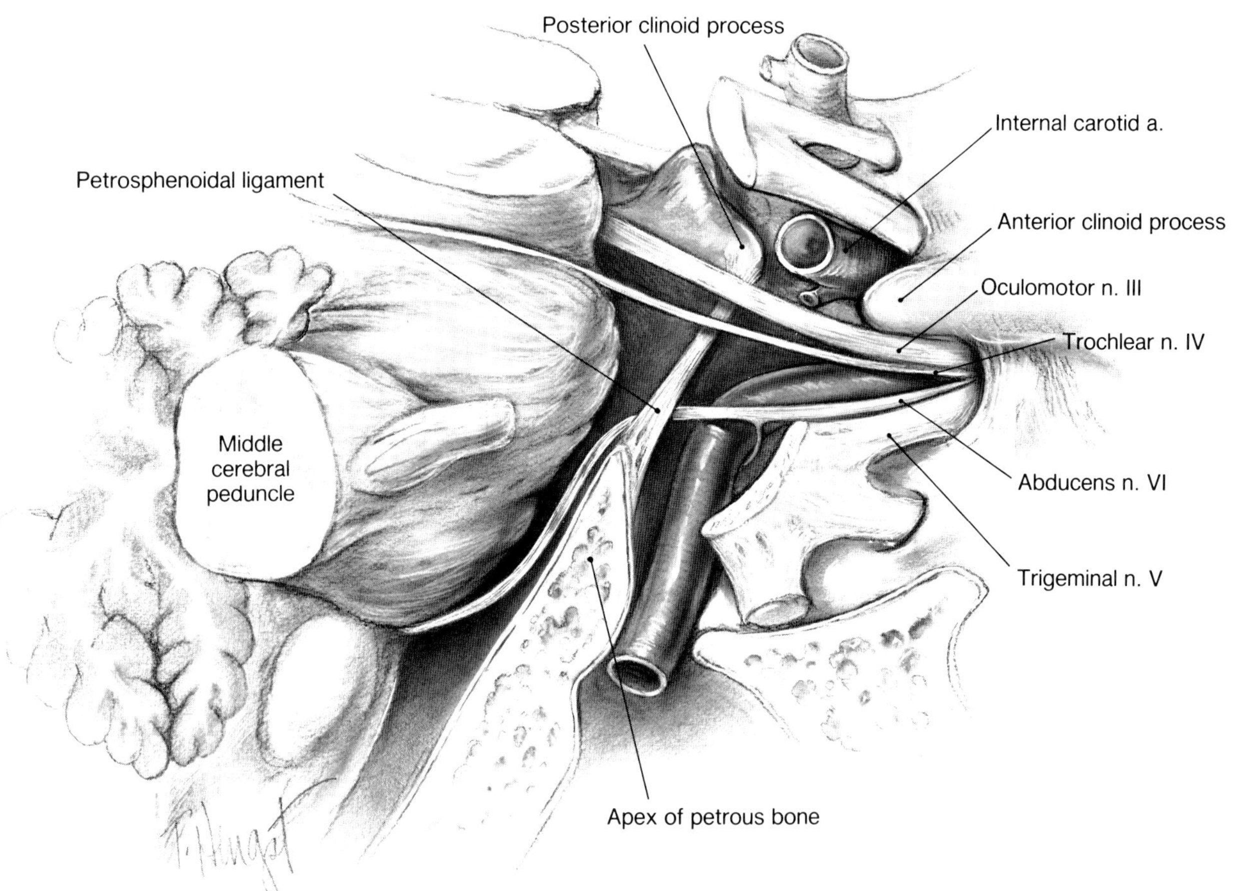

Figure 8.5. Relationship of brainstem, cranial nerves, and basal structures. The internal carotid extends through the carotid canal and foramen lacerum to enter the cavernous sinus with the abducens nerve at its lateral border. The abducens nerve originates from the brainstem at the medullary-pontine sulcus, to travel along the clivus and petrous portion of the temporal bone. The abducens nerve then bends sharply over the petrous ridge beneath the petrosphenoidal ligament prior to entering the cavernous sinus.

then extends toward and lies on the lateral wall of the sinus, prior to entering the superior orbital fissure within the annulus of Zinn. The nerve then continues to the lateral rectus to insert on its conal surface at the junction of the posterior one-third and anterior two-thirds of the muscle.

CLINICAL NOTE

The long course of the abducens nerve predisposes it to a variety of pathologic conditions. Tumors, vascular disease, infection, trauma, diabetes, and multiple sclerosis may present with an abducens palsy (Rucker, 1966; Robertson, et al, 1970; Nathan, et al, 1974; Bajandas, 1977). It is helpful to distinguish an isolated abducens palsy from those associated with additional neurologic signs. The associated neurologic signs help identify the anatomic location of abducens palsies.

Isolated lateral rectus weakness is not always secondary to abducens palsy. Systemic disorders, such as myasthenia gravis and dysthyroid myopathy, or localized orbital inflammation, may produce lateral rectus muscle abnormalities. These conditions should be excluded prior to submitting the patient to an extensive neurologic diagnostic evaluation.

Isolated abducens palsy may be secondary to basilar skull fractures. The long, intracranial course of the

abducens nerve, and its confinement by the dura, and the petrosphenoidal (petroclinoid) ligament predispose it to injury with cranial trauma. In addition, the abrupt angle the abducens nerve makes as it crosses over the petrous ridge places it in a precarious position in cases of basilar skull fractures and increased intracranial pressure.

Abducens palsies may also be associated with additional neurologic signs which are of localizing value. Pontine syndromes produced by infarction, demyelination, or tumor may produce contralateral hemiplegia, and ipsilateral abducens and facial palsy. The proximity of the abducens nerve to the petrous portion of the temporal bone will account for abnormalities in the nerve that are associated with chronic mastoiditis, or suppurative otitis media (Gradenigo, 1907; Wheeler, 1918; Meltzer, 1931). The combination of abducens palsy, pain in the trigeminal area, and suppurative otitis media are recognized as Gradenigo's syndrome. After the abducens nerve enters the cavernous sinus, it may be affected by conditions within the sinus or superior orbital fissure.

Control of Ocular Motility

It is not the purpose of this text to exhaustively describe the neuroanatomical foundation of ocular motility. A superficial examination of the systems involved in the coordination of ocular motility will be presented herein and the reader will be referred to more comprehensive texts, such as Miller's revision of *Walsh and Hoyt's Clinical Neuro-Ophthalmology* (1983)

The brainstem systems which coordinate ocular motility are the medial longitudinal fasciculus (MLF), paramedian pontine reticular formation (PPRF), and the rostral interstitial nucleus of the medial longitudinal fasciculus (Fig 8.1). The MLF is a medial fiber tract extending from the oculomotor nucleus to the spinal cord. The bulk of fibers of the MLF extend from the vestibular nuclei which ascend and influence the oculomotor nuclei. The MLF coordinates ocular motility.

The paramedian pontine reticular formation is also called the horizontal gaze center. It is located adjacent to the abducens nucleus within the pons. Localized lesions to this area produce a paralysis of conjugate gaze to the ipsilateral side. This is to be distinguished from lesions of the abducens nucleus or nerve which produces a paralysis of abduction of the ipsilateral eye only (Carpenter and McMasters, 1963; Keller, 1974).

The rostral interstitial nucleus of the medial longitudinal fasciculus is recognized as the verticle gaze center. A small group of cells at the level of the thalamic nuclei has been reputed to control vertical gaze, based on efferent fiber projections and retrograde tracer studies (Leichnetz, 1982; B'auuttner-Ennever, et al, 1982). The existence of the vertical gaze center will require further validation.

SENSORY NERVE OF THE ORBIT TRIGEMINAL NERVE

The trigeminal nerve has two major components: sensory and motor. The sensory component of the nerve is by far the most extensive. It supplies sensation to the face, the mucus membranes, and the internal structures of the head. The sensory system of the trigeminal nerve is extraordinarily complex because of the diffuse array of stimuli to which it responds, and its need to process these multitudinous impulses. The smaller motor component of the nerve receives impulses from the sensory root and proprioceptive stimuli from the mesencephalic root of the nerve. The motor component of the trigeminal nerve innervates the muscles of mastication.

The combined sensory and motor components of the trigeminal nerve complicate its evaluation. The ocular motor nerves are somatic motor nerves, except for the autonomic portion of the oculomotor nerve, which originate at the brainstem and extend to the end organs. The trigeminal nerve differs since it consists primarily of sensory afferent fibers. The brainstem portion of the trigeminal nerve will first be discussed, followed by the trigeminal ganglion. The three main divisions of the trigeminal nerve, namely the ophthalmic, maxillary, and mandibular nerves will then be discussed.

Since the main component of the trigeminal nerve is sensory, the bulk of the nerve consists of afferent fibers. The fibers synapse at the trigeminal (gasserian, semilunar) ganglion from its three main divisions: the ophthalmic, maxillary, and mandibular nerves. Separate sensory and motor roots of the trigeminal ganglion extend approximately 18 26 mm from its posterior surface to the pons (Gudmundson, et al, 1971). The trigeminal root is contiguous with the superior cerebellar artery in approximately 50% of instances (Hardy and Rhoton, 1978; Hardy, et al, 1980; Haines, et al, 1980), which may be significant in the etiology of trigeminal neuralgia. Upon entering the brainstem, the sensory root divides into ascending and descending tracts which terminate in the main sensory nucleus, and the nucleus of the spinal tract of the trigeminal nerve respectively.

The terminal sensory nucleus of the trigeminal nerve has an enlarged rostral end (the main sensory nucleus), and a long, slender caudal portion (the trigeminal spinal tract). The main sensory nucleus, at the level of the superior cerebellar peduncle, deals with discriminative sense. Fibers are received from the ascending tract derived from the sensory root of the trigeminal ganglion. Fibers extending from the main sensory nucleus proceed to the nucleus ventralis posteromedialis of the thalamus, and are then relayed to the postcentral cerebral cortex.

The descending branches of the sensory root fibers course through the pons and medulla to form the trigeminal spinal tract. The nucleus of the spinal tract is elongated, extending to the second cervical segment of the spinal cord. At this level, the fibers are continuous with the substantia gelatinosa of the dorsal horn of the spinal cord. This serves as the sensory nucleus for pain perception and temperature. Second-order neurons extend to the lateral spinothalamic tract and terminate on the nucleus ventralis posteromedialis of the thalamus.

The motor nucleus of the trigeminal nerve is at the level of the pons, and is the most caudal of a line of motor nuclei, which innervate the brachial musculature (motor nerve of trigeminal, motor nucleus of the facial, and the nucleus ambiguous of the glossopharyngeal and the vagus nerves). Fibers from the motor nucleus of the trigeminal pass through the pons in a ventral direction, and meet with sensory fibers. These fibers are associated with the mandibular division of the trigeminal nerve and are distributed to the muscles of mastication (masseter, temporalis, and pterygoid muscles).

The motor component of the trigeminal nerve also receives proprioceptive impulses from a mesencephalic root. Proprioceptive fibers are associated with the mandibular division and extend to the trigeminal ganglion. The fibers then enter the brainstem. The rostral extent of the mesencephalic nucleus is the level of the fourth ventricle, while caudally it extends to the sensory nucleus of the trigeminal nerve. The mesencephalic root supplies proprioceptive stimuli for the muscles of mastication and the extraocular muscles.

The trigeminal nerve, with its large sensory and small motor root, extends from the brainstem at the lateral portion of the ventral surface of the pons, passing anteriorly in the posterior cranial fossa, to an opening in the dura. This opening leads into Meckel's cave, a structure that is formed by a split in the dura at the petrous portion of the temporal bone in the middle cranial fossa. Meckel's cave extends forward to the cavernous sinus, and houses the trigeminal ganglion. This ganglion is flat, semilunar in shape, and measures 1 × 2 cm. The ganglion is lateral to the posterior portions of the cavernous sinus and internal carotid artery. The motor component of the trigeminal nerve passes beneath the ganglion and proceeds with the mandibular nerve through the foramen ovale.

Three divisions or peripheral extensions of the trigeminal nerve extend from the anterior aspect of the trigeminal ganglion. The ophthalmic nerve is embedded in the deep layer of the lateral wall of the cavernous sinus, which is formed by sheaths of the oculomotor, trochlear, ophthalmic, and maxillary nerves, and a reticular membrane which extends between the sheaths. While in the cavernous sinus the ophthalmic nerve forms a recurrent (tentorial) branch to supply the dura of the posterior pole of the brain. In addition, fine nerves extend to the oculomotor, trochlear, and abducens nerves. Toward the front of the cavernous sinus the ophthalmic nerve divides into its three main branches: lacrimal, frontal, and nasociliary (Fig. 8.6 *A* and *B*).

Separating from the main trunk of the nerve in the cavernous sinus, the lacrimal is the smallest division of the ophthalmic nerve. The lacrimal nerve enters the orbit through the superior orbital fissure and extends anteriorly along the superior border of the lateral rectus muscle. Prior to entering the posterior aspect of the lacrimal gland, the lacrimal nerve is joined by a branch of the zygomaticotemporal nerve. This contribution provides postganglionic parasympathetics containing secretomotor fibers for the lacrimal gland. These fibers originally traveled with the facial nerve and the greater superficial petrosal nerve. The pathway for parasympathetic secretomotor fibers to the lacrimal gland is reviewed in Chapter 5, "The Lacrimal System." The lacrimal nerve pierces the orbital septum to provide sensation to the superior lateral aspect of the eyelid (Fig. 8.7).

The frontal is the largest branch of the ophthalmic nerve and enters the orbit through the superior orbital fissure. The frontal nerve travels anteriorly, between the levator muscle and the periorbita. About halfway to the superior orbital margin, the nerve splits into a small supratrochlear branch and a larger, supraorbital branch. The supratrochlear nerve extends medially, to the trochlear region. The supratrochlear nerve supplies the lower portion of the forehead and the medial aspect of the upper eyelid. The supraorbital nerve exits the orbit through the supraorbital notch or foramen. The supraorbital nerve innervates the forehead. It lies in a plane immediately beneath the frontalis muscle.

The third division of the ophthalmic nerve the nasociliary nerve, has a long, complicated course. The nasociliary nerve enters the orbit through the annulus of Zinn in the oculomotor foramen. The nasociliary nerve crosses over the optic nerve and sends branches to the medial orbital wall. As the nasociliary nerve enters the orbit, a small branch extends along the lateral side of the optic nerve, and enters the ciliary ganglion. This sensory contribution to the ciliary ganglion travels through the ganglion without synapsing, and continues as the short ciliary nerves. These nerves penetrate the sclera near the optic nerve.

The long ciliary nerves, two or three in number, extend from the nasociliary nerve as it crosses the optic nerve. These nerves pass by the ciliary ganglion and progress with the short posterior ciliary nerves to the globe. The long

ciliary nerves enter the sclera and extend anteriorly along the medial and lateral aspects of the globe to supply the iris, ciliary muscle, and cornea. In addition to afferent fibers, the long ciliary nerves carry sympathetic fibers from the superior cervical ganglion to the dilator muscle of the iris. The sympathetic contribution to the nasociliary nerve probably occurs in the cavernous sinus.

The nasociliary nerve continues to the medial orbital wall where it branches to form the posterior ethmoidal, anterior ethmoidal, and infratrochlear nerves. The ethmoidal nerves supply the mucous membranes of the ethmoidal sinuses and nasal mucosa.

After forming the anterior ethmoidal nerve, the nasociliary nerve extends anteriorly as the infratrochlear nerve. The nerve runs along the superior border of the medial rectus muscle and penetrates the orbital septum. The infratrochlear nerve supplies the medial skin of the eyelids, side of the nose, medial conjunctiva, lacrimal sac, and caruncle.

CLINICAL NOTE

The varicella-zoster virus has a tendency to affect the ophthalmic division of the trigeminal nerve to produce herpes zoster ophthalmicus (HZO). The predilection for HZO to occur in the elderly is poorly understood, but is probably related to a deficient cell-mediated immunity (delayed hypersensitivity) (Schlaegel, 1982).

Patients with HZO develop a vesicular eruption in the distribution of the ophthalmic division of the trigeminal nerve. The vesicular eruption may be preceded by paresthesias in the affected area, or nasal mucosa. Hutchinson (1866) recognized the association of ocular involvement with zoster affecting the tip of the nose (Fig. 8.8). This association reflects the distribution of the nasociliary nerve.

The oculocardiac reflex is recognized as a triad of bradycardia, nausea, and faintness evoked by stretching the extraocular muscles, particularly the medial rectus muscle (Apt, et al, 1973; Isenberg and Blechman, 1982), pressure on the globe, and stretching the retractors of the upper and lower eyelid (Roy, 1975; Anderson, 1978). The ophthalmic division of the trigeminal nerve is the afferent limb of this reflex. As noted previously, the ophthalmic nerve proceeds to the trigeminal ganglion and brainstem via the sensory root of the trigeminal nerve. Upon entering the brainstem, fibers descend forming the spinal trigeminal tract. Denny-Brown and Yanagisawa (1973) demonstrated in macaque monkeys that the spinal trigeminal tract acts to converge sensory stimuli from the trigeminal, facial, and vagus nerves, and the second and third cervical sensory roots. Therefore stimulation of the trigeminal nerve passes through polysynaptic pathways in the reticular formation to the visceral motor nucleus of the vagus nerve (Walsh and Hoyt, 1969). Atropine may help block this reflex arc and reverse signs of vagus stimulation.

The maxillary nerve is a pure sensory nerve which extends from the middle of the trigeminal ganglion. The nerve provides sensation for the midface, lower eyelid, side of the nose, and the upper lip; plus the mucous membranes of the nasopharynx, maxillary sinus, soft palate, tonsils, roof of

ganglion. In addition, parasympathetic secretomotor fibers from the ganglion extend to the maxillary nerve. These parasympathetic fibers are destined for the lacrimal gland via the zygomaticotemporal nerve.

Additional branches of the maxillary nerve within the pterygopalatine fossa are the zygomatic nerve and the posterior-superior alveolar nerve (Fig 8.6*A*). The zygomatic nerve enters the orbit through the inferior orbital fissure and divides into the zygomaticotemporal and zygomaticofacial nerves. The zygomaticotemporal nerve runs along a groove in the zygomatic bone, and passes through a small fissure to enter the temporal fossa, thus supplying sensation to the lateral aspect of the forehead. Prior to departing the orbit, however, the zygomaticotemporal nerve provides secretomotor parasympathetic branches to the lacrimal nerve (see Fig. 5.8). The zygomaticofacial branch extends along the inferior-lateral angle of the orbit, through foramina in the zygomatic bone, and supplies the skin of the cheek.

The superior alveolar nerves primarily provide sensory innervation for teeth, but small branches also supply the mucous membranes of the cheek and gums. The posterior-superior alveolar nerve arises from the maxillary nerve just before it enters the infraorbital groove within the pterygopalatine fossa. The middle and anterior-superior alveolar nerves are the only branches of the maxillary nerve within the infraorbital canal. The middle and anterior-superior alveolar nerves run downward in the lateral wall of the antrum to supply premolar, incisor, and canine teeth, respectively.

The maxillary nerve enters the infraorbital canal as the infraorbital nerve and exits approximately 10 mm from the inferior orbital margin on the anterior wall of the maxilla (Fig. 8.6*A*). The terminal extensions of the infraorbital nerve following its exit from the infraorbital foramen are the inferior palpebral, external nasal, and the superior labial branches. The inferior palpebral branch supplies the skin and conjuctiva of the lower eyelid. The external nasal branches, joining terminal branches of the nasociliary nerve, provide sensation to the skin and nasal septum. The superior labial branch provides sensation to the skin of the upper lip and the mucous membrane.

The mandibular division of the trigeminal nerve is a mixed cranial nerve with a sensory and motor root. The mandibular nerves extend from the inferior region of the trigeminal nerve. A sensory and motor root of the nerve exits the cranial cavity through the foramen ovale. The sensory fibers supply the skin below the cheek, excluding the upper lip, and the underlying mucous membranes. The motor root supplies the muscles of mastication: temporalis, masseter, pterygoideus medialis and pterygoideus lateralis. The two roots of the nerve exit the foramen ovale and extend to their designated areas.

CLINICAL NOTE

Trigeminal neuralgia (tic douloureux) is a disorder characterized by brief attacks of severe pain along the distribution of the trigeminal nerve. This disorder may affect all age groups, although there is a predilection for it to occur in middle and old age (Harris, 1943).

The attacks of pain are intense, and may start in one area of the trigeminal nerve, and then spread to adjacent areas. The mandibular division of the trigeminal nerve is most commonly affected (Dandy, 1936), with the ophthalmic division affected in only 1–2% of instances (Jefferson, 1931; Sperling and Stender, 1935). The attacks are occasionally elicited by stimulation of trigger points, frequently in the snout area. As such, the patients may be unable to speak, brush their teeth, eat or drink without suffering spasms of pain. In rare cases, facial pain will become unremitting.

Trigeminal neuralgia may be secondary to a cerebral tumor (Dandy, 1936), trauma, dental problems, or vascular anomalies adjacent to the sensory root of the trigeminal nerve (Goldstein, et al, 1963). Multiple sclerosis is associated with trigeminal neuralgia in 2–4% of instances (Harris, 1943; Rushton and Olafson, 1965). Dandy (1936) postulated the sensory root of the trigeminal nerve as the lesion site based upon procedures to treat trigeminal neuralgia. However, Kerr and Miller (1966) provided evidence for a peripheral etiology for trigeminal neuralgia. Those authors recognized ultrastructural changes in trigeminal ganglia removed as treatment for the disorder.

FACIAL NERVE

The facial nerve (CN VII) has two parts: a larger motor component which supplies the facial musculature (except the mastication muscles); and a smaller component, the nervus intermedius, which contains sensory and parasympathetic fibers. This smaller sensory component is responsible for taste sensation on the anterior two-thirds of the tongue. The parasympathetic portion of the nervus intermedius supplies secretomotor fibers for the submandibular, sublingual, lacrimal, nasal, and palatine glands. The nervus intermedius has distinct features and can be considered a separate cranial nerve.

The motor nucleus of the facial nerve is located deep within the reticular formation of the pons, near the nucleus ambiguous and the lateral horn of the spinal cord (Fig. 8.1). The motor nucleus is approximately 4 mm long. Attempts to topographically organize the facial nucleus in a manner similar to that of the oculomotor nucleus (Warwick, 1953) have been difficult. Van Buskirk (1945) could not consistently organize the facial nucleus into cellular groups that supply specific facial muscles. However, Crosby and De-Jonge (1963) were able to divide the facial nucleus into an upper and a lower segment, with each supplying specific structures. The upper segment supplies the frontalis, occipitalis, upper orbicularis, and corrugator superciliaris, while the lower segment supplies the remaining musculature of facial expression.

Axons from the facial nucleus proceed from its dorsal surface to arch around the abducens nucleus and form the genu of the facial nerve. The genu of the facial nerve produces an elevation, the facial colliculus, in the rhomboid fossa on the dorsal surface of the brainstem. Once the facial nerve root arches around the abducens nucleus, it extends ventrally, slightly laterally, and caudally to emerge from the brainstem between the olive and inferior cerebellar peduncle at the caudal border of the pons. The sensory and parasympathetic fibers of the facial nerve are not present at the genu but join distally to that point. Parasympathetic fibers arise from the superior salivatory nucleus alongside the descending limb of the facial nerve within the pons. Sensory fibers, with diffuse brainstem connections, diverge from the facial root shortly after it enters the pons.

Supranuclear pathways affect the facial nucleus. Volitional activity arises from the lower portion of the precentral gyrus. A disproportionately large area of the precentral gyrus is devoted to the muscles of facial expression, which reflects the intricacy of this type of motor activity. Fibers extend from the precentral gyrus to the lower portion of the internal capsule to reach the contralateral facial nucleus and form the corticobulbar tract. Some uncrossed fibers of the corticobulbar tract terminate in the upper segment of the facial nucleus. Peripheral neurons, which originate from the upper segment of the facial nerve and supply the forehead and upper orbicularis muscle, receive bilateral cortical impulses. (Fig. 8.10). Fibers controlling facial expression, mediated by

fibers of the reticular formation, are also contributed by the thalamus and globus pallidus.

The facial nerve exits from the ventrolateral portion of the brainstem at the cerebellopontine angle as a motor root and the nervus intermedius. The fine nervus intermedius is sandwiched between the motor root and the acoustic nerve. These fibers extend laterally to the internal auditory canal, where the facial nerve separates from the acoustic nerve to enter the facial (fallopian) canal within the petrous portion of the temporal bone. Within the facial (fallopian) canal, the facial nerve follows a serpiginous course. This canal is approximately 30 mm long, and is divided into four sections: meatal, labyrinthine, tympanic, and mastoid (Fig. 8.11). The meatal segment of the facial nerve is that portion prior to the separation of the acoustic nerve from the facial and nervus intermedius. The labyrinthine segment of the facial nerve begins as the nerve enters the facial or fallopian canal. This segment of the nerve is 3–6 mm long, and extends to the geniculate ganglion. The geniculate ganglion is the sen-

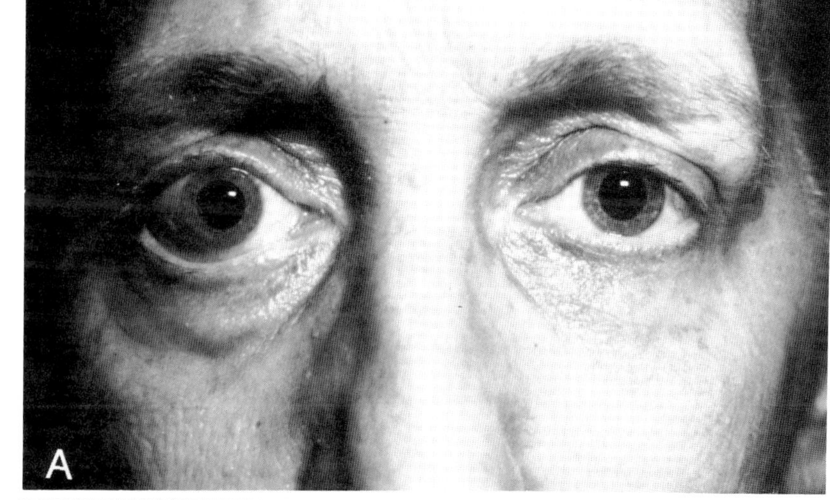
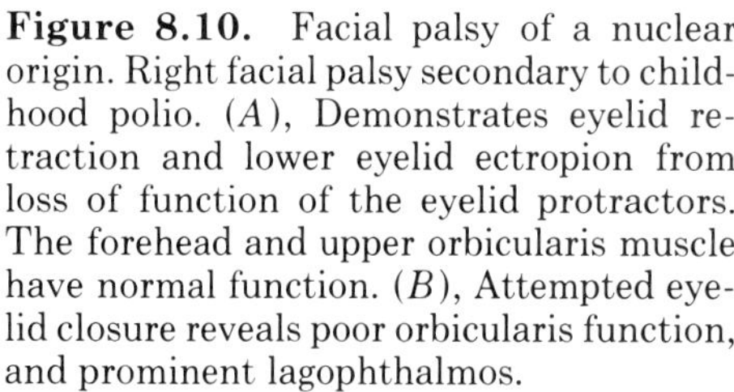

Figure 8.10. Facial palsy of a nuclear origin. Right facial palsy secondary to childhood polio. (*A*), Demonstrates eyelid retraction and lower eyelid ectropion from loss of function of the eyelid protractors. The forehead and upper orbicularis muscle have normal function. (*B*), Attempted eyelid closure reveals poor orbicularis function, and prominent lagophthalmos.

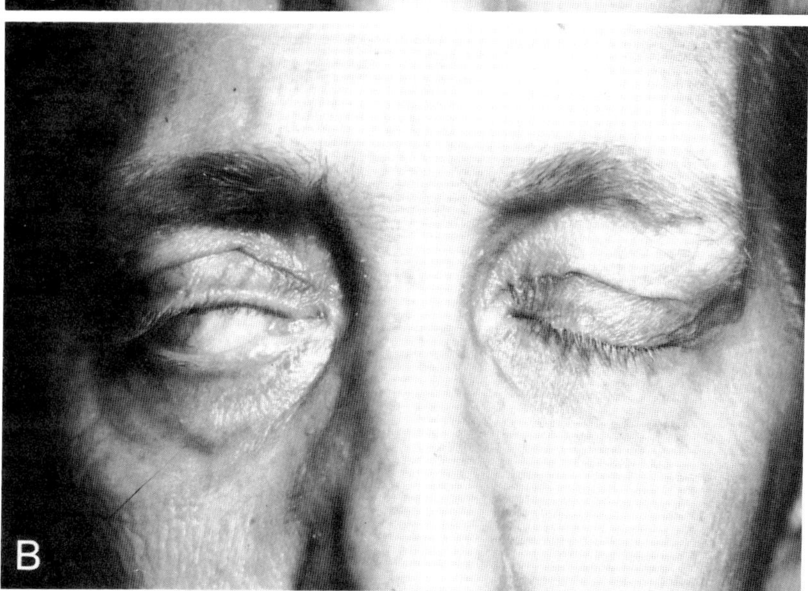

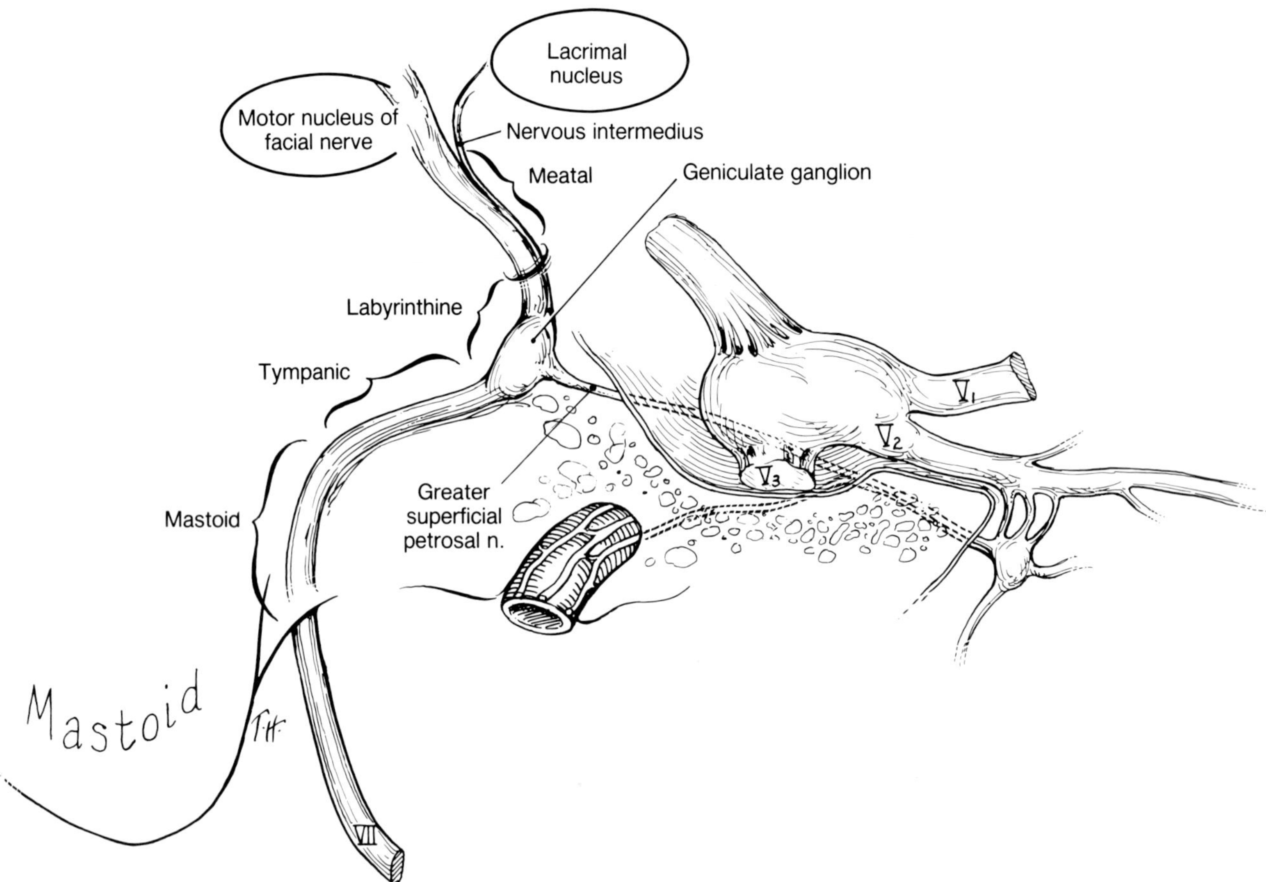

Figure 8.11. The facial nerve. As the facial nerve travels through the petrous and mastoid portion of the temporal bone, it is divided into four segments: meatal, labyrinthine, tympanic, and mastoid.

sory ganglion of the facial nerve, and is primarily composed of taste fibers from the chordi tympani. The greater superficial petrosal nerve originates from the geniculate ganglion to supply parasympathetic secretomotor fibers to the lacrimal gland (see Chapter 5, "The Lacrimal System"). The labyrinthine segment is laterally oriented between the cochlea and semicircular canals. At the geniculate ganglion, the facial canal makes a sharp bend posteriorly, and forms the tympanic or horizontal segment, which measures 8–11 mm. This segment is lateral to the semicircular canals. The mastoid segment of the facial nerve then extends straight downward for 9–12 mm to emerge from the stylomastoid foramen. The stapedius nerve and the chordi tympani, which supply taste fibers to the anterior two-thirds of the tongue, arise from the mastoid segment.

The facial nerve exits from the styloid canal in the mastoid process of the temporal bone and emerges from behind the sternocleidomastoid muscle. It passes beneath the external auditory canal to enter the posterior substance of the parotid gland. Here, it divides into an upper (temporofacial) branch and a lower (cervicofacial) branch. The upper branch further divides into temporal and zygomatic branches; and the lower branch divides into buccal, mandibular, and cervical branches. They divide further, and anastomose as they course foward, leaving the parotid gland on its superior and anterior surfaces, and passing to the facial muscles which are entered on their deep surfaces. Davis and co-workers (1956) described six branching patterns of the facial nerve, ranging from direct extensions to complex plexiform arrangements of the nerve (Fig. 8.12).

CLINICAL NOTE

Uncontrolled spasm or contracture of the facial musculature may be produced by a variety of disorders and may be distinguished by the symmetry (or lack of) and presentation of the spasm. Bilateral involuntary progressive spasm of the eyelids, with or without contraction of other facial musculature, is most often called essential blepharospasm. The asymmetric, mild, moving face may be secondary to facial myokymia, facial tic, aberrant regeneration of the facial nerve, or a spastic paretic facial nerve. Hemifacial spasm is the marked, unilateral, uncontrolled spasm of facial musculature.

Essential blepharospasm is characterized by chronic, progressive bilateral eyelid spasm, resulting in severe visual disability (Fig. 8.13). When the lower facial musculature, particularly the orofacial region, is involved the condition is frequently referred to as Meiges' disease. Patients with essential blepharospasm may be functionally blind due to their inability to open their eyes voluntarily (Henderson, 1956; Frueh, et al, 1976; Gillum and Anderson, 1981). This

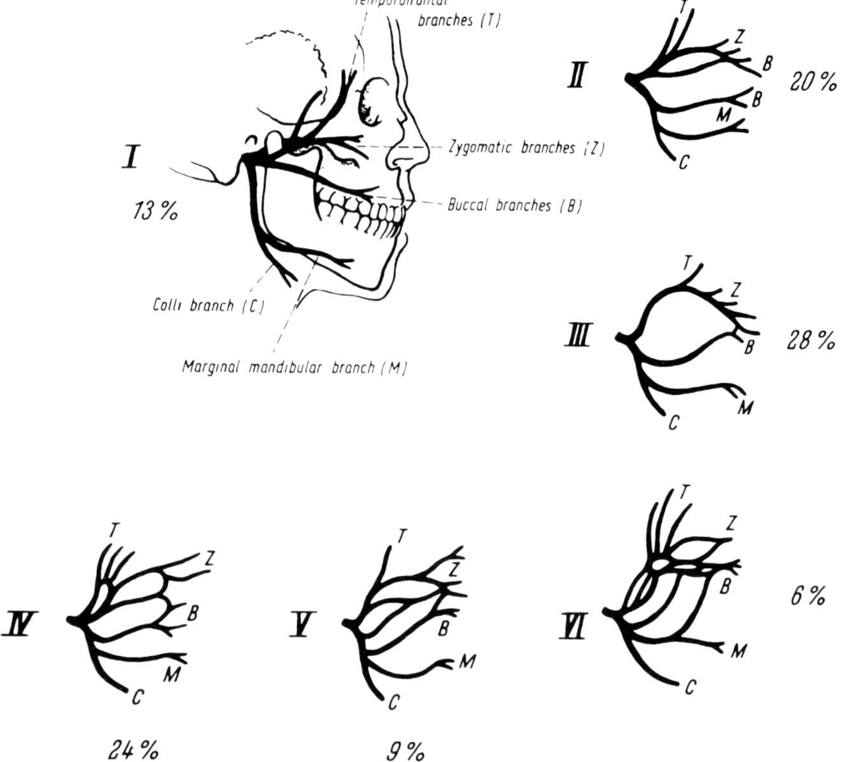

Figure 8.12.   Peripheral course of the facial nerve. Six branching patterns have been recognized (Davis, 1956). (Reproduced from Miehlke A: *Surgery of the Facial Nerve*. Philadelphia, W.B. Saunders, 1973.)

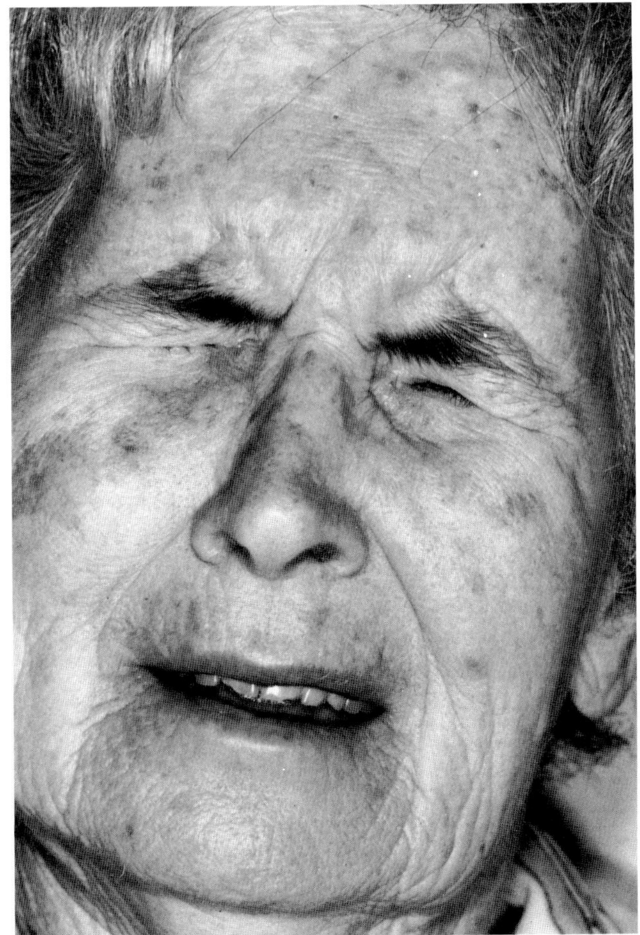

Figure 8.13. Essential blepharospasm. Uncontrolled spasm of the orbicularis muscle produces functional blindness.

disorder generally occurs in older patients. Coles (1977), using a motion picture analysis, described multiple cranial nerve involvement associated with essential blepharospasm. Signs that Coles recognized as indicative of multiple cranial nerve involvement were the impersistence of gaze, lid retraction, tongue thrust, head tilts and jerks, vertical gaze spasm, and assymetric eyelid involvement.

Diagnostic evaluation of blepharospasm has failed to reveal a pathologic foci. Spector and co-workers (1981) postulated a lesion in the motor inhibitory system of the facial nuclei. However, this has not been substantiated. Numerous medical therapies have been advocated, but none has proven successful in the treatment of this disorder (Merikangas and Reynolds, 1979; Miller, 1973; Charkravorty, 1974). Selective section of the facial nerve was recognized as the surgical treatment of choice for blepharospasm (Callahan, 1963). Selective neurectomy, however, produces facial palsy and may exaggerate brow ptosis and upper eyelid dermatochalasis and result in ectropion, tearing, and drooling. In addition, approximately 50% of patients who have undergone facial neurectomy have a recurrence of blepharospasm (Frueh, 1976). Gillum and Anderson (1981) described an anatomic approach to blepharospasm surgery in which removal of the squeezing muscle protractors in the extended eyelid and brow regions effectively eliminates the blepharospasm.

The asymmetric moving face is a distinct pathologic entity separate from essential blepharospasm. Hemifacial spasm occurs predominantly in middle-to-older aged females, and begins with a unilateral eyelid twitch which progresses to involve the entire facial musculature (Fig. 8.14). The synkinetic twitching and facial spasms are generally the only neurologic manifestations of hemifacial spasm. While generally hemifacial spasm is idiopathic, it may be secondary to compression of the facial nerve in the facial canal, or involvement of the proximal intracranial portion of the nerve. Intracranial compression of the facial nerve may be of a vascular etiology (Maroon, 1978), or the result of tumors such as acoustic neuromas, meningiomas, and cholestatomas (Revilla, 1948). Neurologic evaluation to rule out a vascular or tumor etiology of hemifacial spasm is warranted.

Facial myokymia is a unilateral, vermilliform movement of the facial musculature. This is in contrast to the squeezing movements of hemifacial spasm. While usually idiopathic, facial myokymia has been associated with pontine gliomas (Waybright, et al, 1979), brainstem multiple sclerosis (Andermann, et al, 1961), or infections (Boghen, et al, 1977). Myokymia isolated to the orbicularis muscle may also occur.

The anatomy of the facial nerve is important in the evaluation of Bell's palsy or idiopathic facial paralysis (Fig. 8.15). The intracranial course of the facial nerve through the long, confining facial or fallopian canal predisposes it to compression. The location of involvement of the facial nerve is reflected by associated signs.

Bell's palsy has a nearly complete recovery in approximately 70% of patients. Of the remaining 30% of patients, one-half will have a partial recovery, while the other half will have minimal to no recovery of facial function (May and Hardin, 1978). One of the

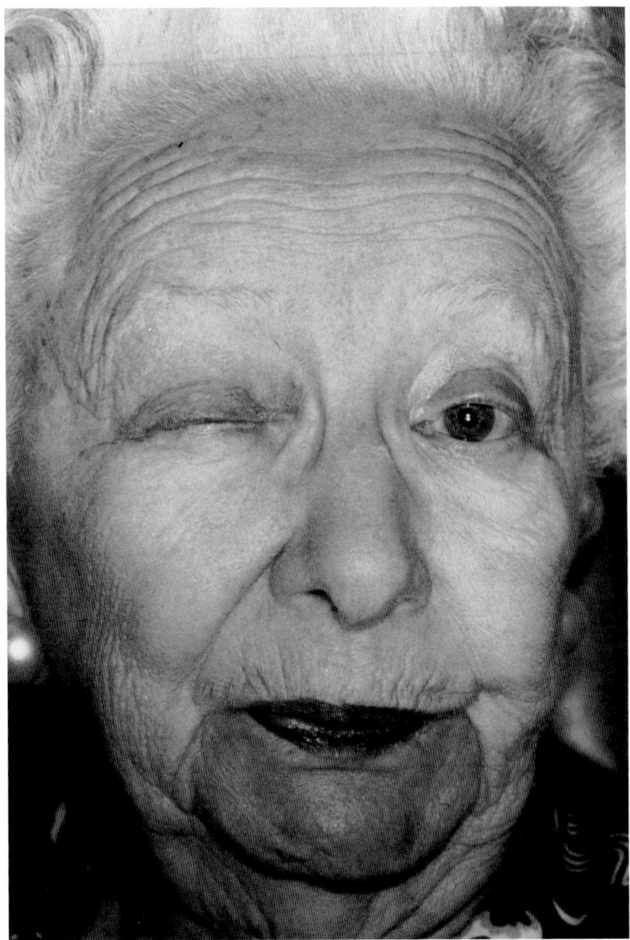

Figure 8.14.   Hemifacial spasm.

goals in the evaluation of Bell's palsy is to determine which patients have a poor prognosis for recovery of function. Prognostic indicators implying a poor return of function include reduction of tears and submandibular secretions. Reduction of tears or submandibular secretions indicate facial nerve compression proximal to or at the level of the geniculate ganglion, the site of origin of the greater superficial petrosal nerve. Dysacousis (loud noises causing pain) and loss of the stapedial reflex indicate proximal cranial nerve compression. Return of facial function occurred in only 10% of patients with reduced lacrimal secetion and in 25% of patients with dysacousis (May and Hardin, 1977). Fisch (1981) used electroneuronography to study patients with facial palsy and recognized that patients with less than 90% degeneration of facial nerve fibers had a good return of function. Patients with degeneration greater than 90% entered a poor prognostic category. It is recommended that those patients with a poor prognosis should have a decompression of the facial canal to eliminate constriction of the facial nerve, and enhance functional return.

AUTONOMIC NERVOUS SYSTEM

The autonomic nervous system is composed of the sympathetic and the parasympathetic systems. The autonomic nervous system is not under direct voluntary command, and

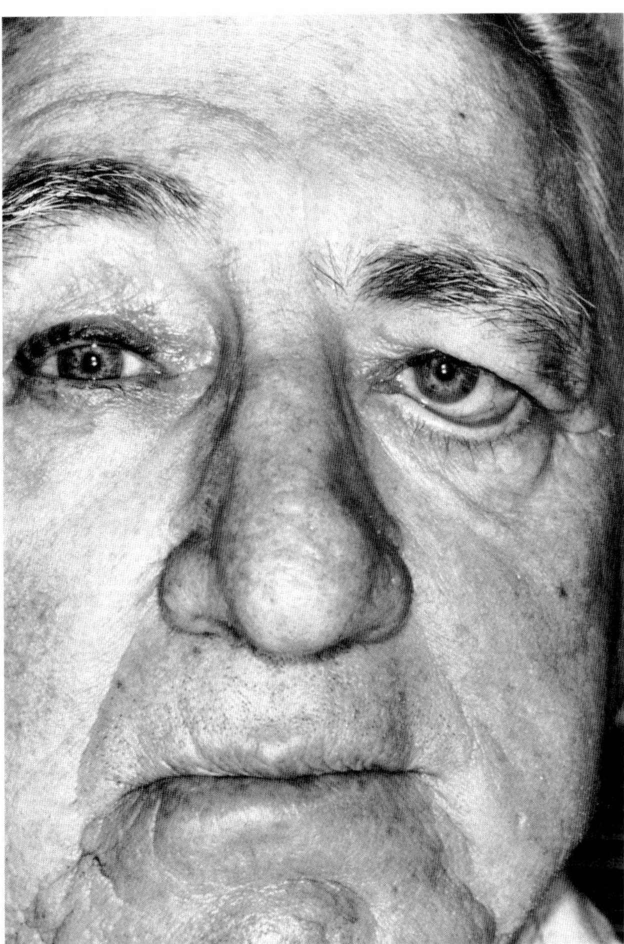

Figure 8.15. Facial palsy. Paralysis of the muscles of facial expression produces a prominent brow ptosis and lower eyelid ectropion.

is controlled by neurons within the central nervous system. The sympathetic and parasympathetic systems are, for the most part, antagonist systems supplying the same organs. In general, the sympathetic system mobilized energy for sudden, quick activities (rage, flight), while the parasympathetic system is geared toward restoring energy reserves.

The sympathetic or thoracolumbar system is connected to the central nervous system through thoracic or upper lumbar segments of the spinal cord. The ganglia or synaptic connections of the neurons are located adjacent to the spinal cord. The parasympathetics have central nervous system connections through the cranial nerves or the craniosacral system. Their ganglia are located near the visceral organ to be supplied. The sympathetic system supplies fibers to the iris dilator muscles and sympathetic muscles of the upper and lower eyelids and lacrimal glands. The parasympathetic system supplies the iris sphincter muscle, ciliary body, and lacrimal glands.

The ocular sympathetics extend from the hypothalamus to the orbit. This is a three-neuron pathway (Fig. 8.16). The first (central) neuron extends from the posterolateral area of the hypothalamus and descends uncrossed in the brainstem and through the lateral column of the cervical spinal cord. The first order neuron terminates in the ciliospinal center of Budge between C-8 and T-1. The second order (intermediate) neurons immediately leave the spinal cord to enter the paravertebral sympathetic chain, passing adjacent to the lower cervical vertebrae. The second order nerves pass over the pulmonary apex, then run through the inferior and middle cervical ganglia, and synapse in the superior cervical ganglion. For every second order (intermediate) neuron, about 15 third-order (postganglion) neurons will exit the superior cervical ganglion (Maloney, et al, 1980). The postganglionic fibers will follow the internal and external carotid arteries providing sympathetic innervation to the head and neck. Postganglionic neurons responsible for facial sweating follow the external carotid artery, while those fibers supplying orbital structures follow the internal carotid artery. A sympathetic branch passes through ciliary ganglion to the globe to supply the iris dilator muscle. The route of the sympathetics to sympathetic muscles of the upper and lower eyelids is diffuse, following either the levator muscle, sensory nerves or arterioles (Collin, et al, 1979).

CLINICAL NOTE

Horner's syndrome is produced by the interruption of the sympathetics, and is characterized by miosis, ptosis, anhidrosis, and apparent enophthalmos (Fig. 8.17). Pupillographic characteristics of Horner's syndrome are the delay of redilation and the absence of dilation with psychosensory stimulation (Maloney, et al, 1980). Patients with congenital Horner's syndrome have hypochromia of the affected iris. Diagnostic evaluation of Horner's syndrome should determine the site of sympathetic interruption as preganglionic or postganglionic. Benign and malignant tumors which may produce Horner's syndrome are generally preganglionic in origin (Malone, et al, 1980; Giles and Henderson, 1958; Thompson and Mensher, 1971; Grimson and Thompson, 1975, 1980). Postganglionic causes of Horner's syndrome are generally benign and do not warrant an extensive diagnostic evaluation.

Pharmacologic evaluation via eye drops is an effective means of localizing the site of ocular sympathetic interruption in Horner's syndrome. Although cocaine will diagnose Horner's syndrome, it provides no information regarding the localization of oculosympathetic interruption. Hydroxyamphetamine (Paradrine) is recognized as the best pharmacologic agent for use in determining the pre- or postganglionic localization of Horner's syndrome (Thompson and Mensher, 1974). In such evaluations, absence of pupillary dilation is characteristic of a postganglionic lesion, and intact pupillary dilation is indicative of a preganglionic lesion. If a preganglionic lesion is found, further extensive evaluation is warranted to rule out the presence of serious underlying pathology.

Raeder's syndrome is severe unilateral facial pain associated with an ipsilateral oculosympathetic palsy (Raeder, 1924). Boniuk and Schlezinger (1962) divided patients with Raeder's syndrome into two groups: those with painful oculosympathetic paresis with parasellar cranial nerve (III, IV, V, and VI) involvement; and another group with the same painful paresis, but no cranial nerve involvement. Raeder's syndrome with associated cranial nerve involvement suggested dis-

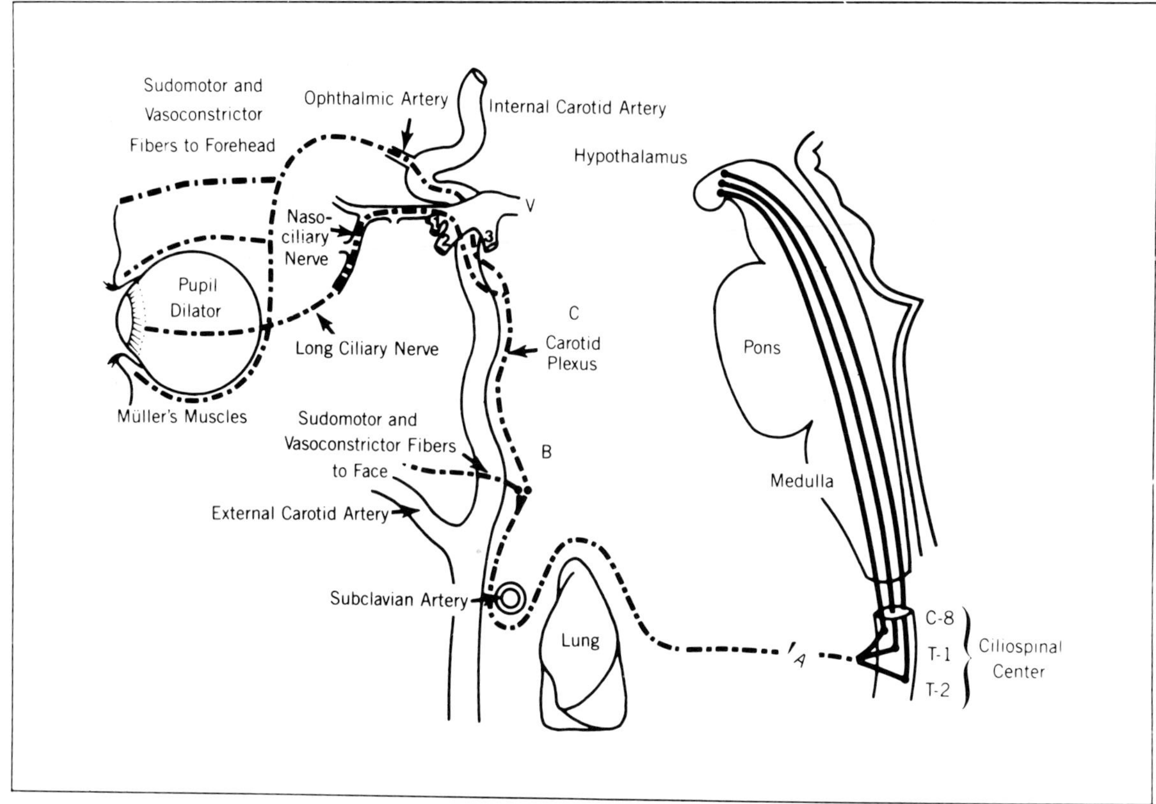

Figure 8.16. Oculosympathetic pathway. (*A*), Preganglionic neuron. (*B*), Superior cervical ganglion. (*C*), Postganglionic neuron. (Reproduced from Weinstein, et al: *Archives of Ophthalmology* 98:1074, 1980. American Medical Association.)

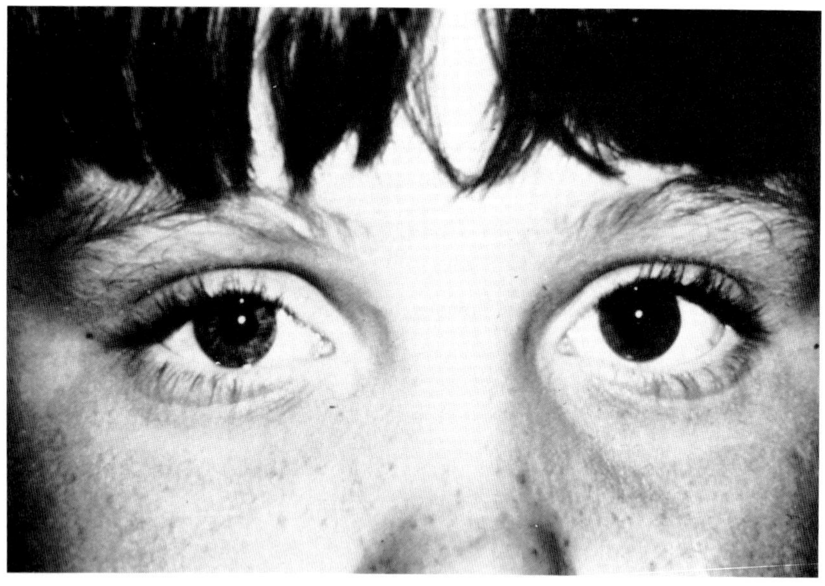

Figure 8.17. Congenital Horner's syndrome. Miosis, ptosis, apparent enophthalmos and heterochromia (lighter pigmented iris in the affected right side). (Reproduced with permission from N. R. Miller.)

ease of the middle cranial fossa, and warranted appropriate diagnostic studies. Patients who did not have cranial nerve involvement were thought to have the paresis as a manifestation of cluster headaches; therefore extensive diagnostic evaluation was not thought to be warranted (Ford and Walsh, 1958; Smith, 1958; Kunkle and Anderson, 1961). However, carotid artery aneurysm (Law and Nelson, 1968) and fibromuscular hyperplasia (Cohen, et al, 1975) have been reported to produce Raeder's syndrome without associated cranial nerve involvement.

The parasympathetic supply to the orbit is closely associated with the oculomotor nerve. The nuclear origin and intracranial course of the ocular autonomic system has been described with the oculomotor nerve. The parasympathetic supply to the lacrimal, submandibular, and sublingual glands is associated with the facial nerve. The ciliary and pterygopalatine ganglia, although discussed briefly in other sections, will be examined here more closely, since they are the parasympathetic ganglia that supply the ocular adnexal area.

The ciliary is a parasympathetic ganglion whose postganglionic fibers innervate the sphincter pupillae and ciliary body. The ciliary ganglion is situated in loose, adipose tissue approximately 1 cm from the orbital apex, and 1.5–2.0 cm posterior to the globe (Fig. 8.18). It measures 1×2 mm and lies on the temporal side of the optic nerve between the optic nerve and the lateral rectus muscle. Three roots extend to the ciliary ganglion; however, only parasympathetic fibers from the inferior division of the oculomotor nerve synapse at the ganglion (Fig. 8.19). A sensory root passes through the ciliary ganglion from the nasociliary branch of the ophthalmic nerve to supply the globe. A sympathetic root passes throughy the ciliary ganglion from the sympathetic plexus in the cavernous sinus.

CLINICAL NOTE

Adies' pupil is an isolated internal ophthalmoplegia that is characterized by a tonic pupil with defective accommodation. The affected pupil dilates in room illumination. However, if viewed in darkness, the pupil on the involved side is smaller than the contralateral normal pupil. The pupil fails to respond normally to light (may produce vermiform movements of pupillary margin), but the pupil does slowly constrict to accomodative effort (light-near dissociation). The site of pathology in Adies' pupil is the ciliary ganglion. Pathologic examination of the ciliary ganglion discloses absent or grossly diminished ganglion cells (Harriman and Garland, 1968).

The pterygopalatine ganglion (Meckel's, sphenopalatine) lies in the upper portion of the pterygopalatine fossa (Fig. 5.7). Similar to the ciliary ganglion, the pterygopalatine ganglion receives parasympathetic, sensory, and sympathetic fibers, but only the parasympathetic secretomotor fibers synapse in the ganglion. The parasympathetic fibers are derived from the greater superficial petrosal nerve. As this nerve enters the pterygoid canal, it is joined by the deep petrosal nerve which contributes sympathetic fibers from the carotid plexus. The union of these nerves forms the vidian nerve or nerve of the pterygoid canal, which contributes directly to the pterygopalatine ganglion. The parasympathetic fibers synapse in the ganglion, then provide secretomotor fibers to the lacrimal, submandibular, and sublingual glands. The sympathetic fibers pass through the ganglion without synapsing, then diffuse from the ganglion. Two or three small twigs extend from the maxillary nerve to the pterygopalatine ganglion. These fibers pass through

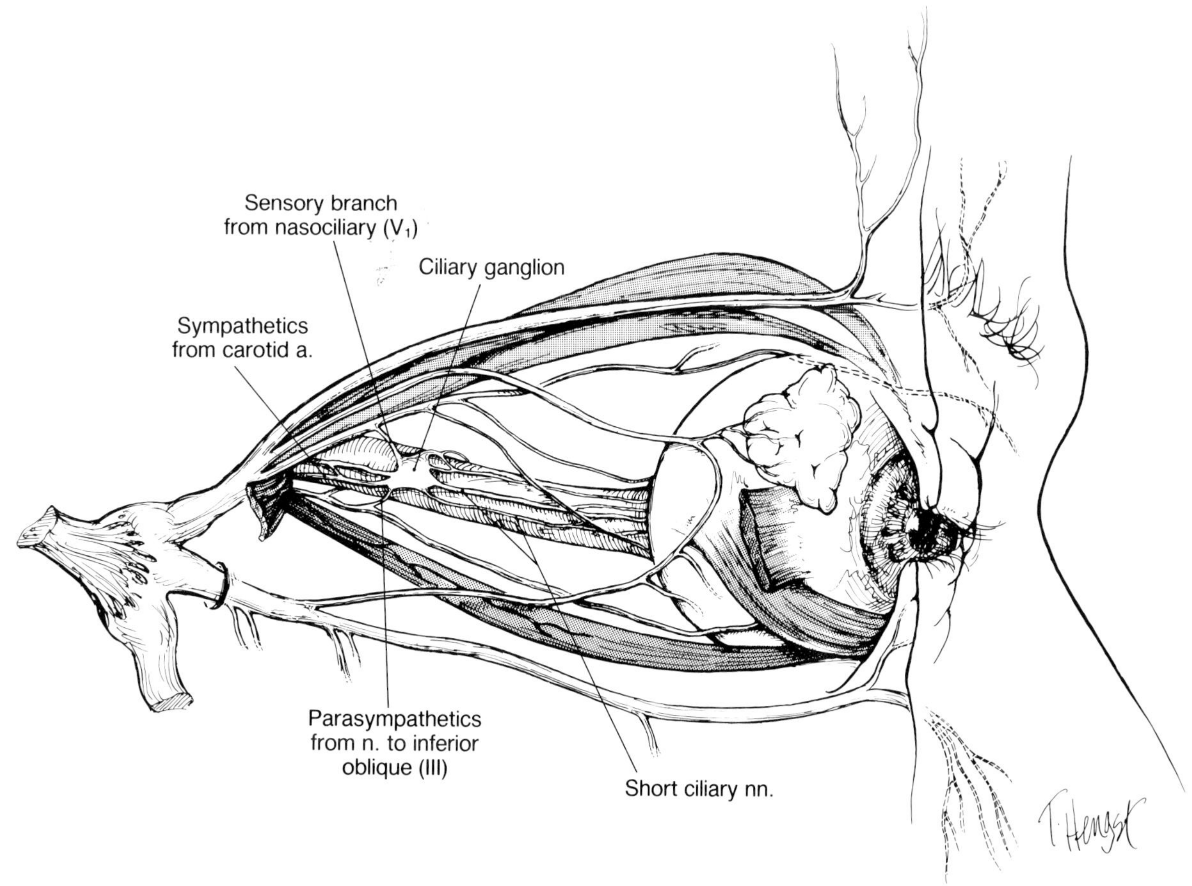

Figure 8.18. Contributions to the ciliary ganglion.

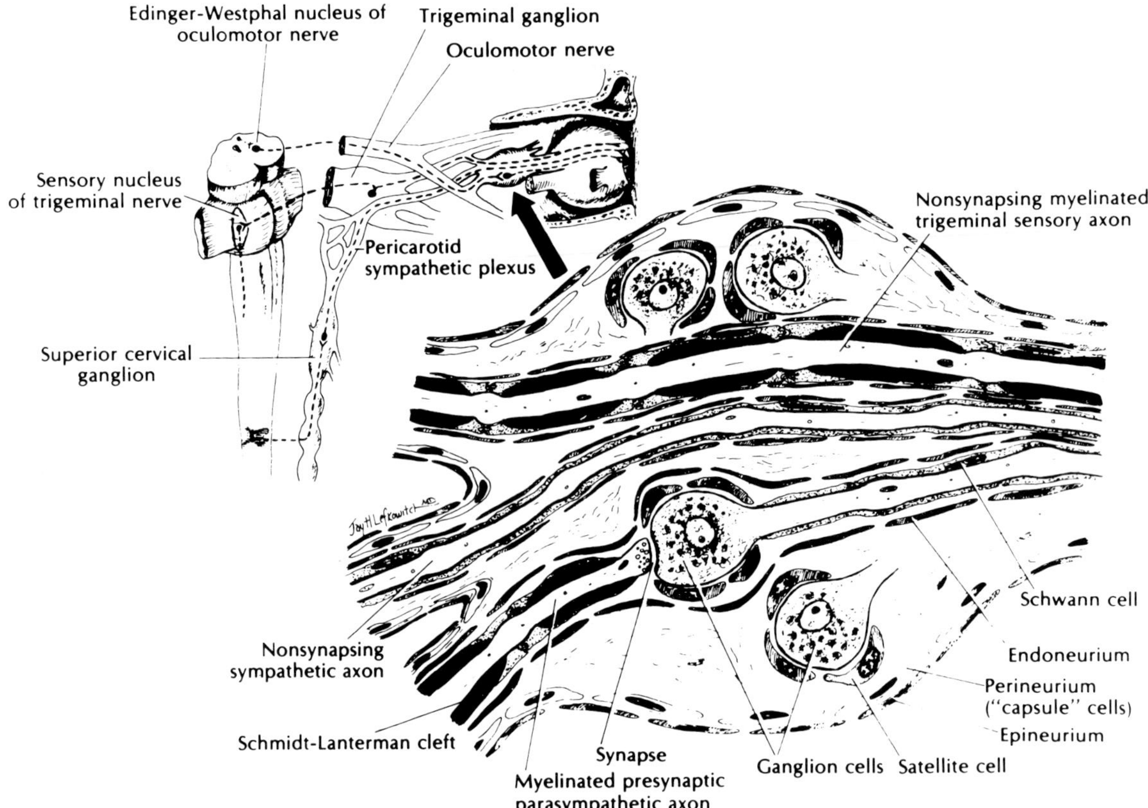

Figure 8.19.   Schematic of the ciliary ganglion. The parasympathetic fibers synapse in the ganglion, while the sympathetic and sensory fibers merely pass through the ganglion without synapsing. (Reproduced from Jakobeic FA, Iwamoto T: Ocular adnexa. Introduction to lids, conjunctiva and orbit. In Duane TD, Jaeger EA: *Biomedical Foundations of Ophthalmology*, Vol. 1, Chapter 28. Philadelphia, Harper & Row, 1982.)

the ganglion without synapsing to provide sensation to the inferior orbit, palate, and nasal cavity.

OPTIC NERVE

The optic nerve is about 50 mm in length and extends from the optic chiasm to the globe. The optic nerve is frequently considered to be an outgrowth from the brain and not a true cranial nerve (Whitnall, 1932). The optic nerve has four parts: intraocular (optic nerve head), intraorbital, intraosseus, and intracranial. The optic nerve will not be discussed in detail. Additional anatomic relationships of the optic nerve can be found in Miller's revision of *Walsh and Hoyt's Clinical Neuro-Ophthalmology.*

The intraocular portion of the nerve is the optic nerve head which is only 1 mm long and corresponds to the scleral thickness at the posterior pole. The anterior surface of the optic nerve head is recognized as the optic disc with a diameter of 1.5–2.0 mm (Straatsma, et al, 1969). At this point, the optic nerve consists of unmyelinated axons from the retinal ganglion cells.

The intraorbital portion of the optic nerve is S-shaped and is 30 mm long. The distance from the posterior globe to the orbital apex is approximately 20 mm, so the S-shaped course of the optic nerve is a result of its excessive length. The course of the optic nerve permits free movement of the globe without compromising its function. The orbital portion

of the optic nerve is 4 mm in diameter, since its myelinated axons add dimension. The optic nerve is surrounded by dura, arachnoid, and pia throughout its orbital course. The dura, or outermost sheath around the optic nerve, is continuous with the sclera anteriorly, and fuses with the periosteum at the orbital apex. The arachnoid is a loose connective tissue bridging the dura and the pia mater. The pia is adjacent to the nerve investing the capillary net on the nerve. The central retinal artery enters the optic nerve on its inferomedial surface, about 10 mm posterior to the globe.

The intraosseus (intracanalicular) optic nerve passes through the optic canal and is generally 5–6 mm long. The optic canal is formed by the union of the two roots of the lesser wing of the sphenoid bone. Each optic canal follows the orientation of the lateral orbital wall. The dura is fused with the periosteum, tightly adherent to the optic nerve with the optic canal. This confinement of the optic nerve predisposes it to compression in a variety of clinical situations.

The intracranial portion of the optic nerve is about 10 mm long, and extends from the optic canal to the optic chiasm. The frontal lobe lies directly above the optic nerve. The optic nerve is closely associated with the anterior cerebral and anterior communicating arteries. In addition, the internal carotid is present at the lateral surface of the optic nerve as it exits the cavernous sinus. After exiting the cavernous sinus, the internal carotid divides into the anterior and middle cerebral arteries. The close anatomic relationships of the optic nerve to these vessels accounts for optic nerve compression as a result of aneurysms in this area.

References

Akagi Y: The localization of motor neurons innervating the extraocular muscles in the oculomotor nuclei of the cat and rabbit, using horseradish preoxidase. *J Comp Neurol* 181:745, 1978.

Andermann F, Cosgrove JBR, Lloyd-Smith DL: Facial myokymia in multiple sclerosis. *Brain* 84:31, 1961.

Anderson RL: The blepharocardiac reflex. *Arch Ophthalmol* 96:1418, 1978.

Apt L, Isenberg SJ, Gaffney WL: The oculocardiac reflex in strabismus surgery. *Am J Ophthalmol* 76:533, 1973.

Asbury AK, Aldredge H, Hershberg R, Fisher CM: Oculomotor palsy in diabetes mellitus: A clinicopathological study. *Brain* 93:555, 1970.

B'auuttner-Ennever JA, Akert K: Medial rectus subgroups of the oculomotor nucleus and their abducens internuclear input in the monkey. *J Comp Neurol* 197:17, 1981.

B'auuttner-Ennever JA, B'auuttner-Ennever U, Cohen B, Baumgartner G: Vertical gaze paralysis and the rostial interstitial nucleus of the medial longitudinal fasciculus (rostral MLF). *Brain* 105:125, 1982.

Bajandas FJ: The six syndromes of the sixth nerve. In Smith JL (ed): *Neuro-Ophthalmology Update.* New York, Masson, 1977, p 49.

Bienfans DC: Crossing axons in the third nerve nucleus. *Invest Ophthalmol* 12:927, 1975.

Boghen D, Filiatrault R, Descarres L: Myokymia and facial contracture in brainstem tuberculoma. *Neurology* 27:270, 1977.

Boniuk M, Schlezinger NS: Raeder's paratrigeminal syndrome. *Amer J Ophthalmol* 54:1074, 1962.

Burger LJ, Kalvin NH, Smith JL: Acquired lesions of the fourth cranial nerve. *Brain* 93:567, 1970.

Callahan A: Blepharospasm with resection of part of orbicularis nerve supply. *Arch Ophthalmol* 70:508, 1963.

Carpenter MB, McMasters RE: Disturbances of conjugate horizontal eye movements in the monkey: II. Physiological effects and anatomical degeneration resulting from lesions in the medial longitudinal fasciculus. *Arch Neurol* 8:347, 1963.

Charkravorty VK: Treatment of blepharospasm with levodopa. *Postgrad Med J* 50:521, 1974.

Cohen DN, Zakov ZN, Salanga VD, Dohn DF: Raeder's paratrigeminal syndrome. *Am J Ophthalmol* 79:1044, 1975.

Coles WH: Signs of essential blepharospasm: A motion-picture analysis. *Arch Ophthalmol* 95:1006, 1977.

Collin JRO, Beard C, Wood I: Terminal course of the nerve supply to Müller's muscle in the rhesus monkey and its clinical significance. *Am J Ophthalmol* 87:234, 1979.

Crosby EC, DeJonge BR: Experimental and clinical studies of the central connections and central relationships of the facial nerve. *Ann Otolaryng* 72:735, 1963.

Dandy WE: Concerning the cause of trigeminal neuralgia. *Am J Surg* 24:447, 1934.

Dandy WE: Trigeminal neuralgia and tic douloureux. In Lewis OD (ed): *Practice of Surgery,* Vol 12. Hagerstown, MD, WF Prior, 1936, p 177.

Daroff RB: Ocular motor manifestations of brainstem and cerebellar dysfunction. In Smith JL (ed): *Neuro-Ophthalmology: Symposium of the University of Miami and the Bascom-Palmer Eye Institute,* Vol 5. Hallendale, FL, Huffman, 1971, p 104.

Davis RA, Anson BJ, Budinger JM, Kurth L: Surgical anatomy of the facial nerve and parotid gland based upon a study of 350 cervicofacial halves. *Surg Gynecol Obstet* 102:385, 1956.

Denny-Brown D, Yanagisawa N: The function of the descending root of the fifth nerve. *Brain* 96:783, 1973.

Fisch U: Surgery for Bell's palsy. *Arch Otolaryng* 107:1, 1981.

Ford FR, Walsh FB: Raeder's paratrigeminal syndrome: A benign disorder, possibly a complication of migraine. *Bull Johns Hopkins Hosp* 103:296, 1958.

Frueh BR, Callahan A, Dortzbach RK: The effects of differential section of the VIIth nerve on patients with intractable blepharospasm. *Trans Am Acad Ophthalmol Otolaryng* 81:595, 1976.

Frueh BR, Callahan A, Dortzbach RK: A profile of patients with intractable blepharospasm. *Trans Amer Acad Ophthalmol Otolaryng* 81:591, 1976.

Gacek RR: Localization of neurons supplying the extraocular muscle in the kitten using horseradish peroxidase. *Exp Neurol* 44:381, 1974.

Giles CL, Henderson JW: Horner's syndrome. An analysis of 216 cases. *Am J Ophthalmol* 46:389, 1958.

Gillum WN, Anderson RL: Blepharospasm surgery: An anatomical approach. *Arch Ophthalmol* 99:1056, 1981.

Goldstein JE, Cogan DG: Diabetic ophthalmoplegia with special reference to the pupil. *Arch Ophthalmol* 64:592, 1960.

Goldstein NP, Gibilisco JA, Rushton JG: Trigeminal neuropathy and neuritis. A study of etiology with emphasis on dental causes. *JAMA* 184:458, 1963.

Gradenigo G: Ueber die paralyse des nervos abducens bei otitis. *Arch f Ohrenh* 74:149, 1907.

Grimson BS, Thompson S: Raeder's syndrome: A clinical review. *Surv Ophthalmol* 24:199, 1980.

Grimson BS, Thompson HS: Drug testing in Horner's syndrome. In Glaser JS, Smith JL (eds): *Neuro-Ophthalmology Symposium of the University of Miami and the Bascom-Palmer Eye Institute.* St. Louis, CV Mosby, 1975, p 265.

Gudmundsson K, Rhoton AL, Jr, Rushton JG: Detailed anatomy of the intracranial portion of the trigeminal nerve. *J Neurosurg* 35:592, 1971.

Haines SJ, Jannetta PJ, Zorub DS: Microvascular relations of the trigeminal nerve. An anatomic study with clinical correlation. *J Neurosurg* 52:381, 1980.

Hardy DG, Peace DA, Rhoton AL Jr: Microscopic anatomy of the superior cerebellar artery. *Neurosurgery* 6:10, 1980.

Hardy DG, Rhoton AL, Jr: Microsurgical relationships of the superior cerebellar artery and the trigeminal nerve. *J Neurosurg* 49:669, 1978.

Harriman DGF, Garland H: The pathology of Adie's syndrome. *Brain* 91:401, 1968.

Harris W: Trigeminal neuralgia at an exceptionally early age: Cured by Gassarian alcohol injection. *Br Med J* 2:49, 1943.

Heinze J: Cranial nerve avulsion and other neural injuries in road accidents. *Med J Aust* 2:1246, 1969.

Henderson JW: Essential blepharospasm. *Trans Am Ophthalmol Soc* 54:453, 1956.

Henderson WR: A note on the relationship of the human maxillary nerve to the cavernous sinus passing through the foramen ovale. *J Anat* 100:109, 1966.

Hutchinson J: A clinical report on herpes frontalis seu ophthalmicus. *Roy London Ophthalmol Hosp Rep* 5:191, 1866.

Hyland HH, Barnett HJM: The pathogenesis of cranial nerve palsies associated with intracranial aneurysms. *Proc Roy Soc Med* 47:141, 1956.

Isenberg SJ, Blechman B: Oculocardiac reflex during postoperative muscle adjustment. *Am J Ophthalmol* 92:422, 1982.

Jampel RS, Mindel J: The nucleus for accommodation in the midbrain of the macaque. *Invest Ophthalmol* 6:40, 1967.

Jefferson G: Observations on trigeminal neuralgia. *Br Med J* 2:879, 1931.

Keller EL: Participation of the medial pontine reticular formation in eye movement generation in monkeys. *J Neurophysiol* 37:316, 1974.

Kerr WL, Miller RH: The pathology of trigeminal neuralgia. Electron microscopic studies. *Arch Neurol* 15:308, 1966.

Kunkle EC, Anderson WB: Significance of minor eye signs in headaches of migraine type. *Arch Ophthalmol* 65:504, 1961.

Law WR, Nelson ER: Internal carotid aneurysm as a cause of Raeder's paratrigeminal syndrome. *Neurology* 18:43, 1968.

Leichnetz GR: A comment on the center for vertical eye movements in the medial prerubral subthalamic region of the monkey considering some of its frontal cortical afferents. *Neurosci Lett* 30:95, 1982.

Lindenberg R: Significance of the tentorium in head injuries from blunt forces. *Clin Neurosurg* 12:129, 1966.

Lubow M: Cited in Duane TD, Jaeger EA (eds): *Clinical Ophthalmology*, Vol 2, Chapter 12. Philadelphia, Harper & Row, 1982, p 10.

Maloney WF, Younge BR, Moyer NJ: Evaluation of the causes and accuracy of pharmacologic localization in Horner's syndrome. *Am J Ophthalmol* 90:394, 1980.

Maroon JC: Hemifacial spasm: A vascular cause. *Arch Neurol* 35:481, 1978.

May M, Hardin WB: Facial palsy: Interpretation of neurologic findings. *Laryngoscope* 88:1352, 1978.

Merikangas JR, Reynolds CF: Blepharospasm: Successful treatment with Clonazepam. *Ann Neurol* 5:401, 1979.

Meltzer PE: Gradenigo's syndrome: Anatomic aspects. *Arch Otolaryng* 26:412, 1931.

Miller E: Dimetryl Aminoethanol in the treatment of blepharospasm. *N Engl J Med* 289:697, 1973.

Miller MT, Urist MJ, Folk ER, et al: Superior oblique palsy presenting in late childhood. *Am J Ophthalmol* 70:212, 1970.

Miller NR: *Walsh and Hoyt's Clinical Neuro-Ophthalmology*, ed 4, Vol 2. Baltimore, Williams & Wilkins, 1983.

Nadeau SE, Trobe JD: Pupil sparing in oculomotor palsy. A brief review. *Ann Neurol* 13:143, 1983.

Nathan H, Ouaknine G, Kosary IZ: The abducens nerve: Anatomic variations in its course. *J Neurosurg* 41:561, 1974.

Raeder JG: "Paratrigeminal" paralysis of oculo-pupillary sympathetic. *Brain* 47:149, 1924.

Revilla AG: Differential diagnosis of tumors at the cerebellopontine recess. *Bull Johns Hopkins Hosp* 83:187, 1948.

Robertson DM, Hines JD, Rucker CW: Acquired sixth nerve paresis in children. *Arch Ophthalmol* 83:574, 1970.

Roy FH: *Ocular Differential Diagnosis*, ed 2. Philadelphia, Lea & Febiger, 1975, p 154.

Rucker CW: The causes of paralysis of the third, fourth and sixth cranial nerves. *Am J Ophthalmol* 61:1294, 1966.

Rushton JG, Olafson RA: Trigeminal neuralgia associated with multiple sclerosis. *Arch Neurol* 13:383, 1965.

Sacks JG: Peripheral innervation of the extraocular muscles. *Am J Ophthalmol* 95:520, 1983.

Schlaegel TF: Uvitis associated with viral infections. In Duane TD, Jaeger EA (eds): *Clinical Ophthalmology*, Vol 4, Chapter 46. Philadelphia, Harper & Row, 1982, p 10.

Smith JL: Raeder's paratrigeminal syndrome. *Am J Ophthalmol* 46:194, 1958.

Spector GJ, Smith PG, Burde RM: Selective facial neurectomy for essential blepharospasm. *Laryngoscope* 91:1896, 1981.

Sperling E, Stender A: "Tic doloureaux" und Gesichtschmerz (Therapeutische and Pathogenetische Betrachtungen). *Deutsch Z Nervenheilk* 173:161, 1955.

Straatsma BR, Foos RY, Spencer LM: The retina: Topography and clinical correlations. *Trans New Orleans Acad Ophthalmol* St. Louis, C.V. Mosby, 1969.

Thompson HS, Mensher JH: Adrenergic mydriasis in Horner's syndrome. Hydroxyamphetamine test for the diagnosis of postganglionic defects. *Am J Ophthalmol* 72:472, 1971.

Thompson HS, Mensher JH: Horner's syndrome. *Amer J Ophthalmol* 78:739, 1974.

Trobe JD, Glaser JS, Quencer RC: Isolated oculomotor paralysis: The product of saccular and fusiform aneurysms of the basilar artery. *Arch Ophthalmol* 96:1236, 1978.

Umansky F, Nathan H: The lateral wall of the cavernous sinus: With special reference to the nerves related to it. *J Neurosurg* 56:228, 1982.

Van Buskirk C: The seventh nerve complex. *J Comp Neurol* 82:303, 1945.

Walsh FB, Hoyt WF: *Clinical Neuro-Ophthalmology*, ed 3. Baltimore, Williams & Wilkins, 1969, p 379.

Warwick J: Representation of the extra-ocular muscles in the oculomotor nucleus of the monkey. *J Comp Neurol* 98:449, 1953.

Waybright EA, Gutmann L, Chou SM: Facial myokymia: Pathologic features. *Arch Neurol* 36:244, 1979.

Weber RB, Darnoff RB, Mackey EA: Pathology of oculomotor nerve palsy in diabetics. *Neurology* 20:835, 1970.

Wheeler JM: Paralysis of the sixth cranial nerve associated with otitis media. *J Am Med Assoc* 71:1713, 1918.

Younge BR, Sutla F: Analysis of trochlear nerve palsies. Diagnosis, etiology and treatment. *Mayo Clin Proc* 52:11, 1977.

CHAPTER 9

Vascular Supply of the Orbit

The orbital vascular system is a diffuse network of anastomosing vessels from internal (deep) and external (superficial) sources. The arterial supply to the orbit is mainly from the ophthalmic branches of the internal carotid artery, with only minor contributions from some of the branches of the external carotid artery. Hayreh (1962) stressed the variability in the course and branching pattern of the ophthalmic artery, which complicates its study.

The venous drainage of the orbit is mainly by the two ophthalmic veins (superior and inferior) extending to the cavernous sinus, which connect with the surrounding veins such as the angular vein, and the pterygoid plexus. The vascular system of the orbit will be described in a systematic manner. Branches of the ophthalmic artery will be described in detail according to a topographical schema: ocular, orbital, and extraorbital branches. The second major area to be analyzed is the venous system of the orbit, which will include discussions of the cavernous sinus and pterygoid plexus. The final area to be discussed is the lymphatics of the eyelids and orbit.

ARTERIAL SYSTEM

The arterial supply of the orbit is essentially from the ophthalmic artery, a branch of the internal carotid artery. A number of orbital branches of the ophthalmic artery anastomose with the adjacent branches from the external carotid artery, so that the ophthalmic artery forms an important anastomotic channel between the internal and external carotid arteries (Hayreh, 1963) (Fig. 9.1).

The internal carotid artery begins at the bifurcation of the common carotid artery at the level of the upper border of the thyroid cartilage (Fig. 9.2). The internal carotid artery runs upward to the petrous portion of the temporal bone, and enters the carotid canal; the artery leaves the canal and enters the cranial cavity through the foramen lacerum. The internal carotid artery proceeds to the posterior clinoid process, then makes a sharp turn to enter the cavernous sinus with the abducens nerve at its lateral border. Within the cavernous sinus, the artery has an S-shaped configuration forming the carotid siphon. The cavernous portion of the internal carotid artery has the abducens nerve adjacent to it in the posterior aspect of the cavernous sinus. The oculomotor, trochlear, and ophthalmic nerves are within a loose fibrillar connective tissue in the deep layer of the

lateral wall of the cavernous sinus (see Fig. 9.11 *A–B*). The artery perforates the dura at the medial side of the anterior clinoid process and turns backwards below the optic nerve. The ophthalmic artery arises from the internal carotid artery as soon as it exits the cavernous sinus and lies under the optic nerve. The internal carotid artery divides into its terminal branches, the middle and anterior cerebral arteries, at the medial end of the lateral cerebral sulcus.

The ophthalmic artery is the first major branch of the internal carotid artery. In rare instances, the ophthalmic artery may either arise from the middle meningeal artery and enter the orbit through the superior orbital fissure, or have two trunks—a small one arising from the internal carotid artery and a big one from the middle meningeal artery (Hayreh and Dass, 1962a).

The course of the ophthalmic artery can be divided into intracranial, intracanalicular, and orbital parts. The intracranial portion of the ophthalmic artery, from its origin to the intracranial dural opening of the optic canal, averages 2.6 mm (range 0–4.8 mm) in length (Hayreh and Dass, 1962a). The intracranial portion of the ophthalmic artery lies under the inferior surface of the optic nerve.

The intracanalicular part lies in the optic canal below the optic nerve and gradually pierces the dura mater. The artery then enters the orbit through the optic canal, inferior and lateral to the optic nerve.

The intraorbital course and branching pattern of the ophthalmic artery are variable (Hayreh and Dass, 1962 a and b, Singh and Dass, 1962). Hayreh (1958) and Hayreh and Dass (1962b) divided the intraorbital course of the ophthalmic artery into three parts (Fig. 9.3). The first part extends from the apex of the orbit, on the inferolateral aspect of the optic nerve, to a point where the artery curves around the optic nerve. In the second part the artery crosses over (in 82.6%) or under (17.4%) the optic nerve to run in a medial direction. The third part extends from a "bend" in the artery at the superomedial portion of the optic nerve to its termination. The artery is anchored to the medial orbital wall by the penetration of the posterior and anterior ethmoidal arteries through their respective foramina.

Hayreh (1962) described, in detail, the branches of the ophthalmic artery. Because of the variable course of the ophthalmic artery, the order of origin of its branches also varies. Whitnall (1932) classified branches of the ophthalmic artery in three ways: (i) by the order of origin; (ii) by the

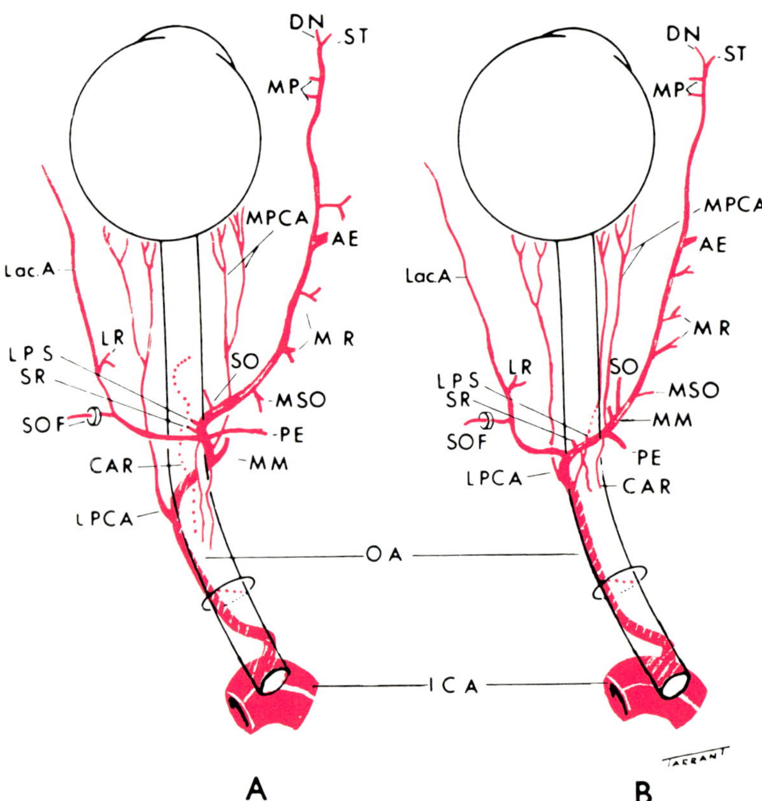

Figure 9.4. Branching pattern of the ophthalmic artery under (*A*) or over (*B*) the optic nerve.

AE	anterior ethmoidal a.	Muscular branches to:	
CAR	central retinal a.	LR	lateral rectus
DN	dorsal nasal a.	LPS	levator palpebrae superioris
ICA	internal carotid a.	MR	medial rectus
lac A	lacrimal a.	MSO	muscular branch to superior oblique
LPCA	lateral posterior ciliary a.	SR	superior rectus
MM	medial muscular a.	OA	ophthalmic a.
MP	medial palpebral a.	PE	posterior ethmoidal a.
MPCA	medial posterior ciliary a.	SO	supra-orbital a.
		SOF	superior orbital fissure

inferior rectus, and the inferior oblique muscles, and less frequently other muscles. The muscular branches to the rectus muscles give out the respective anterior ciliary arteries, with each rectus muscle usually having two anterior ciliary arteries, except the lateral rectus which has only one.

The lacrimal artery arises from the second part of the ophthalmic artery and, along with the lacrimal nerve, extends to the lateral wall of the orbit above the superior border of the lateral rectus muscle. Branches of the lacrimal artery are the recurrent meningeal, zygomatic, glandular (to the lacrimal gland), and lateral palpebral arteries. Smaller branches include the muscular and, sometimes, posterior ciliary arteries, although the latter usually arises directly from the ophthalmic artery. The recurrent meningeal artery passes through the lateral portion of the superior orbital fissure, and anastomoses with the anterior branch of the middle meningeal artery. The lacrimal artery supplies the lacrimal gland, and then pierces the orbital septum to give the two lateral palpebral arteries, which anastomose with the corresponding medial palpebral arteries to form arterial arcades in the eyelids. Prior to its reaching the lacrimal gland, the lacrimal artery gives out the zygomaticotemporal and zygomaticofacial arteries. These branches with their associated nerves pass through the foramina in the zygomatic bone and anastomose with the anterior deep temporal and the transverse facial arteries, respectively.

The supraorbital artery arises from the ophthalmic artery as it crosses the optic nerve. Hayreh (1962) noted that the supraorbital artery was absent in 18.2% of the specimens in which the ophthalmic artery crossed over the optic nerve and in 7.1% of those in which it crossed under. The supraorbital artery extends superiorly near the medial border of the superior rectus and levator muscles. The supraorbital artery and accompanying nerve extend to the supraorbital foramen to supply the muscles of the eyebrows and forehead. The supraorbital artery has branches which partially supply the superior rectus, superior oblique, and levator muscles.

The extraorbital branches of the ophthalmic artery include the posterior ethmoidal, anterior ethmoidal, medial palpebral, supratrochlear, and dorsal-nasal arteries. In Hayreh's (1962) series, the posterior ethmoidal artery branch was absent in 15% of the specimens when the ophthalmic artery crossed under the optic nerve, and in 19% where it crossed over. It extends through a foramen of the same name to supply the posterior ethmoidal air cells.

The anterior ethmoidal artery accompanies the anterior ethmoidal nerve through the anterior ethmoidal foramen. This artery, larger than the posterior ethmoidal artery, is consistently present. The anterior ethmoidal artery supplies the anterior and middle ethmoidal air cells, the frontal sinus, and the dura of the anterior cranial fossa (via a meningeal branch). The nasal branches of the anterior ethmoidal artery supply the lateral wall of the nose and septum.

The medial palpebral arteries are two in number and arise from the ophthalmic artery below the trochlea. The medial palpebral arteries pass above and below the medial canthal tendon to enter the upper and lower lids respectively, where they anastomose with the corresponding lateral palpebral arteries to form the marginal arcades. In the upper eyelid the marginal arcade lies in, or just anterior to, the tarsus, approximately 4 mm from the eyelid margin, and in the lower eyelid 2 mm from the margin.

The supratrochlear (frontal) artery is one of the two terminal branches of the ophthalmic artery. It pierces the orbital septum and proceeds superiorly, terminating in the scalp.

The dorsal nasal artery is the other terminal branch of the ophthalmic artery. It pierces the orbital septum above the medial canthal tendon and anastomoses with the angular artery, thereby establishing anastomosis between the internal and external carotid artery systems. It supplies the scalp and forehead near the midline, and sometimes the lacrimal sac.

The external carotid artery arises at the bifurcation of the common carotid artery, at the level of the upper border of the thyroid cartilage, and extends to the back of the neck of the mandible, where it divides into superficial temporal and maxillary arteries. The external carotid artery gives out the following branches: superior thyroid, ascending pharyngeal, lingual, facial, occipital, posterior auricular, superficial temporal, and maxillary arteries (Fig. 9.5). The branches which supply the orbital structures are the facial, superficial temporal, and maxillary arteries.

The facial (external maxillary) artery arises near the angle of the mandible, extends deeply beneath the posterior belly of the digastric and stylohyoid muscles, and enters a groove in the submandibular gland. The artery then runs downward and forward to the lower border of the mandible. The facial artery then proceeds superiorly, lateral to the nose, to the medial canthal region. At the medial canthus, it anastomoses with the dorsal nasal artery.

The superficial temporal artery is one of the two terminal branches of the external carotid artery. The superficial temporal artery gives out the transverse facial, middle temporal, and zygomatic arteries. The transverse facial artery supplies the parotid gland and duct, and the masseter mus-

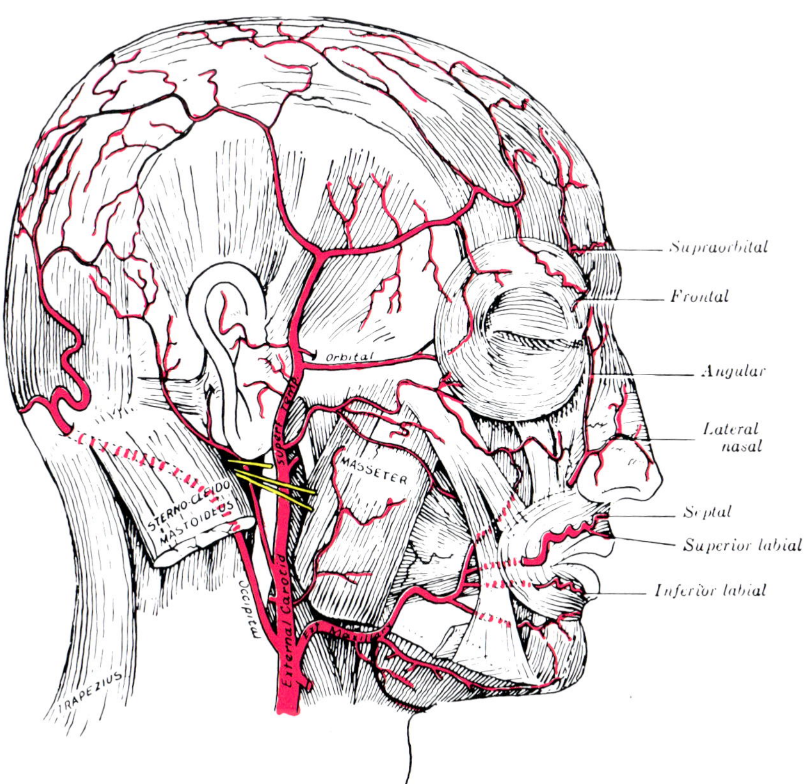

Figure 9.5. Branches of the external carotid artery. Terminal branches of the external carotid artery are the superificial temporal and maxillary arteries. (Reproduced from Goss CM (ed): *Gray's Anatomy of the Human Body* (ed 28). © 1966, Philadelphia, Lea & Febiger.)

cle, and ends in branches which anastomose with the infraorbital artery and branches of the facial artery in that region. The zygomatic artery is another branch of the superficial temporal artery which extends over the zygomatic arch to supply the orbicularis oculi and anastomoses with the lacrimal and palpebral branches of the ophthalmic artery. The frontal branch of the superficial temporal artery supplies the eyebrows and forehead, and anastomoses with the supraorbital and frontal arteries. Clinically, the superficial temporal artery is important, since this is biopsied to confirm the diagnosis of giant cell (temporal) arteritis.

The maxillary (internal maxillary) artery is the other terminal branch of the external carotid artery. The artery has an extensive supply (Fig. 9.6). It gives intracranial branches (accessory and middle meningeal arteries) to supply the trigeminal ganglion and the dura. The branch of ophthalmic interest is the infraorbital artery, which arises in the pterygopalatine fossa. The infraorbital artery enters the orbit through the inferior orbital fissure and runs forward in the infraorbital groove and canal, to emerge on the face from the infraorbital foramen. In the orbit the infraorbital artery gives branches to the lacrimal gland, and the inferior rectus and inferior oblique muscles. The facial branches of the infraorbital artery supply the lacrimal sac and the lower eyelid, and anastomose with the branches of the facial artery (Fig. 9.7).

VENOUS SYSTEM

Discussion of the orbital venous system first addresses the venous patterns in the orbit, and then in the cavernous sinus and pterygoid plexus.

Historically, two views exist concerning the venous system of the orbit. One view states that the branching pattern and course of the superior ophthalmic vein correspond to that of the ophthalmic artery. (Duke-Elder, 1962: Wolff, 1968). This venous pattern, in which a vein accompanies an artery, is generally present throughout the body. A second and completely opposing view is that orbital veins have a course and direction that are totally independent from that of the arterial system (Gurwitsch, 1883). Radiographic evaluation using venograms of the orbit confirms the independent nature of the orbital venous system, and the rather constant course of the superior ophthalmic vein (Boudit, 1955; Brismar, 1974). The orbital venous system is, therefore, unique as a result of its independence from the arterial system.

Blood is drained from the orbit by three major systems. Primary venous drainage is directed posteriorly by the superior ophthalmic vein to the cavernous sinus. The inferior ophthalmic vein usually fuses with the superior ophthalmic vein prior to entering the cavernous sinus, but a branch may extend to the pterygoid plexus. Blood also flows anteriorly, through continuity of the superior ophthalmic vein and the angular vein of the facial system. The direction of blood flow through the valveless venous system is determined by pressure gradients.

The principal vein in the orbit is the superior ophthalmic vein, which is formed by two roots: superior (supraorbital) and inferior (Fig. 9.8). The superior root is an extension of the supraorbital vein and extends from the superior nasal aspect of the orbital rim along the roof of the orbit to the medial aspect of the levator muscle. There, it unites with the inferior root. The inferior root is the orbital extension

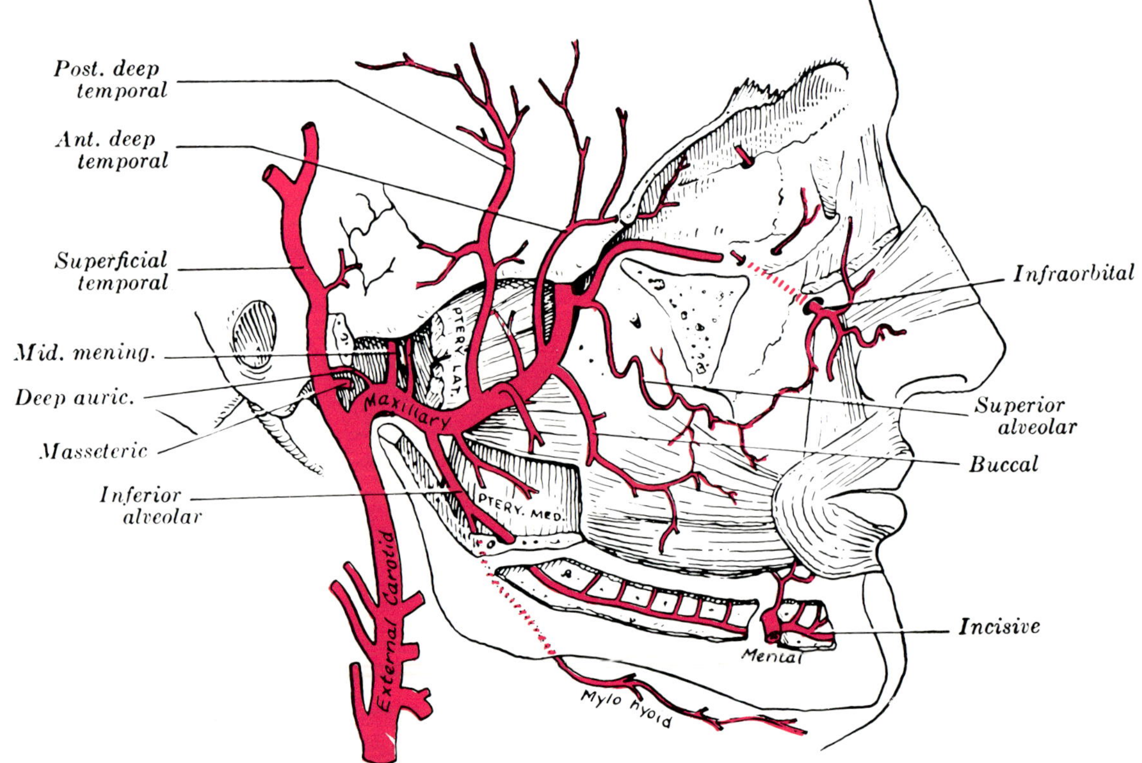

Figure 9.6. Branches of the internal maxillary artery. (Reproduced from Goss CM (ed): *Gray's Anatomy of the Human Body* (ed 28). © 1966, Philadelphia, Lea & Febiger.)

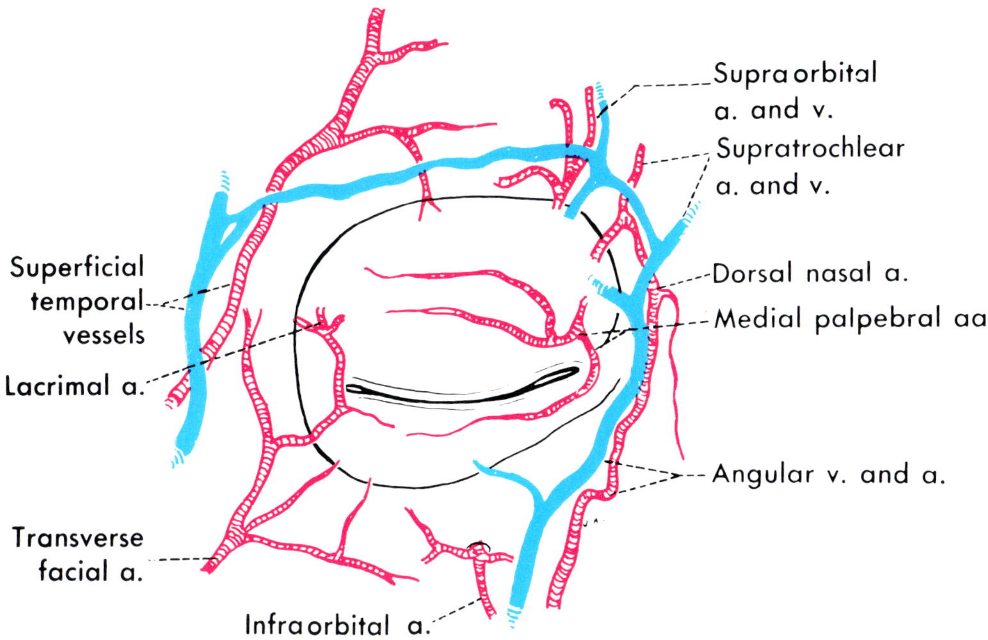

Figure 9.7. Vessels near the orbital margin and eyelids. (Reproduced from Hollinshead WH: *Textbook of Anatomy* (ed 3). © 1974; Philadelphia, Harper & Row Publishers.)

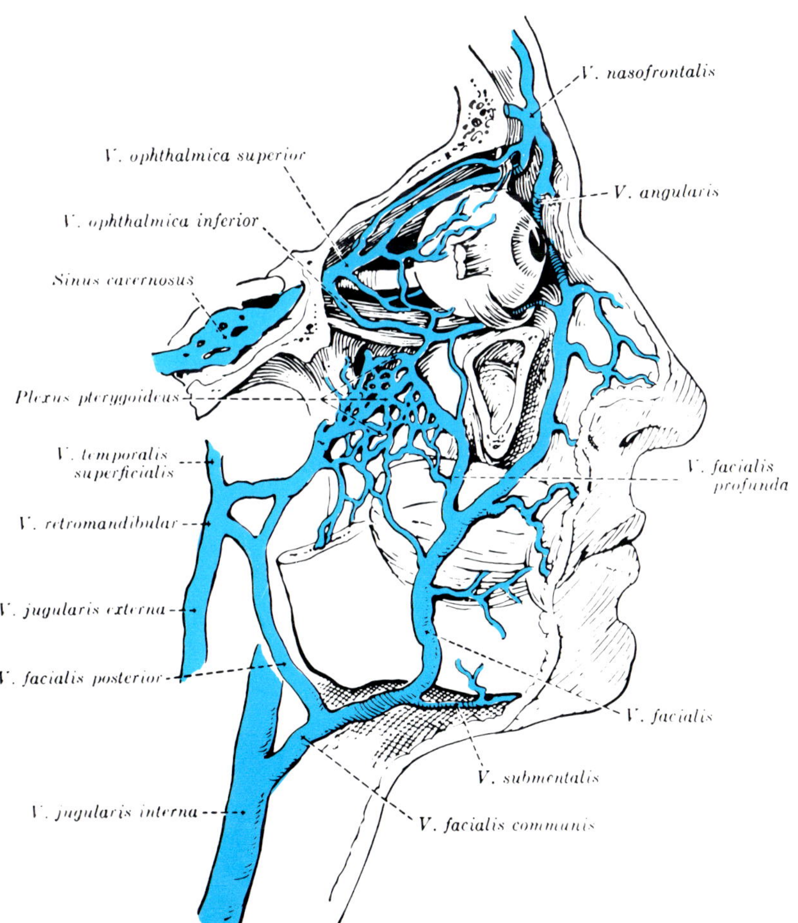

Figure 9.8. Principle venous system of the face. Venous drainage from the orbit primarily extends posteriorly to the cavernous sinus and pterygoid plexus, but may extend anteriorly to the angular vein and facial system. (Reproduced from Goss CM (ed): *Gray's Anatomy of the Human Body* (ed 28). © 1966, Philadelphia, Lea & Febiger.)

of the angular vein, which pierces the orbital septum then extends in a posterosuperior direction to join the superior root (Henry, 1959). The superior ophthalmic vein extends posteriorly to the medial border of the superior rectus muscle. The vein, supported by a hammock of connective tissue, then descends beneath the superior rectus muscle and progresses to the lateral border of the superior rectus muscle (Koornneef, 1977). The superior ophthalmic vein travels along the lateral border of the superior rectus to the superior orbital fissure. Numerous branches feed into the vein as it proceeds posteriorly: ciliary veins, superior vortex (vorticose) veins, lacrimal veins, and posterior and anterior ethmoidal veins. After the superior ophthalmic vein passes through the superior orbital fissure, it enters the cavernous sinus.

The inferior ophthalmic vein is more variable than the superior ophthalmic vein. The inferior ophthalmic vein originates as a diffuse plexus on the orbital floor. The plexus receives venules from the lateral rectus, the inferior rectus, and oblique muscles. In addition, branches are supplied by the inferior vortex veins, inferior conjunctiva, and lacrimal sac. Posteriorly, the inferior ophthalmic vein has two branches. The primary extension of the inferior ophthalmic vein is to the superior ophthalmic vein prior to its entering the cavernous sinus. A smaller branch extends inferiorly at the posterior aspect of the infraorbital fissure to the pterygoid plexus.

The central retinal vein travels in close relationship with the central retinal artery. The vein emerges from the optic nerve, 8–15 mm from the posterior portion of the globe. The central retinal vein usually passes directly to the cavernous sinus, without joining the superior or inferior ophthalmic veins (Whitnall, 1932). Prior to its reaching the cavernous sinus, the central retinal vein passes a small plexus of veins. This additional connection may provide an ancillary route for venous drainage in cases of cavernous sinus thrombosis.

Two additional named orbital veins have been recognized. The "veine ophtalmique moyenne," a posterior muscular vein, was first described by Henry (1959). This vessel arises in the retrobulbar space and drains the muscular branches from the inferior, medial, and lateral rectus. The vein extends posteriorly to the superior ophthalmic vein or directly to the cavernous sinus. Utilizing orbital phlebography, Brismor (1974) recognized a medial ophthalmic vein in 40% of the orbits he examined. This vein arises from the angular vein, or branches from the superior ophthalmic vein, and runs along the medial orbital wall to the cavernous sinus.

Although orbital venous drainage is primarily posterior to the cavernous sinus, anterior drainage does exist. The most prominent drainage vessel is the angular vein in the nasal orbital region (Fig. 9.9). The angular vein is continuous with the facial vein. In addition, diffuse anastomic channels arise from the inferior orbital vein, and cross over the inferior orbital rim to reach the facial vein to provide anterior venous drainage.

Cavernous Sinus

The cavernous sinus is an endothelially-lined, venous channel formed by a splitting of the dura at the body of the sphenoid bone. The sinus receives major tributaries from the superior and inferior ophthalmic veins, central retinal vein, the middle and inferior cerebral veins, and the middle meningeal veins. The sinus also communicates with the superior and inferior petrosal sinus and with the pterygoid plexus, inferiorly.

The cavernous sinus lies within the middle cranial fossa, adjacent to the temporal lobe of the brain. The pituitary gland is above and medial to the cavernous sinus. The cavernous sinus extends from the medial, or widest portion of the superior orbital fissure, to the apex of the petrous portion of the temporal bone. The inferior border of the cavernous sinus is contiguous with the gassarian (or semilunar) ganglion of the trigeminal nerve.

The cavernous sinus is an extensive venous channel, representing an anastomotic array of venules. Traditionally the sinus was thought to have two basic sections: an outer, or superficial, reticular section and an inner, spacious section, closer to the bone. The sinus was visualized as an unbroken, trabeculated structure, continuous with the contributing veins. However, anatomic studies by Parkinson (1972) and Browder and Kaplan (1976) demonstrated that the cavernous sinus is composed of various sized venules. The diffusely anastomotic venules surround the intracavernous portion of the internal carotid artery. The previously described trabecular structure of the cavernous sinus is a result of the postmortem collapse of these venules (Kline, et al, 1982).

The primary arterial structure within the cavernous sinus is the internal carotid artery (Figs. 9.10 and 9.11). The intracavernous portion of the internal carotid artery forms an "S" shaped bend. The carotid, attached at its entrance and exit to the cavernous sinus, is freely movable within the sinus itself. Closely associated with the internal carotid artery, the sympathetic plexus proceeds with the ophthalmic artery to provide sympathetic innervation to orbital structures. In addition, the abducens nerve is closely adherent to the lateral border of the internal carotid artery after entering the cavernous sinus through Dorello's canal. Within the cavernous sinus, the abducens nerve extends to the lateral wall of the sinus (Fig. 9.11 *B*).

The neural structures within the cavernous sinus are the oculomotor, trochlear, abducens, and the ophthalmic division of the trigeminal nerve. The maxillary nerve lies in the inferior portion of the cavernous sinus prior to exiting at the foramen rotundum. (Henderson, 1966). The nerves change their position slightly in their posterior to anterior excursion through the sinus (Figs. 9.11 *A* and *B*). In the posterior extent of the sinus, the oculomotor nerve is superior, entering the sinus after it has dipped beneath the anterior clinoid process. The abducens nerve, as noted previously, enters the cavernous sinus with the internal carotid artery, then extends to the lateral wall of the sinus as it extends anteriorly. In addition, posteriorly, the cavernous sinus envelops Meckel's cave to contain the proximal portion of the three branches of the trigeminal nerve (ophthalmic, maxillary, and mandibular). As the nerves proceed anteriorly, the superior to inferior order of the nerves are oculomotor, trochlear, ophthalmic, and abducens. The oculomotor and trochlear nerves lie within a common dural tunnel in the superior apex of the sinus. At the anterior tip of the

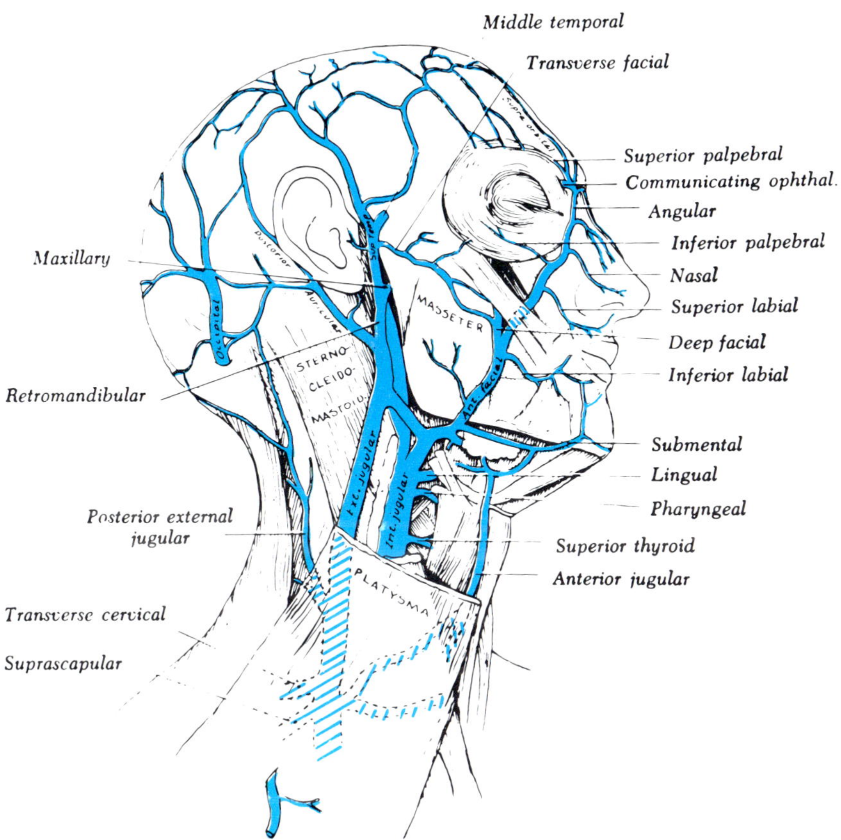

Figure 9.9. Superficial venous systems of the orbit and face. (Reproduced from Goss CM (ed): *Gray's Anatomy of the Human Body* (ed 28). © 1966, Philadelphia, Lea & Febiger.)

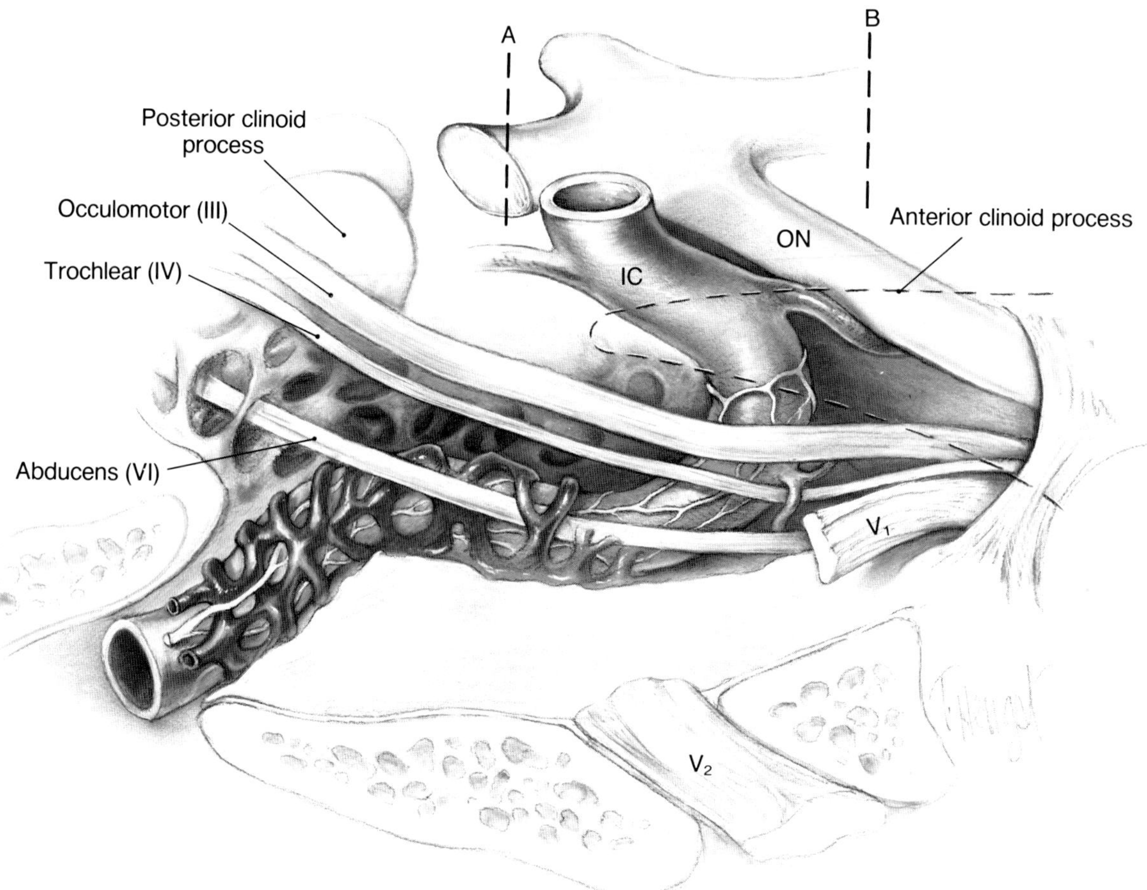

Figure 9.10. Internal carotid artery within the cavernous sinus. Within the cavernous sinus, the internal carotid artery is S-shaped, forming the carotid siphon. The ophthalmic artery arises from the internal carotid artery as it exits the cavernous sinus.

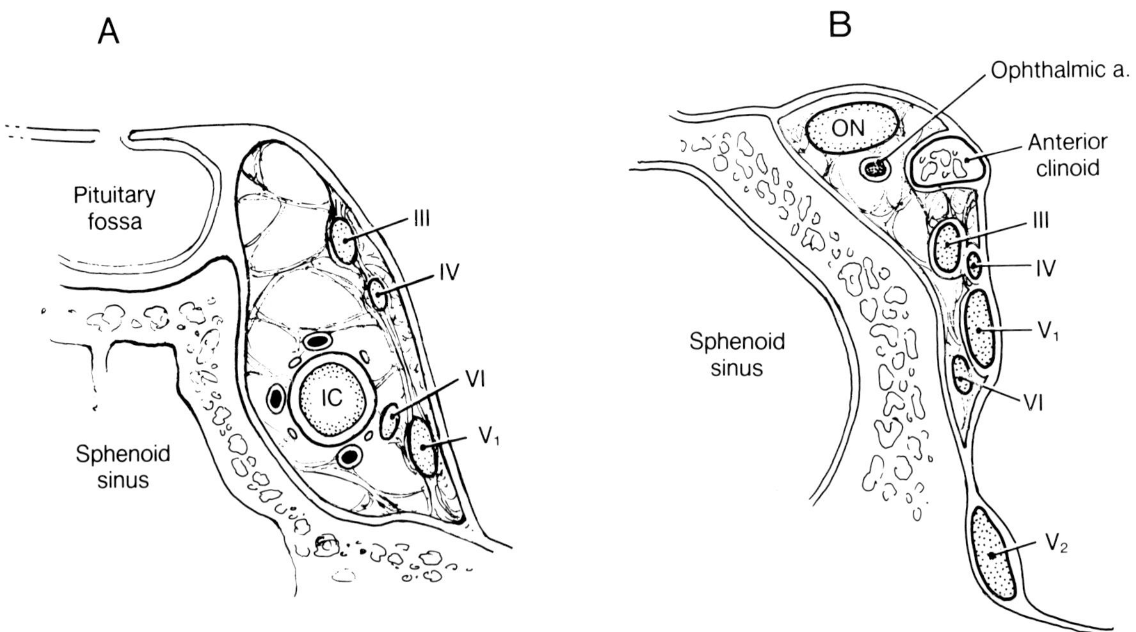

Figure 9.11. Coronal sections of the cavernous sinus. (*A*), at the level of the pituitary fossa, the abducens nerve is still adjacent to the internal carotid artery while the oculomotor, trochlear, and ophthalmic nerves are within the loose layer of the lateral wall of the cavernous sinus. (*B*), At the level of the anterior clinoid process the abducens nerve is within the loose fibrillar tissue of the lateral wall of the cavernous sinus.

sinus, the trochlear nerve lies superior to the oculomotor. The oculomotor nerve enters the orbit through the annulus of Zinn, while the trochlear nerve enters the orbit outside the annulus (Renn and Rhoton, 1975; Kaplan, et al, 1976; Harris and Rhoton, 1976).

Umansky and Nathan (1982) studied the lateral wall of the cavernous sinus and identified two layers: an outer thick dural (superficial) layer, and an inner loose fibrillar (deep) layer (Fig. 9.11 *A* and *B*). The oculomotor, trochlear, and ophthalmic and maxillary divisions of the trigeminal nerve lie within the loose fibrillar connective tissue of the inner layer of the cavernous sinus. The inner layer is poorly defined and is formed by sleeves or sheaths of dura mater which accompany the adjacent nerves. It was apparent from Umansky and Nathan's study that the nerves are not contained within a single dural wall of the cavernous sinus (Warwick and William, 1973). In addition, a distinct septum containing the nerves and dividing the cavernous sinus into two cavities was not identified.

CLINICAL NOTE

The clinical presentation of carotid cavernous sinus fistulas varies, depending on the pressure gradient of the system. The typical carotid cavernous fistula occurs following severe head trauma, and is clinically evident by a bruit, pulsating exophthalmos, conjunctival chemosis, engorgement of epibulbar veins, restriction of ocular motility, unilateral elevation of intraocular pressure, and ocular ischemia (Fig. 9.12 *A* and *B*). The typical carotid cavernous fistula is secondary to continuity of internal carotid artery and the cavernous sinus venous plexus (Fig. 9.13). Studies report visual deficit in 60% of cases (Henderson and Schneider, 1959). The reduced visual acuity is probably secondary to ocular ischemia produced by the altered arteriovenous gradient (Sanders and Hoyt, 1969).

Surgical management of carotid cavernous fistulas by clipping the internal carotid artery may account for additional visual loss. Palestine, et al (1981) reviewed the Mayo Clinic experience with carotid cavernous fistulas and reported a 28% incidence of further visual loss and a 10% incidence of cerebral ischemia. Harris and Rhoton (1976) also noted a tendency for visual loss following clipping carotid aneurysms. They stressed the need for selective internal and external angiography to precisely locate the fistula. Dandy and Follis (1941) reported proptosis occurring on the side opposite the fistula, which further illustrates the need for selective angiography.

An atypical form of cavernous sinus fistula has recently been recognized (Phelps, et al, 1982; Grove, 1984). Atypical fistulas occur in older patients who have no history of trauma. These patients have a lower pressure gradient and produce milder symptoms. Clinical features of the atypical fistulas are episcleral venous congestion, slightly elevated intraocular pressure, and mild exophthalmos. Carotid bruits are rarely present, and proptosis of a pulsatile nature is not encountered. Vision was not affected in 19 patients reviewed by Phelps and coworkers.

Radiographic evaluation of patients with atypical carotid cavernous fistulas confirms the distinction of this syndrome from the classic features of the syndrome. Selective carotid angiography discloses the fistula to be a dural-cavernous fistula. In addition, due to the lower pressure gradient involved in the atypical syndrome, spontaneous closure is encountered in approximately 50% of patients. Low pressure fistulas may produce chronic open angle glaucoma requiring appropriate treatment.

Lesions in and about the sella tursica may involve the cranial nerves within the cavernous sinus. Pituitary tumors, particularly eosinophilic adenomas, may produce a clinical syndrome of multiple cranial nerve palsies, severe headache, and variable visual disturbances, including blindness. Although rare, intracavernous aneurysms may produce a parasellar-cavernous sinus syndrome (Meadows, 1962). Intracavernous carotid aneurysms have a tendency to occur in middle to older aged women, who may or may not have headaches. Occasionally, these aneurysms will invade the sphenoid sinus and produce intractable nose bleeds (Keane and Talalla, 1959). Rarely, intracavernous aneurysms will rupture and produce the characteristics of a carotid-cavernous sinus fistula.

The cavernous sinus and superior orbital fissure may be involved with a granulomatous inflammation producing the Tolosa-Hunt syndrome (Tolosa, 1954; Hunt et al, 1961). These patients present with a painful ophthalmoplegia, which is exquisitely responsive to steroids (Smith and Taxdru, 1966), although this is not pathognomic, since tumors may also respond dramatically to steroids (Thomas and Yoss, 1970). The diagnosis of Tolosa-Hunt syndrome is one of exclusion, and is made only after an exhaustive evaluation rules out additional causes of painful ophthalmoplegia.

Cavernous sinus thrombosis may result from hematogenous extension of infection from the face, nasal cavity, paranasal sinuses, and the ear. Cavernous sinus thrombosis may be a rare complication of orbital cellulitis (Schramm, et al, 1982). Signs of cavernous sinus thrombosis are bluish-purple eyelid coloration, proptosis, chemosis, and ocular motor palsies. The absence of ocular pain and decreased sensation in the area supplied by the maxillary division of the trigeminal nerve are characteristic findings on physical examination. Bilateral involvement is common in cavernous sinus thrombosis. Bilateral involvement results from the normal tributaries connecting the sinuses.

The pterygoid plexus is buried within the fatty tissue of the pterygoid fossa. The pterygoid plexus has diffuse connections with the cavernous sinus and inferior ophthalmic vein through the inferior orbital fissure. Communication to the facial (superficial) system is through the deep facial vein. The plexus also receives blood from the area supplied by the internal maxillary artery. The plexus, therefore, receives venous drainage from the sphenopalatine, middle meningeal, deep temporal, pterygoid, masseteric, buccinator, alveolar, and some palatine veins.

Using orbital dissection techniques similar to those of Koornneef (1977), Bergen (1982) studied the relationship of orbital connective tissue to arteries and veins. Bergen recognized differences between the arterial and venous patterns within the orbit (Fig. 9.14 *A* and *B*). The major arteries are embedded in connective tissue compartments, bordered by connective tissue septa (Fig. 9.15). As arteries branch anteriorly, they make contact with, but are not embedded within,

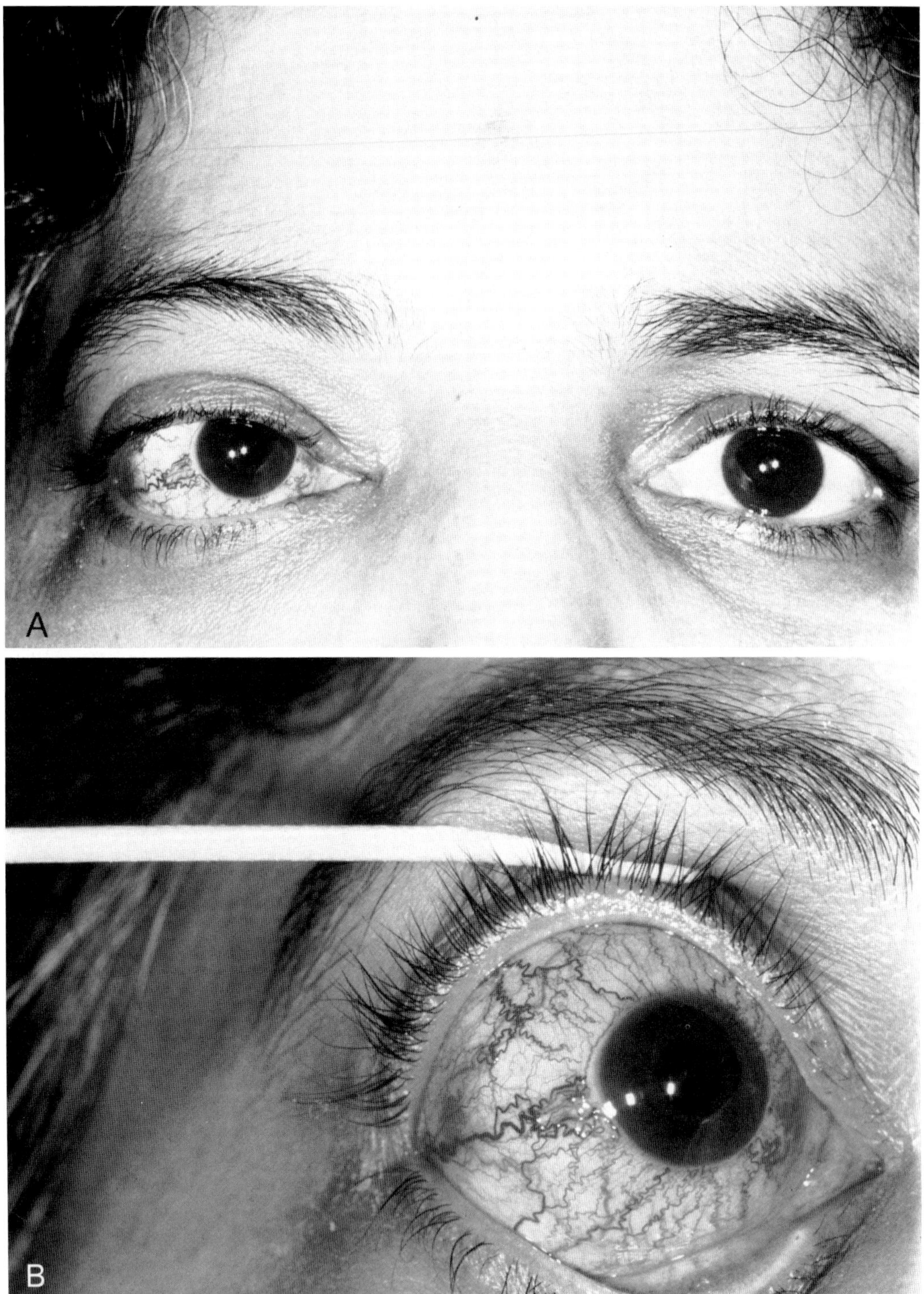

Figure 9.12.   Clinical appearance of carotid-cavernous sinus fistula. (A), Full face photographs demonstrating proptosis of the right orbit. (B), Appearance of the right eye "caput medussa" or episcleral vascular congestion.

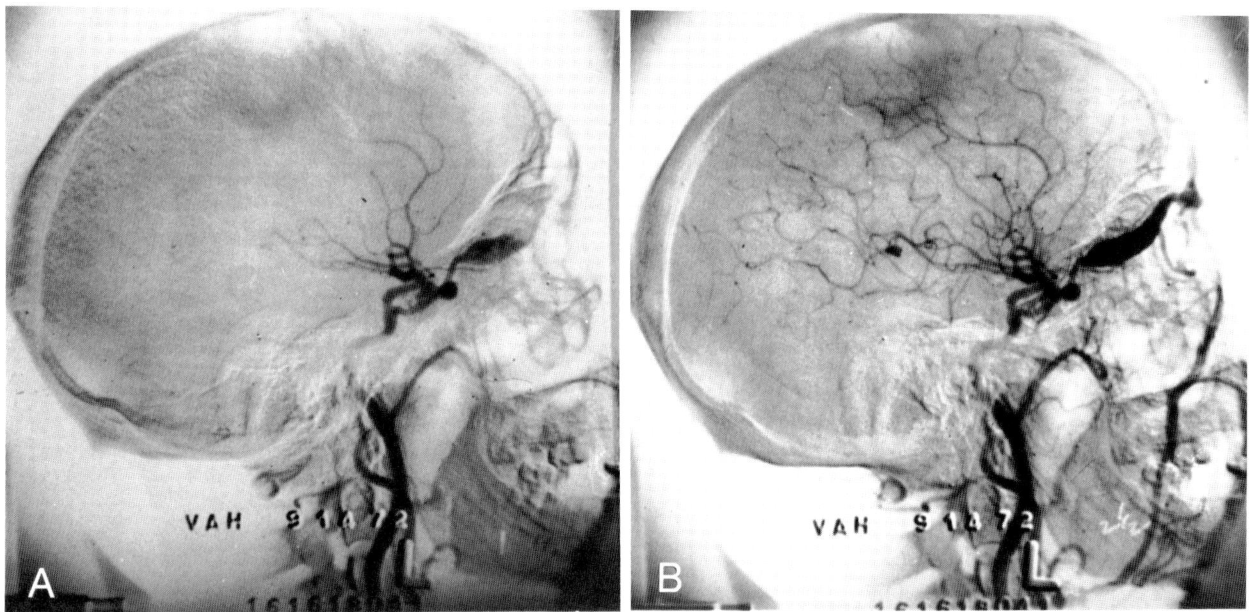

Figure 9.13. Arteriogram of a carotid cavernous sinus fistula. (A), Early phase of arteriogram demonstrating carotd fistula. (B), Marked dilation of superior ophthalmic vein in the superior orbit.

Figure 9.14. Spatial drawing demonstrating the septal architecture of the orbital connective tissue system. (*A*), Overview of the system. Abbreviations: *SOV*, superior ophthalmic vein; *ON*, optic nerve. ■ Cavity in which the superior rectus muscle should be located. ☆ Cavity in which the levator palpebrae superioris should be located. ● Cavity in which the inferior rectus muscle should lie. □ Cavity in which the lateral rectus muscle should lie. (*B*), Detailed drawing demonstrating the superior ophthalmic vein in its hammock beneath the superior rectus muscle. (Reproduced from Bergin MP: *Vascular Architecture in the Human Orbit.* © 1982, Lisse, Netherlands, Swets and Zeitlinger BV.)

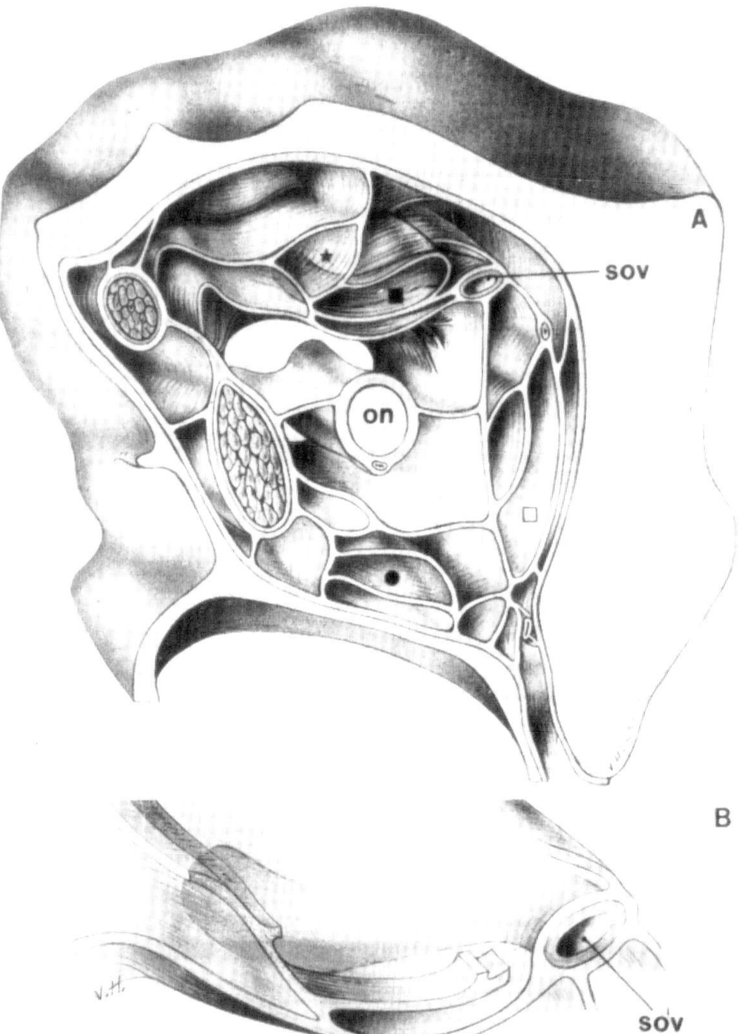

Figure 9.15. Spatial drawing of the arterial and venous relationship to connective tissue septa within the orbit. The arteries radiate toward their target organs and perforate the connective tissue septa. The veins run circularly are confined to the connective tissue septa. Abbreviations: *SOV*, superior ophthalmic vein; *LA*, lacrimal artery; *MA*, muscular artery. (Reproduced from Bergin MP: *Vascular Architecture in the Human Orbit.* © 1982, Lisse, Netherlands, Swets and Zeitlinger BV.)

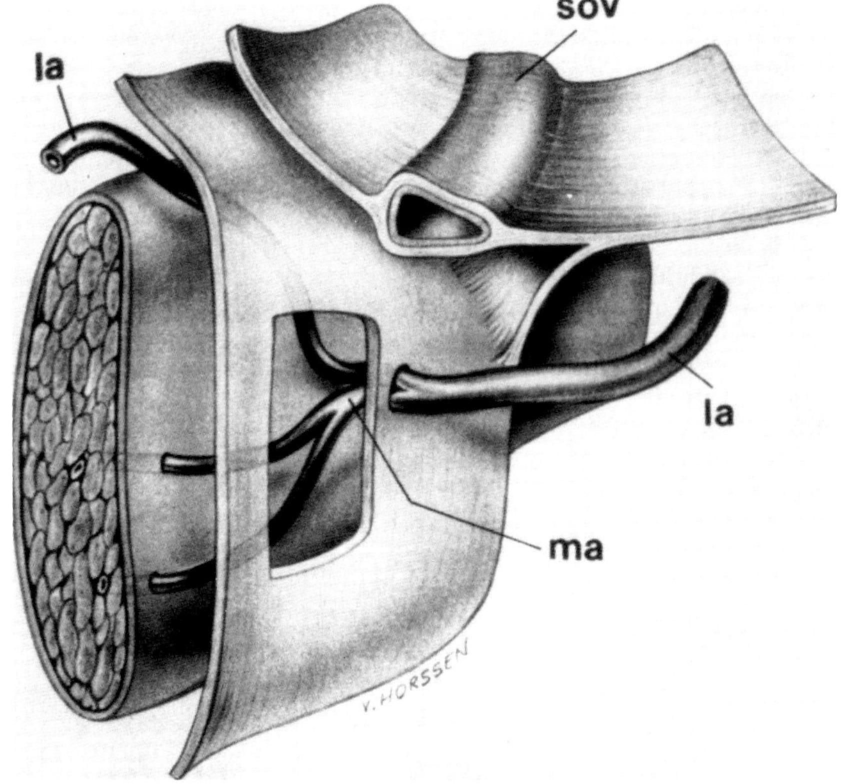

the connective tissue walls. The ophthalmic artery is relatively free of connective tissue attachments.

The relationship of connective tissue to venous structures is different from that of arteries. The major veins are embedded within walls of fibrous connective tissue (Fig. 9.15). This limiting attachment of veins to the connective tissue septa produces a circumferential arrangement of vessels inside and outside of the muscle cone. The microvascular bed is segmented by adipose tissue compartments and few microvascular communications extend across these adipose compartments.

LYMPHATIC SYSTEM

The lymphatics of the eyelids form a deep and a superficial system. The deep vessels form plexuses along the tarsal borders and drain the tarsal and conjunctival regions. The superficial system drains the skin of the lids and orbicularis. The lymphatics of the lids drain into the parotid nodes and submaxillary nodes (Fig. 9.16). The drainage of most of the upper lid, lateral one-half of the lower lid, and lateral one-half of the conjunctiva is to the parotid nodes. The medial portion of the upper and lower lids, as well as the medial one-half of the conjunctiva, drain into the submaxillary nodes by channels which follow the angular and facial arteries. In the orbit itself, and in the globe, there are said to be no lymph nodes or vessels.

CLINICAL NOTE

An interesting article by Corwin and Weiter (1981) which deals with the immunology of chorioretinal disorders brought to light important concepts. The eye does not drain to a specific lymph node, but, in fact, has unique features: (1) avascularity of the cornea and lens; (2) the presence of blood-aqueous and blood-retina barriers; and (3) the absence of lymphatic drainage. It is the absence of lymphatics in the eye that allows antigens to be processed in an extraocular site. This extraocular processing of antigens may allow for the production of suppressor T-cells that enhance antigenic activity. The best example is toxoplasma gondii, which may remain dormant in the eye and be reactivated periodically.

We have found that the orbit as well is very unique. It has no lymphatics and yet acts much like its own lymphatic or immunologic organ. Despite the absence of lymphatics, lymphangiomas, pseudotumors, and lymphomas are common in the orbit. The orbit also has unusual immunologically related responses to systemic diseases, such as Sjögren's syndrome, Wegener's, and other collagen-vascular related disorders.

The most common, most dramatic, and most unique immunologically related orbital disease is thyroid orbitopathy. This common, but very poorly understood condition accounts for nearly one-half of the cases of proptosis in adults. The greatest reaction and swelling

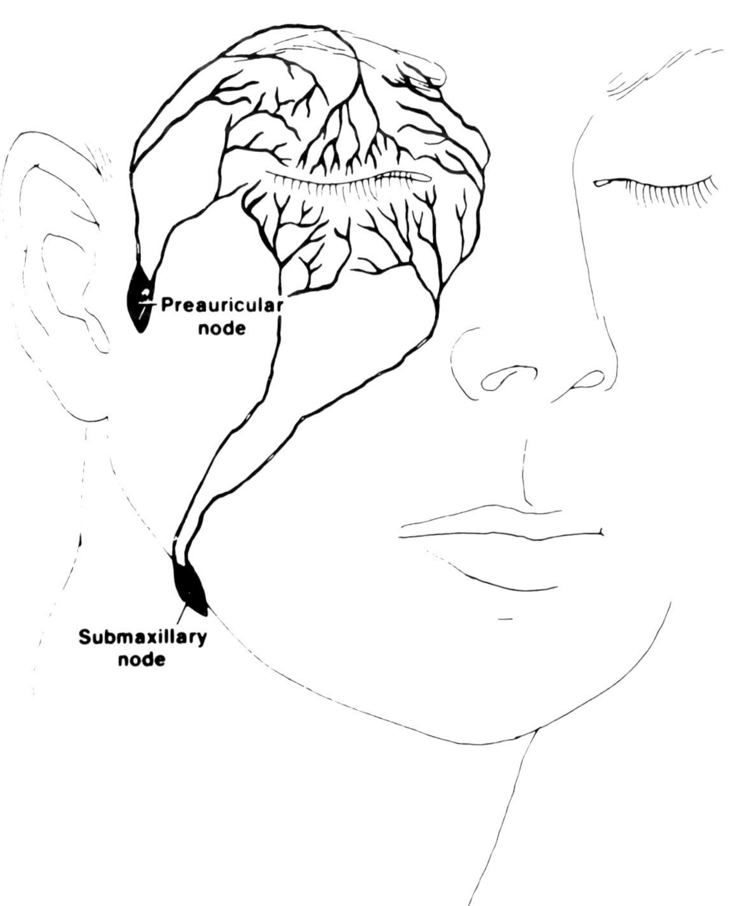

Figure 9.16. Lymphatic drainage of the eye lids. (Reproduced from Meltzer M: *Ophthalmic Plastic Surgery for the General Ophthalmologist.* © 1979, Baltimore, Williams & Wilkins Co.)

in thyroid orbital disease occurs in the extraocular muscles. Hopefully, when the unique immunology of the orbit is better understood, these common but very difficult orbital conditions will be better managed.

References

Bergen MP: Spatial aspects of the orbital vascular system. In Duane TD, Jaeger EA (eds): *Biomedical Foundations of Ophthalmology*, Vol 1, Chapter 32. Philadelphia, Harper & Row, 1982.

Boudeit C: La place du phlebogramme orbitaire dans le diagnostic des tumeurs de l'orbite. *Arch Ophthalmol* 15:597, 1955.

Brismar J: Orbital phlebography. II. Anatomy of the superior ophthalmic vein and its tributaries. *Acta Radiol [Diagn] [Stockh]* 15:481, 1974.

Brismar J: Orbital phlebography. III. Topography of orbital veins. *Acta Radiol [Diagn] [Stockh]* 15:577, 1974.

Browder J, Kaplan HA: *Cerebral Dural Sinuses and Their Tributaries*. Springfield, IL, Charles C Thomas, 1976, p 85.

Corwin JM, Weiter JJ: Immunology of chorioretinal disorders. *Surv Ophthalmol* 25:287, 1981.

Dandy WE, Follis RH Jr: On pathology of the carotid-cavernous aneuryms (pulsating exophthalmos). *Am J Ophthalmol* 24:365, 1941.

Duke-Elder S: *System of Ophthalmology*, Vol VII: The Foundations of Ophthalmology. London, Kimpton, 1962, p 517.

Grove AS: The dural shunt syndrome. Pathophysiology and clinical course. *Ophthalmology* 91:31, 1984.

Gurwitsch M: Ueber die anastomosen zwischen den Gesichts-und orbitalvenen. A *von Graefes Arch Ophthalmol* 29(4):31,1883.

Harris FS, Rhoton AL Jr: Anatomy of the cavernous sinus. A microsurgical study. *J Neurosurg* 45:169, 1976.

Hayreh SS: A study of the central artery of the retina in human beings in its intra-orbital and intraneural course. Thesis, Panjab University, 1958.

Hayreh SS: The ophthalmic artery. III. Branches. *Br J Ophthalmol* 46:212, 1962.

Hayreh SS: Arteries of the orbit in the human being. *Br J Surg* 50:938, 1963.

Hayreh SS: Blood supply of the optic nerve head and its role in optic atrophy, glaucoma and oedema of the optic disc. *Br J Ophthalmol* 53:721, 1969.

Hayreh SS, Dass R: The ophthalmic artery. I. Origin and intracranial and intra-canalicular course. *Br J Ophthalmol* 46:65, 1962a.

Hayreh SS, Dass R: The ophthalmic artery. II. Intra-orbital course. *Br J Ophthalmol* 46:165, 1962b.

Henderson JW, Schneider RC: The ocular findings in carotid-cavernous fistula in a series of 17 cases. *Am J Ophthalmol* 48:585, 1959.

Henderson WR: A note on the relationship of the human maxillary nerve to the cavernous sinus and to an emissary sinus passing through the foramen ovale. *J Anat* 100:905, 1966.

Henry JGM: Contribution a l'etude de l'anatomie des vaisseaux de l'orbite et de la loge caverneuse-pas injection de matieres plastiques du tendon de Zinn et de la capsule de Tenon. These de Paris, 1959. Cited by Bergen MP: *Vascular Architecture in the Human Orbit*. Lisse, Netherlands, Swets and Zeitlinger BV, 1982.

Hunt WE, Meagher JN, LeFever HE, et al: Painful ophthalmoplegia: Its relation to indolent inflammation of the cavernous sinus. *Neurol* 11:56, 1961.

Kaplan HA, Browder J, Krieger AJ: Intercavernous connections of the cavernous sinus. The superior and inferior circular sinuses. *J Neurosurg* 45:166, 1976.

Keane JR, Talalla A: Posttraumatic intracavernous aneurysm: Epistaxis with monocular blindness preceeded by chromatopsia. *Arch Ophthalmol* 87:701, 1972.

Kline LB, Acker JD, Post MJD: Computed tomographic evaluation of the cavernous sinus. *Ophthalmology* 89:374, 1982.

Koornneef L: New insights in the human orbital connective tissue: Result of a new anatomical approach. *Arch Ophthalmol* 95:1269, 1977.

Meadows JSP: Intracavernous aneurysms of the internal carotid artery. *Arch Ophthalmol* 62:566, 1959.

Palestine AG, Younge BR, Piepgras DG: Visual prognosis in carotid-cavernous fistula. *Arch Ophthalmol* 499:1600, 1981.

Parkinson D: Anatomy of the "cavernous sinus." In Smith JL (ed): *Neuroophthalmology Symposium of the University of Miami and the Bascom Palmer Eye Institute*, Vol 6. St. Louis, CV Mosby, 1972, p 73.

Paturet G: Traite d'Anatomie Humaine. In *Systeme Nerveux*, Vol 4. Paris, Masson, 1964, p 721.

Phelps CD, Thompson HS, Ossoinig KC: The diagnosis and prognosis of atypical carotid-cavernous sinus fistula (red-eye shunt syndrome). *Am J Ophthalmol* 93:423, 1982.

Renn WH, Rhoton AL, Jr: Microsurgical anatomy of the sellar region. *J Neurosurg* 43:288, 1975.

Sanders MD, and Hoyt WF: Hypoxic ocular sequelae of carotid-cavernous fistula: Study of the causes of visual failure before and after neurosurgical treatment in a series of 25 cases. *Br J Ophthalmol* 53:82, 1969.

Schramm VL, Curtin HD, Kennerdell JS: Evaluation of orbital cellulitis and results of treatment. *Laryngoscope* 92:732, 1982.

Singh S, Dass R: The central artery of the retina: I. Origin and cause. *Br J Ophthalmol* 44:193, 1960.

Smith JL, Taxdal DSR: Painful ophthalmoplegia: The Tolosa-Hunt syndrome. *Am J Ophthalmol* 61:1466, 1966.

Thomas JE, Yoss RE: The parasellar syndrome: Problems in determining etiology. *Mayo Clin Proc* 45:617, 1970.

Tolosa EJ: Periaortic lesions of the carotid siphon with clinical features of carotid infraclinoidal aneurysm. *J Neurol Neurosurg Psychiat* 17:300, 1954.

Umansky F, Hilel N: The lateral wall of the cavernous sinus: With special reference to the nerves related to it. *J Neurosurg* 56:228, 1982.

Wallace S, Goldberg HI, Leed NE, Mishkin MM: The cavernous branches of the internal carotid artery. *Am J Roentgenol* 101:34, 1967.

Warwick R, William PL (eds): *Gray's Anatomy*, ed 35. Edinburgh, Longman, 1973, p 695.

Whitnall SE: *Anatomy of the Human Orbit and Accessory Organs of Vision*, ed 2. London, Oxford Medical Publications, 1932, pp 173, 303.

Wolff E: *The Anatomy of the Eye and Orbit*, ed 6. London, Lewis, 1968.

CHAPTER 10

Orbital Dissection

The previous chapters have been designed to complement this chapter on orbital dissection. The chapters on eyelid and orbital anatomy and physiology should be read before or in conjunction with the dissection of these anatomical structures. Detailed information on these structures is provided in the appropriate chapter. There are always some minor anatomical variabilities from orbit to orbit, but the anatomical relationships noted in this chapter are in general consistent.

The chapter on osteology, should be studied prior to orbital dissection and a skull and sphenoid bone, are very useful for reference and orientation during orbital dissection.

The following instruments are recommended for orbital dissection:

Group Instruments to be shared by all:
1. Oscillating saws (electric).
2. Heavy rongeurs (orthopaedic or neurosurgical).
3. Large and small chisels (wood or carpenter chisels are adequate).
4. Mallets (hammers adequate).

Individual Dissection Instruments (1 set for each cadaver):
1. Smooth forceps.
2. Tooth forceps.
3. Blunt dissection probes (large lacrimal probes adequate).
4. Scissors.
5. Scalpel with blades.
6. Lacrimal probes, medium to large.
7. Muscle hook.
8. A set of orbital surgery instruments including Kerrison rongeurs are very useful, but not mandatory.

Whether one begins by dissecting the orbit from above or anteriorly is somewhat arbitrary. If complete cadaver heads are being dissected and the brain is to be removed, collected, and preserved, then we recommend continuing with the orbital dissection from above. If complete heads are not available and only a single isolated orbit is being dissected, then we recommend that the orbit be approached from the anterior first, and followed by the orbit from above.

THE ORBIT FROM ABOVE

The scalp is incised circumferentially, approximately 2 cm above the supraorbital rim, and the scalp is removed. The frontalis, procerus, and corrugator superciliaris muscles

should be identified. The calvarium is then removed with an oscillating saw. The brain is removed carefully if it is to be preserved for dissection at a later time. The attachments of the cranial nerves are noted and the cranial cavity is studied to determine the origin and intracranial course of the nerves and vessels (Fig. 10.1). The optic nerve is followed to the chiasm to determine its relationship with the pituitary gland. The nerves to the extraocular muscles are followed through their course. The oculomotor nerve is closely associated with the posterior communicating artery of the circle of Willis, following its exit from the brainstem. The trochlear nerve arises from the dorsal surface of the brainstem, and has a long intracranial course, which accounts for its susceptibility to injury. The abducens nerve is closely aligned to the sphenoid ridge and is confined by Gruber's (petroclinoidal) ligament, which predisposes it to involvement in cranial trauma and increased intracranial pressure. The greater superficial petrosal nerve may be identified as it runs along the petrous ridge. The dura overlying the cavernous sinus should be removed and the relationship of structures appreciated (Fig. 10.2).

The dura covering the roof of the orbit is now removed. The frontal bone forming the roof of the orbit is thin and

The color photographic plates accompanying this chapter are reproduced from David L. Bassett, MD's *A Stereoscopic Atlas of Human Anatomy*, published by Sawyer's Inc. Portland, Oregon, 1954, color photographs by Wm. B. Gruber. All of these figures are from Section II on the Head and Neck. Dr. Bassett's elegant anatomical dissections are the finest of which we are aware, and we doubt they will ever be excelled.

We are very grateful to Mrs. David L. Bassett for allowing us to reproduce these photographs to illustrate this chapter.

I know you will find these photographs as beneficial as we have in learning orbital anatomy. The figures used in this chapter from *A Stereoscopic Atlas of Human Anatomy* are: 49-3 (Fig. 10.29), 50-2 (Fig. 10.1), 51-5 (Fig. 10.2), 52-3 (Fig. 10.13), 52-4 (Fig. 10.14), 53-1 (Fig. 10.15), 53-4 (Fig. 10.16), 53-6 (Fig. 10.17), 53-7 (Fig. 10.19), 54-1 (Fig. 10.4), 54-2 (Fig. 10.5), 54-3 (Fig. 10.7), 54-4 (Fig. 10.8), 54-5 (Fig. 10.9), 54-6 (Fig. 10.10), 54-7 (Fig. 10.11), 55-4 (Fig. 10.20), 55-7 (Fig. 10.21), 56-1 (Fig. 10.22), 56-2 (Fig. 10.23), 56-7 (Fig. 10.28), 57-1 (Fig. 10.18), 57-2 (Fig. 10.25), 57-5 (Fig. 10.26), and 57-6 (Fig. 10.27). These photographs and their legends should be studied in detail. Generally, each photograph is referenced only once in the text, even though it may show many structures and have applications in many parts of the dissection, as well as in preceding chapters. Considerable knowledge regarding orbital anatomy can be gained by carefully studying these photographs in conjunction with the orbital dissection.

easily fractured with a chisel, and then removed with a rongeur (Fig. 10.3). The orbital roof is removed to the thickened superior orbital rim, anteriorly; to the optic canal, posteriorly; and to the ethmoids, nasally. One should attempt to preserve periorbita while removing the orbital roof.

The periorbita is then opened and the frontal nerve, which may adhere to the periorbita, is the first prominent structure noted. As the nerve extends anteriorly it divides into the supratrochlear and supraorbital nerves (Fig. 10.4). Cautious removal of orbital fat in a piecemeal fashion exposes the lacrimal nerve and its accompanying artery in the superior lateral orbit. The lacrimal nerve extends anteriorly in the orbit along the superior border of the lateral rectus muscle, and enters the posterior portion of the orbital lobe of the lacrimal gland (Fig. 10.5).

Prior to studying the levator muscle, one should remove the superior orbital margin. The skin and tissue overlying the rim is cleared with a periosteal elevator. A saw is utilized to make vertical cuts through the rim. Nasally, the incision should extend to a point just lateral to the trochlea, to avoid disruption of this area (Fig. 10.6). Laterally, the bone is incised just medial to the lateral orbital rim. The superior orbital rim is then grasped firmly and rotated to free it from its underlying attachments. Inspection of the orbital rim's inner surface prior to discarding it reveals the lacrimal gland fossa.

The full extent of the levator and superior rectus muscles can now be appreciated (Fig. 10.7). The levator muscle originates at the orbital apex, superior to the annulus of Zinn. The muscle extends anteriorly with fascial attachments to the underlying superior rectus muscle. After the muscle reaches Whitnall's ligament, it fans out in medial and lateral directions, which greatly enhances its horizontal dimension.

After passing beneath the superior transverse ligament of Whitnall, the levator gradually transforms to its aponeurosis, as previously described. The medial and lateral expanses of the aponeurosis (the levator horns) may be traced to the medial and lateral retinaculum.

After studying the levator muscle it should be transected at the orbital apex and reflected anteriorly (Fig. 10.8). The superior rectus muscle, as noted previously, has fascial attachments to the overlying levator muscle. Careful dissection of the posterior one-third of the superior rectus and levator muscles exposes the superior division of the oculomotor nerve as it enters their conal surfaces. The branch of the superior division of the oculomotor nerve which supplies the levator muscle extends through or around the superior rectus prior to innervating the levator on its inferior surface. Like all the rectus muscles the superior rectus is innervated by its nerve on the conal surface in the posterior one-third of the muscle belly. As the superior rectus approaches the globe, intermuscular fibrous tissue becomes more prominent. This tissue unites the superior rectus with the medial and lateral rectus. The intermuscular fibrous septa helps form the muscle cone but is not well-developed, posteriorly.

The superior oblique muscle traverses the superior medial portion of the orbit parallel to the medial orbital wall and can be followed to the trochlea anteriorly, where it becomes a tendon. Its tendon passes through the trochlea and then angles posterior and laterally to its insertion under the superior rectus muscle. The superior rectus should be cut near its insertion and reflected posteriorly to better visualize the superior oblique and its insertion. The trochlear nerve (CN IV) which innervates the superior oblique enters the muscle in its posterior one-third, but on its superior medial surface. This is the only extraocular muscle which is innervated on its extraconal surface.

Incising the superior oblique at the trochlea and retracting the levator, superior rectus, and superior oblique muscles posteriorly permits visualization of deeper orbital structures. A greater appreciation of the orbital nerves and vessels is effected by exposure of the superior orbital fissure. Removal of the lesser wing of sphenoid "unroofs" the fissure to the optic canal. After careful exposure of the superior orbital fissure, dissection may proceed posteriorly into the cavernous sinus.

Careful, piecemeal removal of the orbital fat permits visualization of the orbital structures (Fig. 10.9). As the orbital apex is cleared of fat, three structures are seen to cross over the optic nerve in a lateral to medial orientation from the orbital apex; ophthalmic artery (only rarely does this cross beneath the nerve), nasociliary nerve, and the superior ophthalmic vein. The very large vascular structure, sometimes a network of vessels in the superior orbital fat, is the superior ophthalmic vein. After it is identified it may need to be transected to better visualize deeper structures. The ophthalmic artery originates at the bifurcation of the internal carotid artery as it exits the cavernous sinus. The ophthalmic artery enters the orbit through the optic canal, as does the optic nerve and the orbital sympathetics. The ophthalmic artery enters the orbit on the temporal aspect of the optic nerve and crosses the nerve superiorly (in the majority of cases) in close association with the nasociliary nerve.

The central retinal artery is, generally, the first major branch of the ophthalmic. The central retinal artery extends anteriorly within the dural sheath of the optic nerve and penetrates the nerve, usually on its inferior nasal border, approximately 8–10 mm from the globe. In addition, the globe is supplied by the posterior ciliary arteries which divide into multiple branches and form the long and short posterior ciliary arteries (Fig. 10.10). The lacrimal artery, a significant branch of the ophthalmic, extends laterally to accompany the corresponding nerve to the orbital lobe of the lacrimal gland. Muscular branches of the ophthalmic artery extend as a lateral (superior) branch and a medial (inferior) muscle branch. Each rectus muscle has two anterior ciliary arteries, except the lateral, which has only one. The ophthalmic artery then gives off the posterior and anterior ethmoidal arteries and the supraorbital artery which supplies the forehead. Terminal branches of the ophthalmic artery are the supratrochlear and the dorsal-nasal artery. The dorsal-nasal artery extends inferiorly to anastomose with the angular artery which is the terminal branch of the facial artery.

Identification of the orbital nerves is greatly facilitated by tracing them from their origin to termination. Dissection of the superior orbital fissure exposes the lacrimal, frontal,

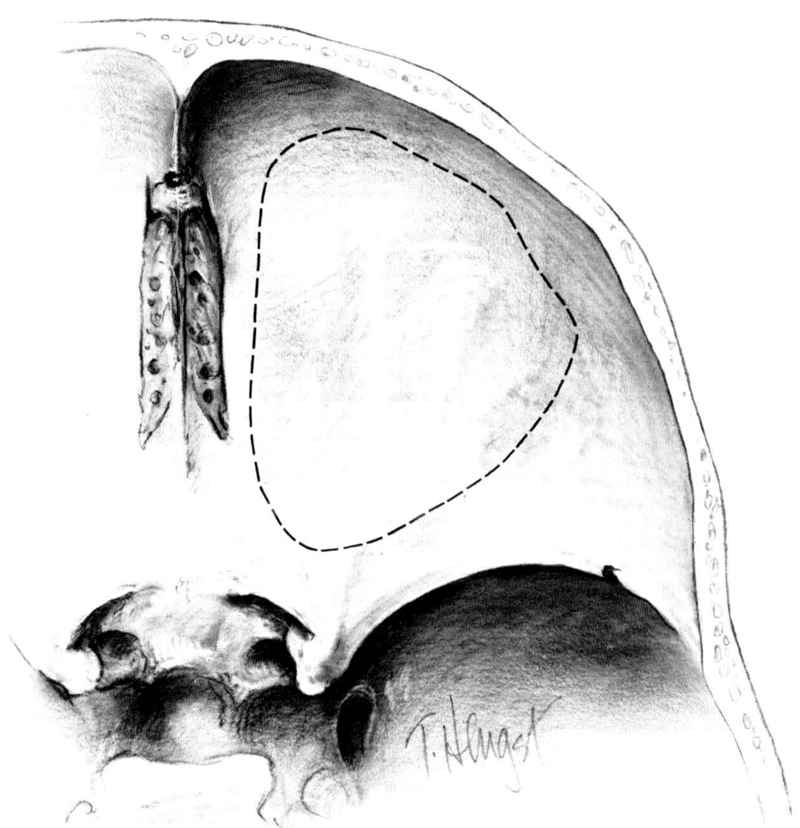

Figure 10.3. Dissection right orbit—superior approach. Bone removal to expose superior orbital structures.

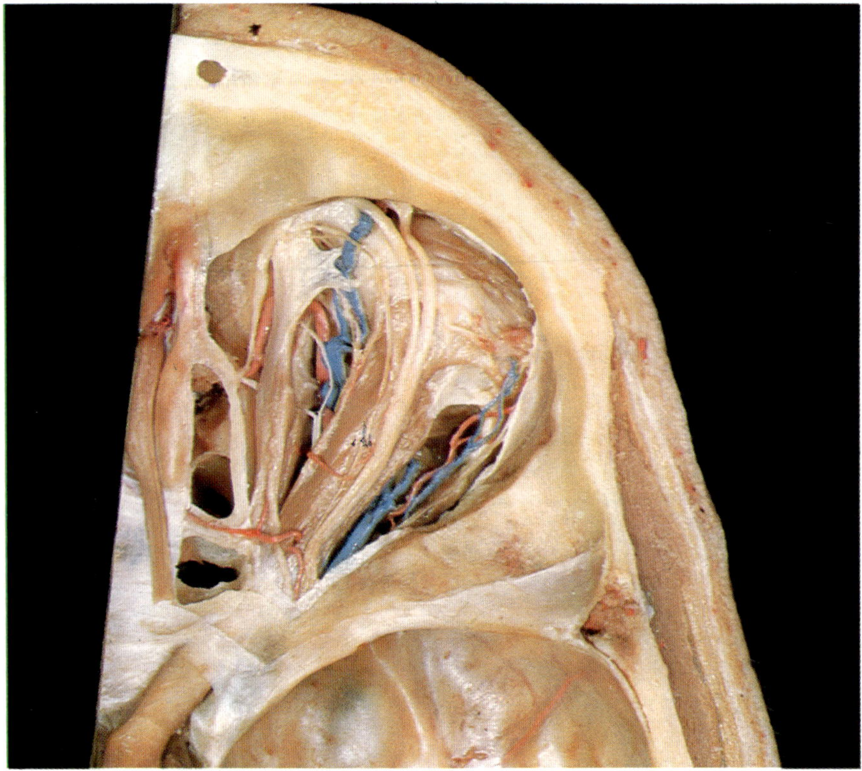

Figure 10.4

(From David L. Bassett, M.D., *Sterescopic Atlas of Human Anatomy*, published by Sawyer's Inc., Portland, Oregon © 1954, color photographs by Wm. B. Gruber.)

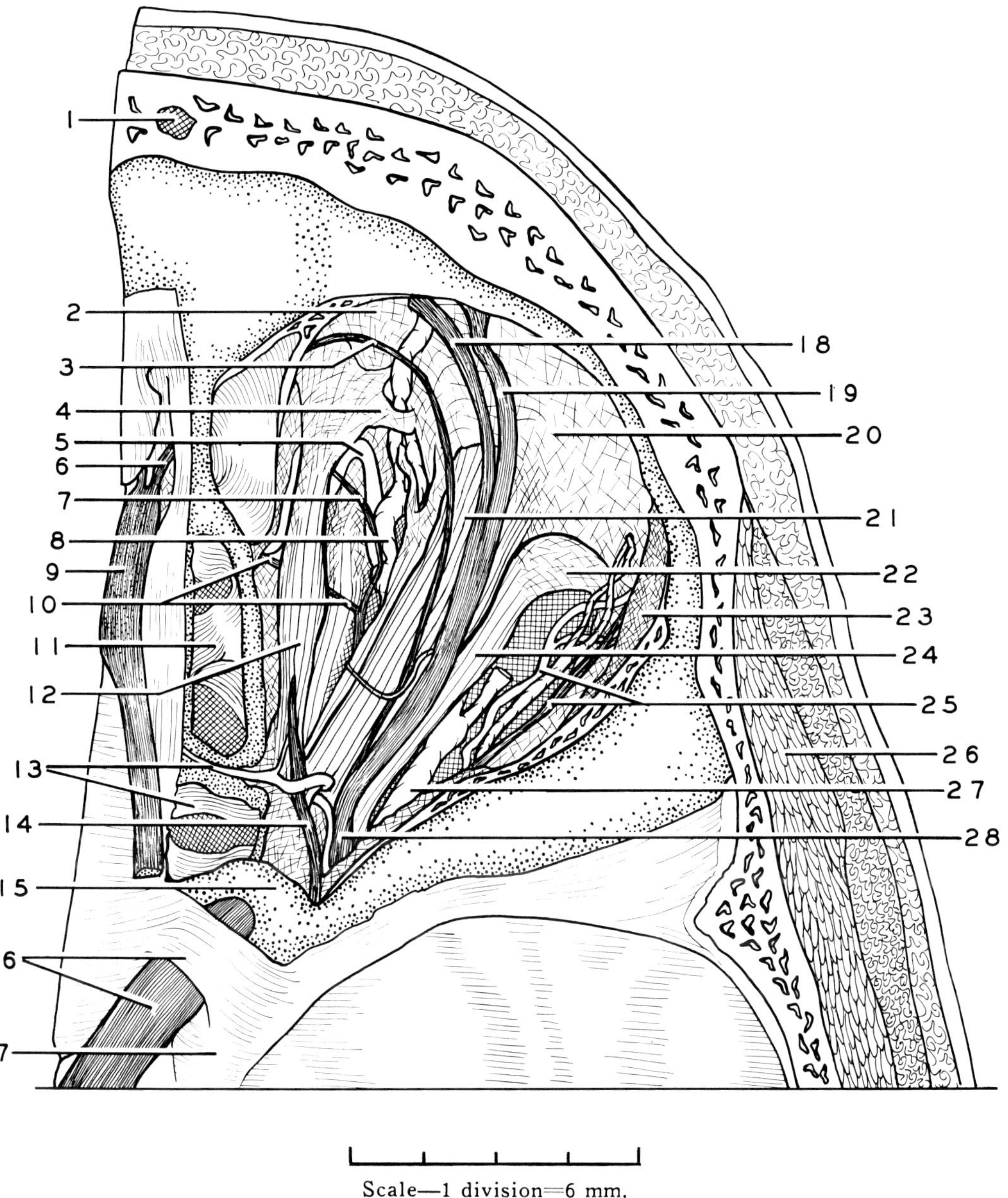

Scale—1 division=6 mm.

DISSECTION RIGHT ORBIT—SUPERIOR APPROACH

1. Frontal sinus (opened)
2. Position of trochlea: Superior oblique muscle
3. Branch of n. frontalis which communicates with n. infratrochlearis
4. Superior oblique muscle (fascia)
5. Ophthalmic a.
6. Olfactory n.
7. Infratrochlear n.
8. Superior ophthalmic v.
9. Olfactory bulb
10. Anterior ethmoidal n.
11. Ethmoidal air cells (middle)
12. Superior oblique m.
13. Upper pointer: Posterior ethmoidal a.
    Lower pointer: Ethmoidal air cells (posterior)
14. Trochlear n. (IV)
15. Roof of optic canal (upper root of lesser wing of sphenoid bone)
16. Dura mater of the optic n. (II)
17. Anterior clinoid process
18. Supratrochlear n.
19. Frontal n. (supraorbital and frontal branches not separated)
20. Fascia of m. levator palpebrae superioris
21. Levator palpebrae superioris m.
22. Fascia of superior rectus m.
23. Periorbita
24. Superior rectus m.
25. Lacrimal a., v., and n.
26. Temporalis m.
27. Superior ophthalmic v.
28. Frontal n.

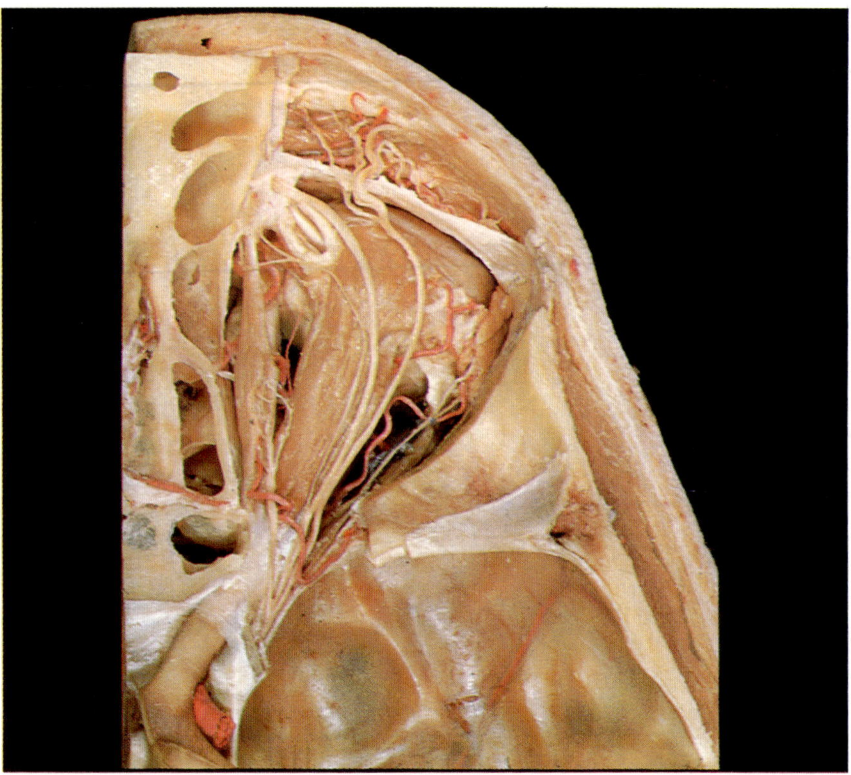

Figure 10.5

(From David L. Bassett, M.D., *Sterescopic Atlas of Human Anatomy*, published by Sawyer's Inc., Portland, Oregon © 1954, color photographs by Wm. B. Gruber.)

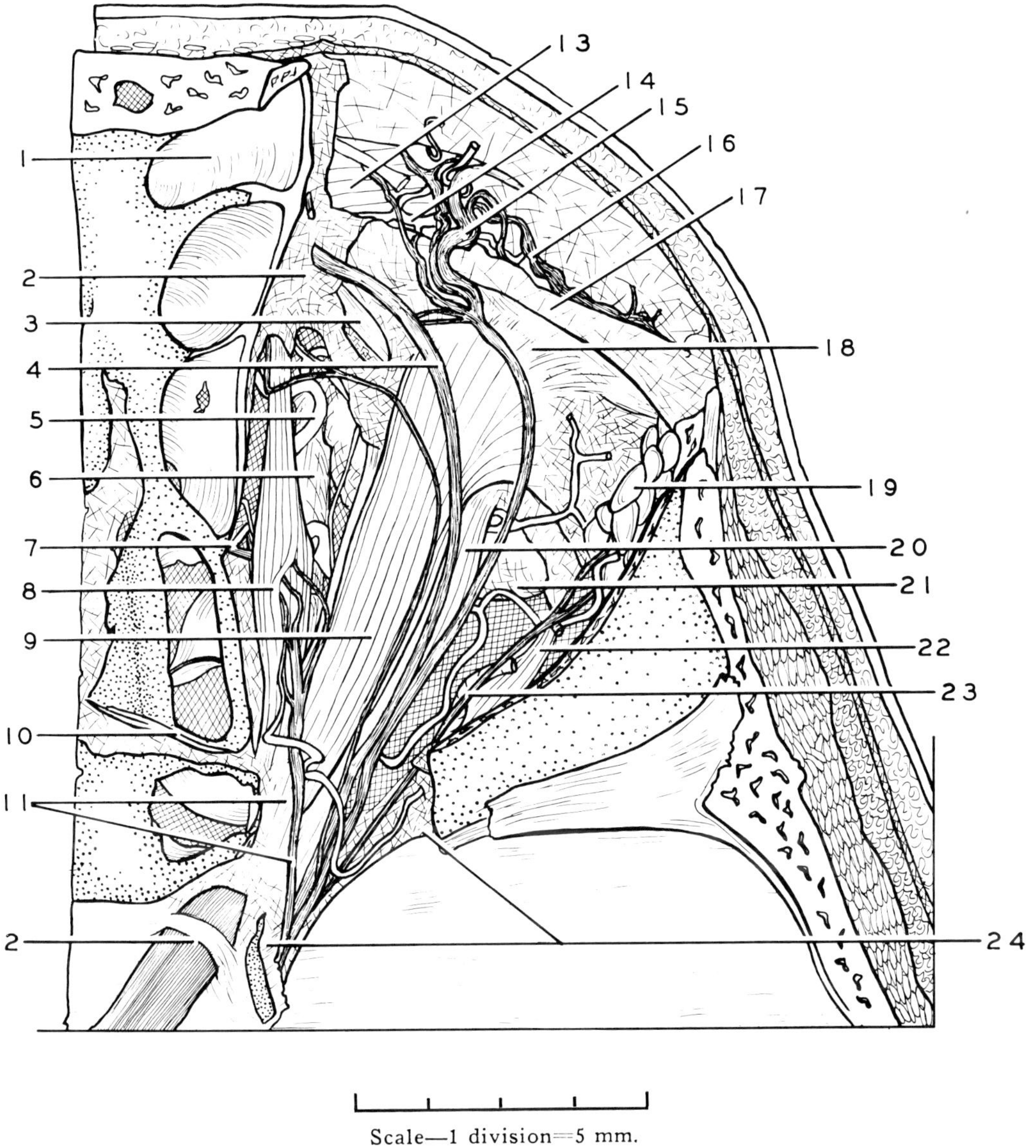

Scale—1 division=5 mm.

DISSECTION RIGHT ORBIT—SUPERIOR APPROACH

1. Frontal sinus
2. Trochlea: Superior oblique muscle
3. Reflected part of tendon of superior oblique m. (the thick fascia which surrounds the tendon blends with fascia of the levator and superior rectus muscles as well as with the fascia bulbi)
4. Supratrochlear n.
5. Ophthalmic a.
6. Fascia covering medial rectus m.
7. Anterior ethmoidal a. and n. (entering the anterior ethmoidal foramen)
8. Superior oblique m.
9. Levator palpebrae superioris m.
10. Posterior ethmoidal a. (posterior ethmoidal nerve absent)
11. Upper pointer: Annulus of Zinn
    Lower pointer: Trochlear n. (IV)
12. Optic n. (II)
13. Corrugator superciliaris m.
14. Supraorbital a. (supraorbital branch of ophthalmic artery absent in this specimen)
15. Supraorbital n.
16. Communicating branch between n. supraorbitalis and n. facialis
17. Periorbita at supraorbital margin
18. Aponeurosis of levator palpebrae superioris m.
19. Superior lacrimal gland
20. Superior rectus m.
21. Bulbus oculi
22. Lateral rectus m.
23. Lacrimal n.
24. Superior orbital fissure (opened)

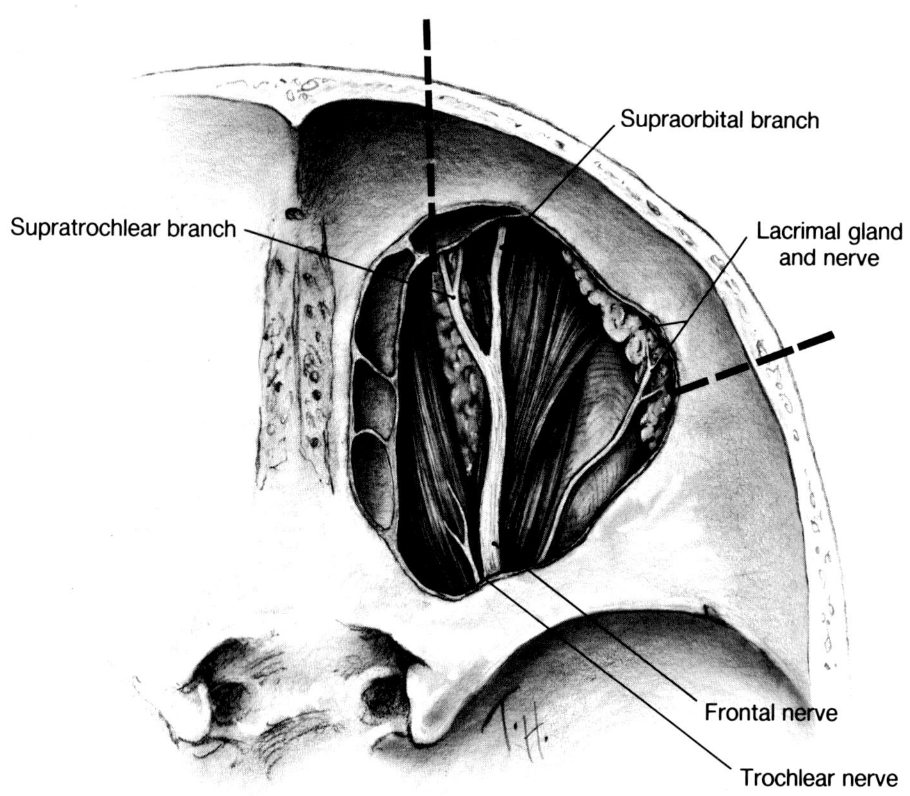

Figure 10.6. Dissection superior orbit. Removal of superior orbital margin.

trochlear, superior, and inferior divisions of the oculomotor, nasociliary, and abducens nerves.

As the optic nerve is exposed, the nasociliary nerve is seen to cross the optic nerve in conjunction with the ophthalmic artery near the orbital apex. Near the orbital apex, a fine sensory branch of the nasociliary nerve extends to the ciliary ganglion, which lies lateral to the optic nerve. The nasociliary nerve then proceeds anterior medially in the orbit and branches into the anterior ethmoid and infratrochlear nerves.

The ciliary ganglion is approximately 2 mm long and is identified on the lateral aspect of the optic nerve approximately 10 mm from the orbital apex (Fig. 10.11). One can identify the ganglion by tracing the nerve to the inferior oblique, which runs along the lateral margin of the inferior rectus, posteriorly. This disproportionately large branch of the inferior division of the oculomotor nerve gives off a small branch superiorly to the ciliary ganglion. This small branch provides the parasympathetic supply of this ganglion. The ciliary ganglion gives off 8–20 short ciliary nerves which surround the optic nerve at its entrance into posterior sclera.

EYELIDS AND ANTERIOR ORBIT

The skin of the orbital region is incised with a scalpel as illustrated in Fig. 10.12. Two skin flaps are thus created, which meet in the horizontal meridian that is formed by the palpebral aperture. These skin flaps should be reflected from their lateral portion. The underlying orbicularis oculi muscle is adherent to the skin and great care should be taken to sharply dissect the subcutaneous attachments of the skin from the underlying orbicularis. A scalpel is used to dissect the attachments between orbicularis and skin which have been placed on tension by a forceps. As the thick skin of the eyebrow is reflected, the firm attachments of the corrugator superciliaris, procerus, and frontalis muscles to the subcutaneous tissues are appreciated.

Once the skin has been reflected, the orbicularis muscle is readily visible. The orbital and palpebral (preseptal and pretarsal) sections of the orbicularis muscle, as well as the procerus and corrugator superciliaris muscles, in the medial portion of the brow are identified (Fig. 10.13). The superficial extension of the sensory nerves, which are accompanied by arteries and veins, should be noted. These include the supratrochlear and infratrochlear nerves and vessels in the superior nasal aspect of the orbit, and the supraorbital and infraorbital nerves and vessels near the junction of the medial one-third and temporal two-thirds of the orbit. The infraorbital nerve and vessels exit a foramen inferior to the rim while the supraorbital vessels exit at the superior rim via either a groove or, less commonly, a foramen.

The orbicularis muscle is continuous but by definition divided into its orbital, preseptal, and pretarsal portions by the structures which it overlies. After studying these portions the orbicularis muscle is disinserted laterally and reflected toward its medial insertion. As the medial canthal region is approached, the superficial and deep insertions of the orbicularis should be studied (Fig. 10.14). The medial canthal tendon, a white glistening structure, is just medial to the palpebral aperture and represents the superficial

insertion of the orbicularis muscle. The preseptal orbicularis has a deep posterior insertion (Jones' muscle) onto the fascia overlying the lacrimal fossa (lacrimal diaphragm). Jones described this muscle and postulated its importance in the lacrimal pump. The pretarsal orbicularis muscle also has a deep insertion (the tensor tarsi muscle of Horner). This deep insertion is anchored in the medial canthus and functions by moving the canalicular portion of the eyelid medially with each blink, thus shortening the canaliculus. The angular artery and vein lie in the medial canthal region, along the side of the nose.

Reflection of the orbicularis muscle exposes the underlying orbital septum, which may be thinned or rarified. In the upper eyelid, the orbital septum fuses with the levator aponeurosis just superior to the upper border of the tarsus. In the lower eyelid, the orbital septum and the lower eyelid aponeurosis fuses near the inferior border of the tarsus. It may be difficult to appreciate all the clinically delineated fat pockets which have been described due to the fixation and retraction of orbital tissues. However, an attempt should be made to identify those fat pockets which are present. These fat pockets are more easily seen after the orbital septum is opened.

The orbital septum should be incised superior to its fusion with the levator aponeurosis and reflected to its origin at superior orbital margin (arcus marginalis). The underlying preaponeurotic fat is identified and removed with forceps, thereby exposing the levator aponeurosis. The transition of the levator muscle to aponeurosis begins at the transverse ligament of Whitnall. In some specimens there is an abrupt transition at Whitnall's ligament and in others it is more gradual and not complete until approximately 10 mm above the superior border of tarsus. The diffuse inferior insertion of the aponeurosis should be appreciated. The anterior fibers of the levator aponeurosis terminate in the subcutaneous tissues of the eyelid at and below the lid crease while the posterior fibers are more firmly attached to the anterior surface of the tarsus. Disinsertion of the levator aponeurosis from its tarsal insertion, and reflection superiorly exposes the superior tarsal muscle (Müller's muscle) which inserts on the superior margin of the tarsus (Fig. 10.15). The smooth muscle fibers comprising this muscle and the peripheral arcade of vessels in Müller's muscle immediately superior to the upper border of tarsus can be seen in most specimens. Blunt dissection between Müller's muscle and the conjunctiva in the superior fornix exposes fine sheath-like fibers densely adherent to the conjunctiva in the superior fornix (the suspensory ligament of the fornix). These fibers extend from the levator muscle and Müller's muscle to the fornix.

The superior transverse ligament of Whitnall lies inferior to the superior orbital margin. The distinct, whitish transverse band is formed from the superior fascia of the levator muscle and suspends the levator near the orbital rim. Whitnall's ligament attaches medially near the trochlea. It extends laterally through the lacrimal gland to help suspend the lacrimal gland, and inserts primarily on the superior lateral orbital wall. A small branch passes inferiorly to the lateral retinaculum.

The orbital septum in the lower lid is incised inferior to its fusion with the retractor aponeurosis near the lower edge

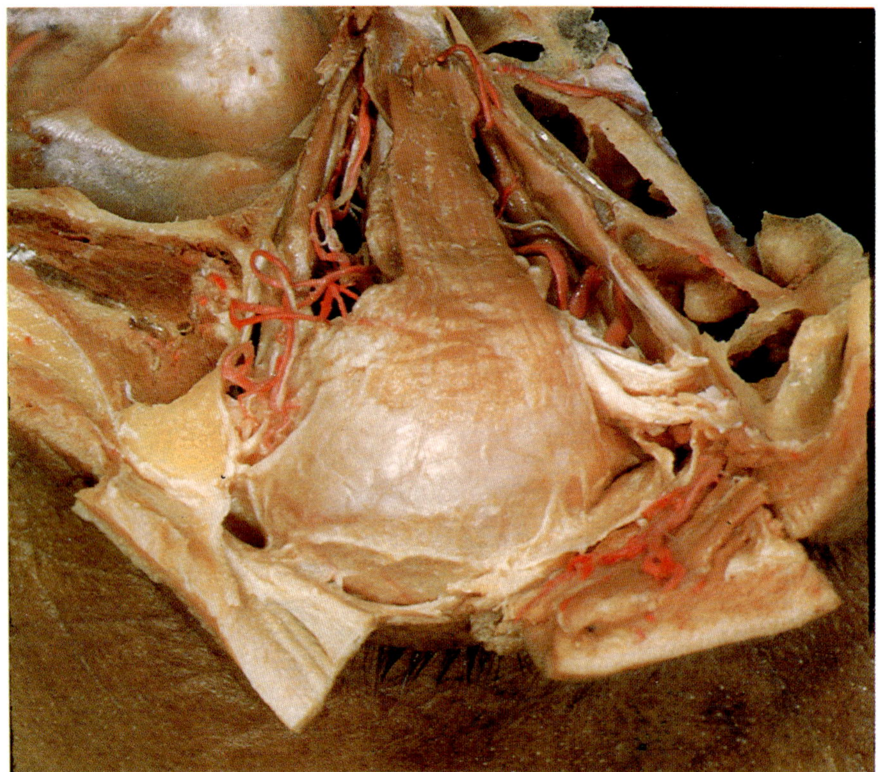

Figure 10.7

(From David L. Bassett, M.D., *Sterescopic Atlas of Human Anatomy*, published by Sawyer's Inc., Portland, Oregon © 1954, color photographs by Wm. B. Gruber.)

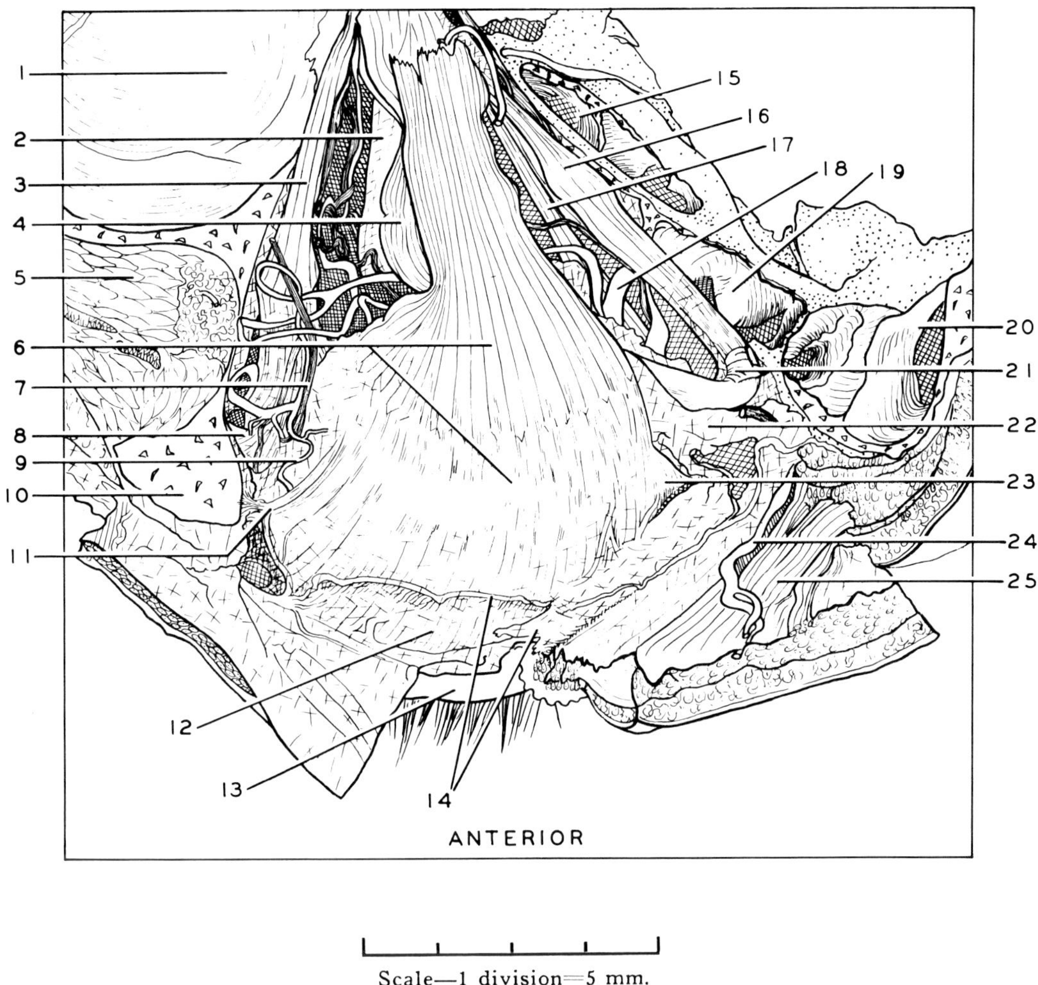

Scale—1 division=5 mm.

DISSECTION RIGHT ORBIT—SUPERIOR APPROACH

Insertion of levator palpebrae superioris muscle

1. Middle cranial fossa
2. Optic n. (II)
3. Lateral rectus m.
4. Superior rectus m.
5. Temporalis m. (within fossa temporalis)
6. Levator palpebrae superioris m. and its aponeurosis
7. Lacrimal n.
8. Lateral check ligament
9. Area for lacrimal gland
10. Zygomatic process of the frontal bone
11. Lateral horn of aponeurosis of levator palpebrae superioris m.
12. Aponeurosis of levator palpebrae superioris m. extending to anterior surface of superior tarsus
13. Upper eyelid
14. Cutaneous insertion of fibers of aponeurosis of levator palpebrae superioris m. (partially cut across)
15. Ethmoidal air cell (middle)
16. Superior oblique m.
17. Medial rectus m.
18. Ophthalmic a.
19. Ethmoidal air cell (anterior)
20. Frontal sinus
21. Trochlea: Superior oblique tendon
22. Fascia surrounding reflected superior oblique tendon
23. Medial horn of levator aponeurosis
24. Frontal a.
25. Corrugator superciliaris m.

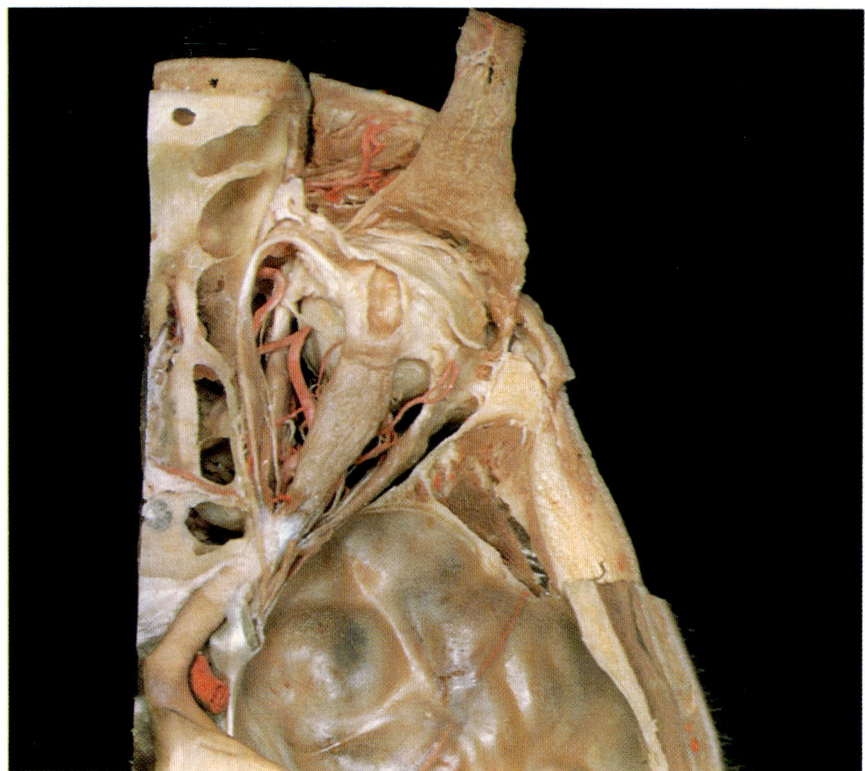

Figure 10.8

(From David L. Bassett, M.D., *Sterescopic Atlas of Human Anatomy*, published by Sawyer's Inc., Portland, Oregon © 1954, color photographs by Wm. B. Gruber.)

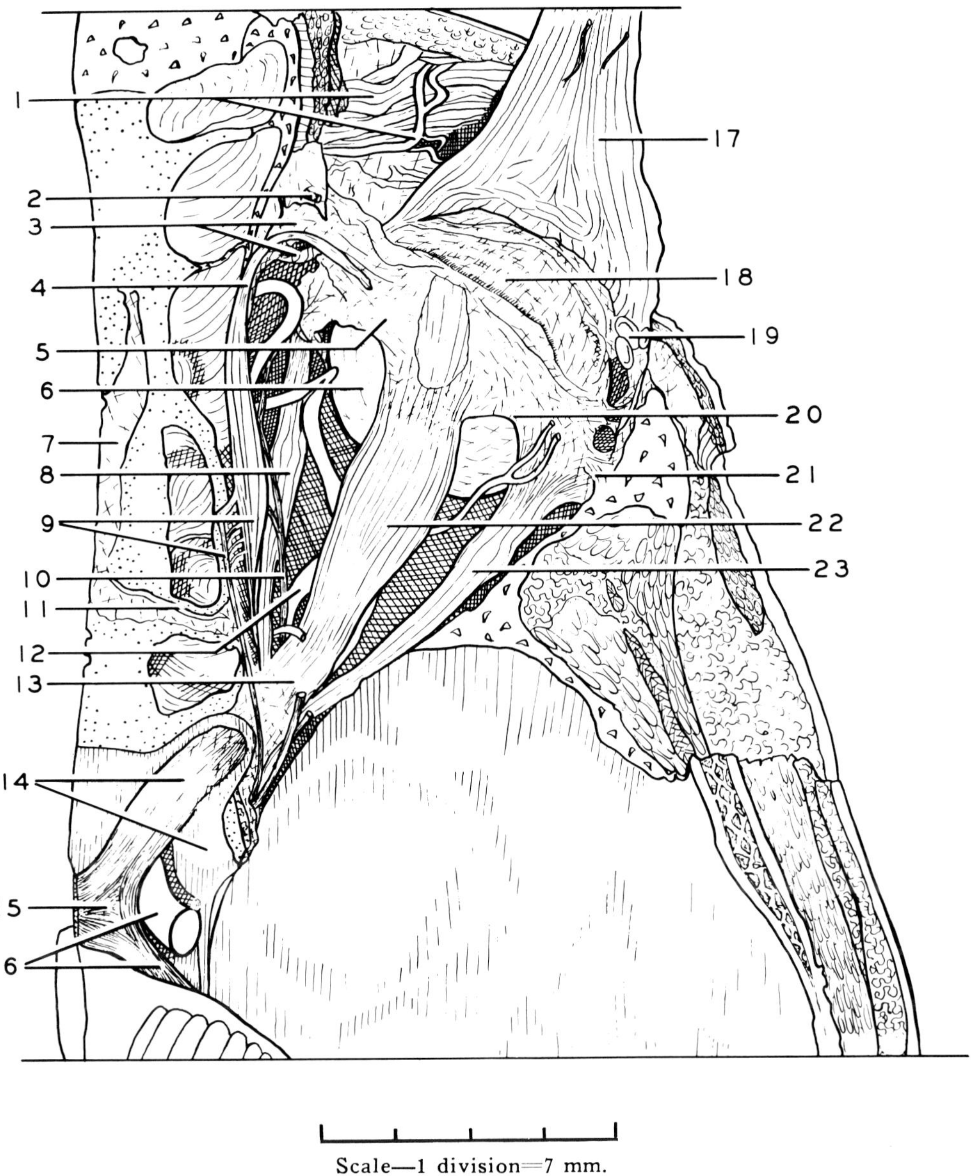

Scale—1 division=7 mm.

DISSECTION RIGHT ORBIT—SUPERIOR APPROACH

Superior rectus muscle and fascial attachments

1. Corrugator superciliaris m. and frontal a.
2. Supratrochlear n. (cut off)
3. Trochlea: Superior oblique tendon
4. Superior oblique tendon
5. Medial expansion of fascia of superior rectus m.
6. Tenon's capsule
7. Lamina cribrosa
8. Medial rectus m.
9. Superior oblique m.; Trochlear n.
10. Nasociliary n.
11. Posterior ethmoidal a.
12. Ophthalmic a.
13. Annulus of Zinn
14. Optic n. (II) and anterior clinoid process
15. Optic chiasm
16. Internal carotid a. and optic n.
17. Levator palpebrae superioris m. (reflected)
18. Fascia beneath levator muscle (17)
19. Lacrimal gland (inferior)
20. Lateral expansion of fascia of superior rectus m.
21. Lateral check ligament
22. Superior rectus m.
23. Lateral rectus m.

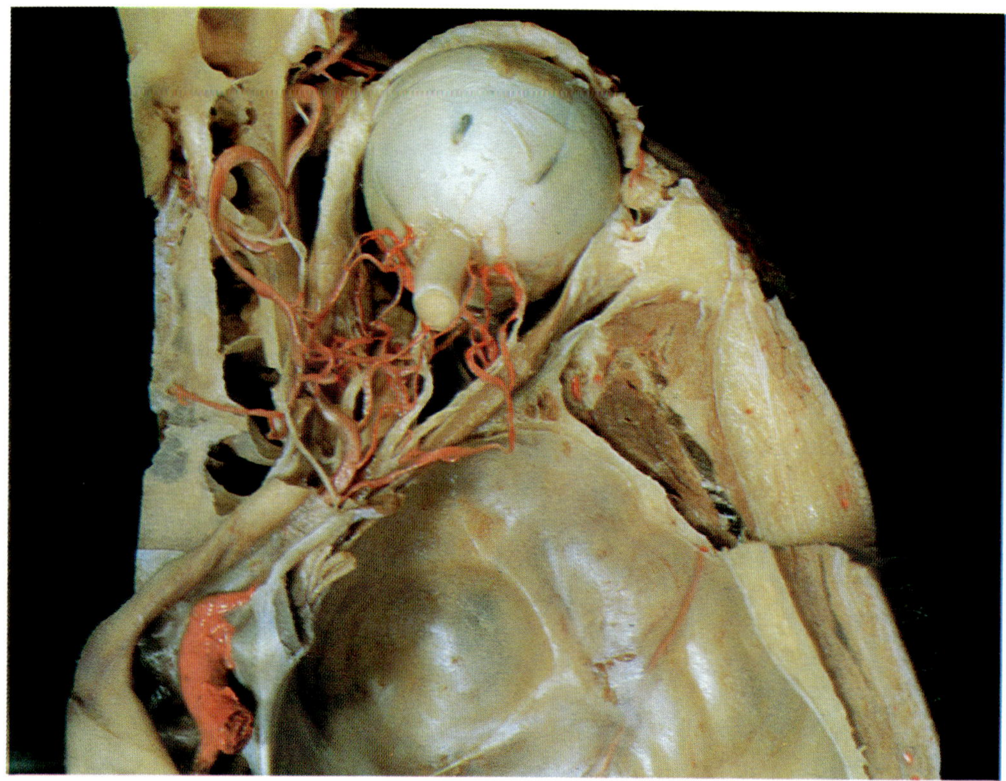

Figure 10.10

(From David L. Bassett, M.D., *Stereoscopic Atlas of Human Anatomy*, published by Sawyer's Inc., Portland, Oregon © 1954, color photographs by Wm. B. Gruber.)

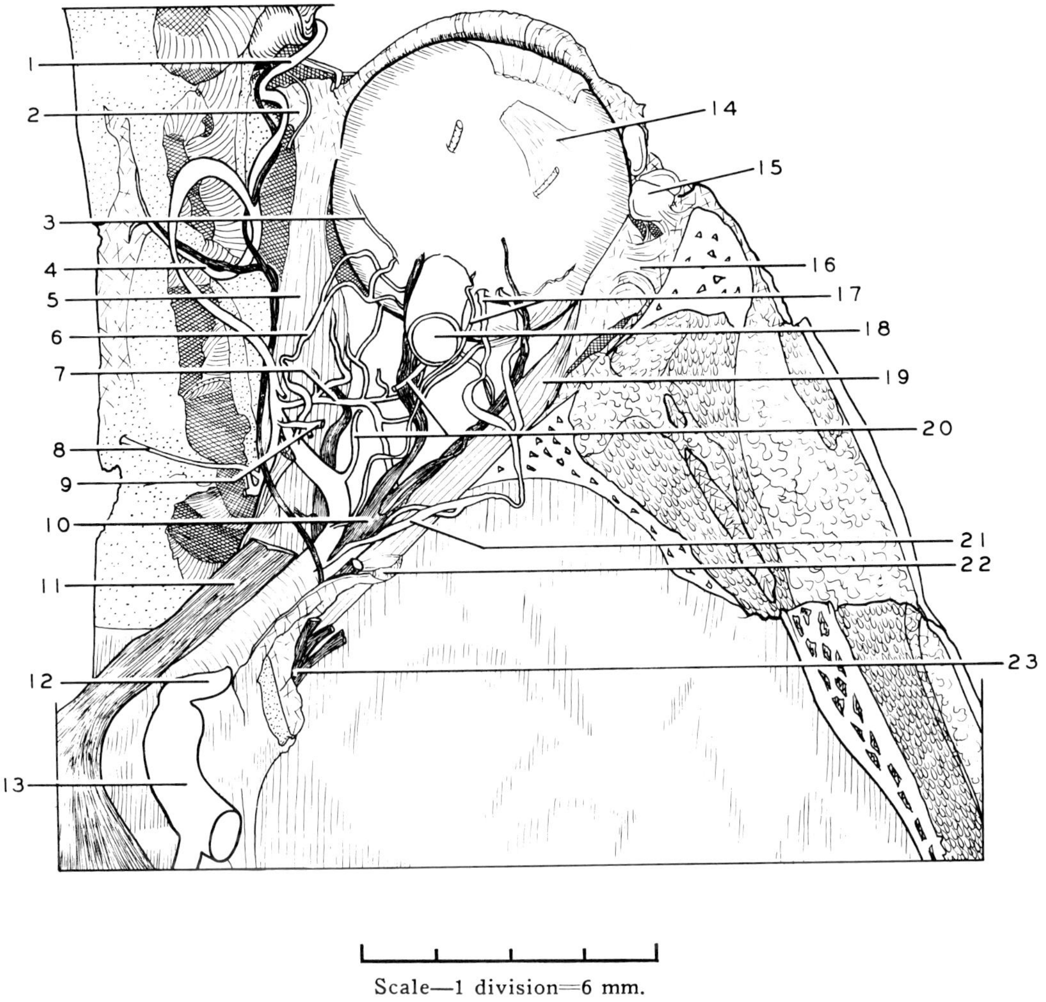

DISSECTION RIGHT ORBIT—SUPERIOR APPROACH

Branches of the ophthalmic artery

1. Frontal a.
2. Medial check ligament
3. Anterior ciliary a. (within sclera)
4. Anterior ethmoidal a.
5. Medial rectus m.
6. Posterior ciliary a.
7. Inferior muscular branch of ophthalmic a. (branches pass to medial, inferior, and lateral rectus muscles and inferior oblique muscle)
8. Posterior ethmoidal a.
9. Lacrimal a. (cut off)
10. Ciliary ganglion
11. Optic n. (II) (elevated)
12. Ophthalmic a. entering optic canal
13. Internal carotid a.
14. Insertion of superior oblique m.
15. Lacrimal gland
16. Lateral check ligament
17. Posterior ciliary aa.
18. Optic n. within sheath
19. Lateral rectus m.
20. Central retinal a. (continuity of artery interrupted in drawing but not in specimen)
21. Posterior ciliary a.
22. Origin of superior rectus m. (cut off)
23. Nerves entering orbit through superior orbital fissure

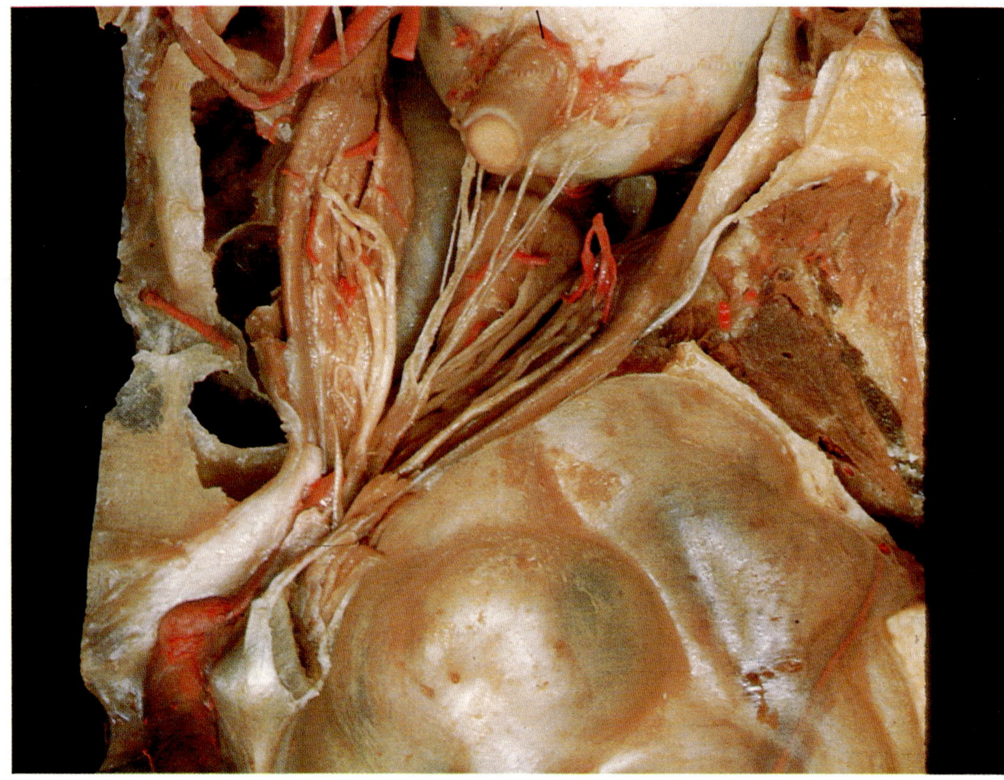

Figure 10.11

(From David L. Bassett, M.D., *Sterescopic Atlas of Human Anatomy*, published by Sawyer's Inc., Portland, Oregon © 1954, color photographs by Wm. B. Gruber.)

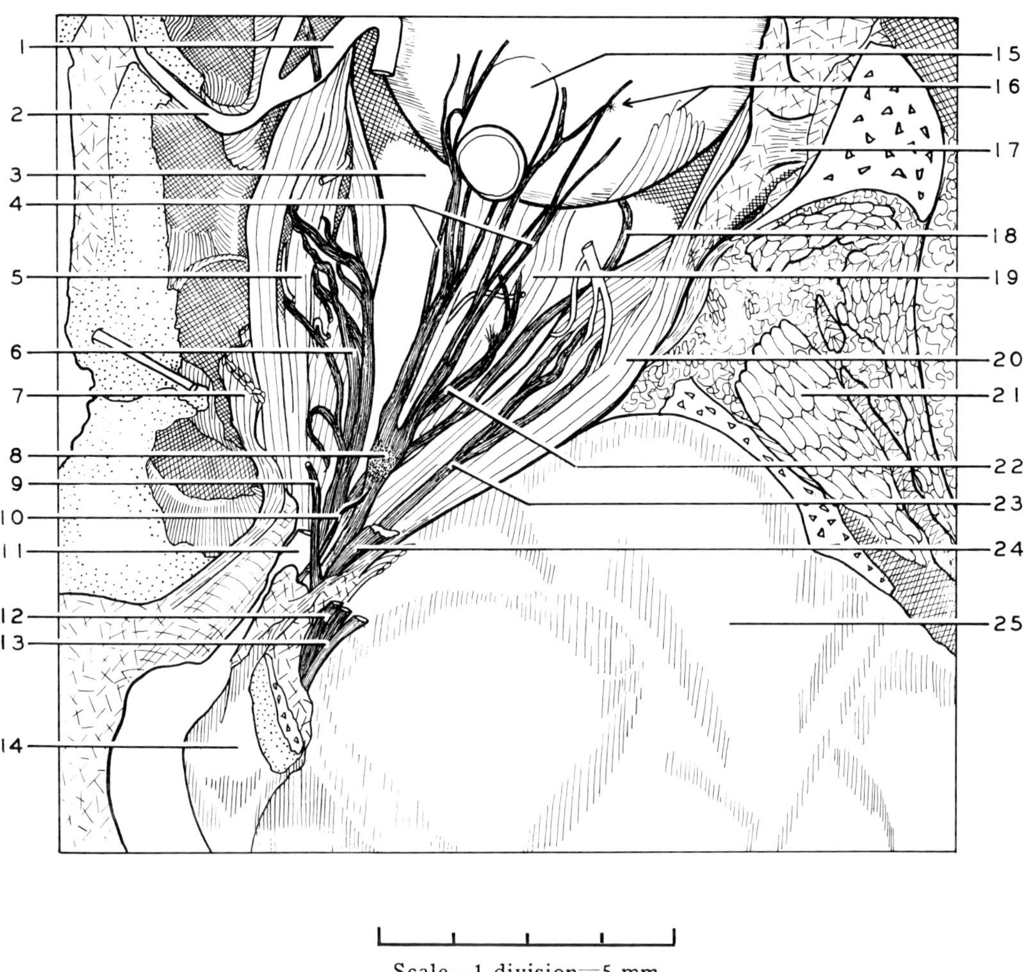

Scale—1 division=5 mm.

DISSECTION RIGHT ORBIT—SUPERIOR APPROACH

Ciliary ganglion

1. Ophthalmic a.
2. Anterior ethmoidal a.
3. Orbital floor
4. Posterior ciliary n.
5. Medial rectus m.
6. Branches of oculomotor n. to medial rectus m.
7. Superior oblique m. (cut off)
8. Ciliary ganglion
9. Nasociliary n. (cut off)
10. Filament from sympathetic plexus to ciliary ganglion
11. Ophthalmic a.
12. Trochlear n. (IV) (cut off)
13. Ophthalmic n. (cut off)
14. Anterior clinoid process
15. Optic n.
16. Upper pointer: Position of macula lutea in interior of eyeball (marked by *)
    Lower pointer: Insertion of inferior oblique m.
17. Lateral check ligament
18. Branch of oculomotor n. (III) to inferior oblique m.
19. Inferior rectus m.
20. Lateral rectus m.
21. Temporalis m.
22. Branch of oculomotor n. (III) to inferior rectus m.
23. Abducens n. (VI)
24. Branch of oculomotor n. (III) to superior rectus m. (cut off)
25. Middle cranial fossa

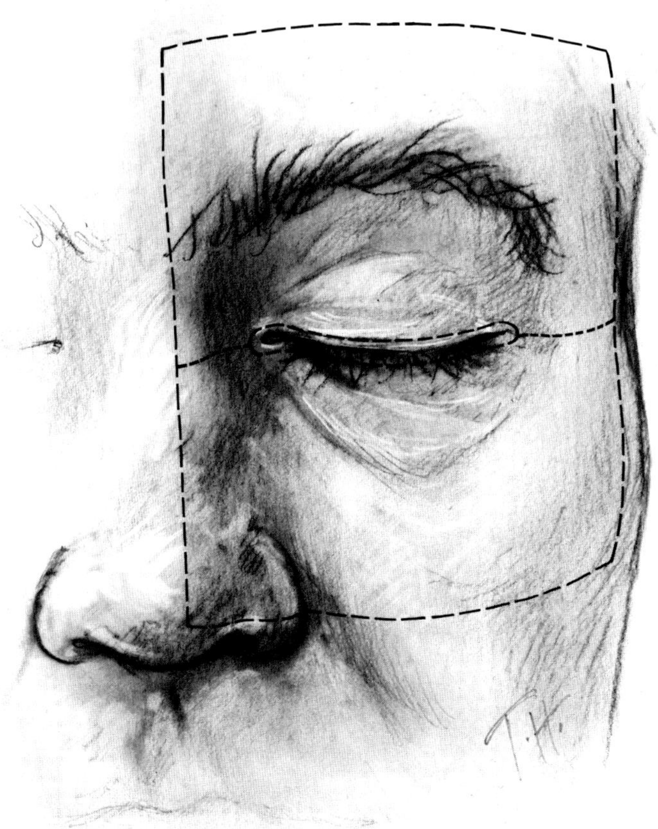

Figure 10.12.　Preparation of skin flaps for anterior orbital dissection.

of tarsus and reflected to its origin at the arcus marginalis. Reflection of the orbital septum and careful removal of preaponeurotic fat reveal the lower lid aponeurosis. The lower eyelid does not have a specific lid retractor like the levator muscle of the upper eyelid. The lower lid retractor is the aponeurotic extension of the inferior rectus muscle, the capsulopalpebral fascia. The capsulopalpebral fascia splits around the inferior oblique muscle and then extends anteriorly to insert on the inferior portion of the tarsus in the lower eyelid. As the lower lid aponeurosis is followed inferiorly a transverse fascial band, the suspensory ligament of Lockwood, is noted (Fig. 10.16). Lockwood's suspensory ligament, which extends from the medial to the lateral retinaculum to support the globe, is formed by the fascial extensions of various structures. These include, principally, the aponeurosis of the lower eyelid or capsulopalpebral fascia, Tenon's capsule, intermuscular fibrous septa, and the fascia surrounding the inferior oblique and inferior rectus muscles. The distinct contribution of each of these structures is rather difficult to isolate, but together they form an important anatomic and structural unit (Fig. 10.17). The inferior oblique muscle can be exposed at the level of Lockwood's ligament and traced to its point of origin in the inferior nasal corner of the orbit, directly posterior to the orbital margin (Fig. 10.18). Immediately posterior to the lower lid aponeurosis and adherent to both the aponeurosis and conjunctiva is the sympathetic muscle of the lower eyelid. It is less well-defined than Müller's muscle in the upper eyelid.

The largest blood vessels supplying the eyelids are the medial and lateral palpebral vessels. The medial and lateral palpebral vessels branch to form a marginal and a more peripheral vessel. The marginal vessels anastomose with each other to form the marginal arcade and the peripheral vessels anastomose to form the peripheral arcade (Fig. 10.19).

THE ORBIT FROM THE LATERAL ASPECT

The periosteum of the lateral orbital rim is incised and elevated with a periosteal elevator and the lateral orbital (Whitnall's) tubercle is identified. The thickened periosteum overlying the lateral orbital tubercle helps form the insertion for the lateral retinaculum. This is made up of portions of the lateral canthal tendon, the lateral horn of the levator aponeurosis, a lateral expanse of the retractor aponeurosis, an inferior slip of Whitnall's ligament, Lockwood's ligament, and check ligaments of the lateral rectus muscle.

The zygomaticofacial branch of the zygomatic nerve is identified in the zygomaticofacial foramen at the inferior temporal corner of the orbit. This nerve exits its foramen to supply sensation to cheek skin. The zygomaticofacial nerve extends anteriorly from the orbital floor in a groove in the zygomatic bone. It then passes through its foramen in the zygomatic bone, and enters the temporalis fossa. Prior to departing the orbit, approximately 15 mm posterior to the orbital rim, the zygomaticotemporal nerve sends a branch superiorly in the periorbita of the lateral orbital wall to provide secretomotor parasympathetic fibers to the lacrimal gland (Fig. 10.20). These fibers communicate with the lacrimal nerve (from the ophthalmic division of the trigeminal nerve) prior to its entering the lacrimal gland.

The lateral wall of the orbit is removed by making a saw cut immediately above the zygomatic arch and another as high as possible on the lateral orbital wall. The lateral wall is fractured temporally with heavy rongeurs and the sphenoid portion of the lateral wall (greater wing of the sphenoid) is rongeured away. Periorbita should be incised, at which time the previously noted branch of the zygomaticotemporal nerve to the lacrimal gland may be seen. Continue to remove the sphenoid, frontal, and temporal bone until the superior orbital fissure and cavernous sinus are exposed. Identify the structures in this area (Fig. 10.21).

The lacrimal gland occupies the superior lateral quadrant of the anterior orbit. Retraction of the orbital lobe of the lacrimal gland defines the separation between the orbital and palpebral lobes of the gland provided by the lateral horn of the levator aponeurosis.

Orbital fat is carefully removed to expose the lateral rectus muscle. Anteriorly, the check ligaments of the lateral rectus make a contribution to the lateral retinaculum. The intermuscular fibrous septa contribute to the muscle cone. The lateral rectus is disinserted and reflected posteriorly and the abducens nerve is apparent as it enters the conal surface in the posterior one-third of the muscle (Fig. 10.22).

By reflecting the lateral rectus muscle posteriorly, and removing orbital fat inferiorly, the broad insertion of the inferior rectus muscle is appreciated (Fig. 10.23). The large nerve which supplies the inferior oblique proceeds anteriorly in the orbit on the lateral border of the inferior rectus, and enters the inferior oblique muscle on its posterior inferior surface. The nerve to the inferior oblique delivers the parasympathetic contribution to the ciliary ganglion posteriorly in the orbit. Study the ciliary ganglion, optic nerve, and nerves and vessels of the muscle cone from this approach. The lateral orbitotomy provides surgical access to this area and the importance of this anatomy to the orbital surgeon is obvious.

LACRIMAL SYSTEM

The lacrimal excretory system begins in the medial canthal angle at the termination of the tarsus. The elevated lacrimal papillae with their central puncta are apparent about 6 mm temporal to the medial canthal angle with the lower punctum extending slightly more temporal than the upper punctum. The puncta should be dilated and the canaliculi probed with a lacrimal probe (Fig. 10.24). In cadavers the rigidity of a medium or large lacrimal probe is required to pass in the fixed tissues. The initial 2 mm of the canalicular system is oriented vertically with an ampulla at the base of the vertical portion (Fig. 10.16). The canaliculus then extends horizontally for approximately 6 mm to pass beneath the medial canthal tendon where the two canaliculi join to form a common canaliculus. After probing approximately 12 mm through the canaliculus and common internal punctum, one notes a hard or firm resistance upon encoun-

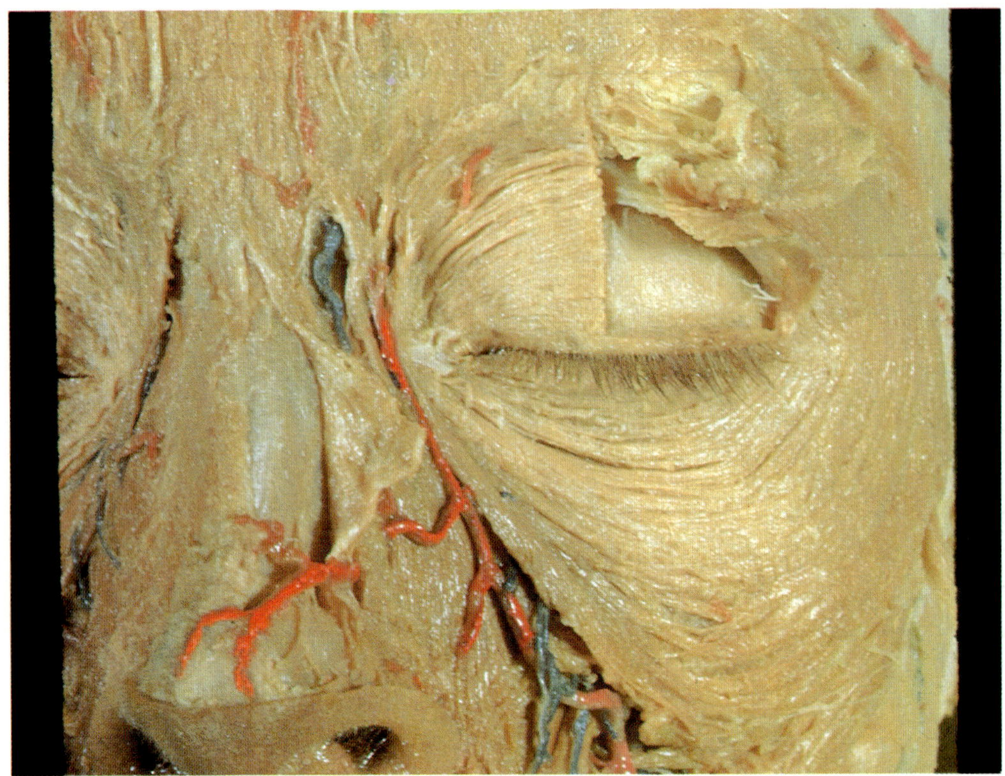

Figure 10.13

(From David L. Bassett, M.D., *Sterescopic Atlas of Human Anatomy*, published by Sawyer's Inc., Portland, Oregon © 1954, color photographs by Wm. B. Gruber.)

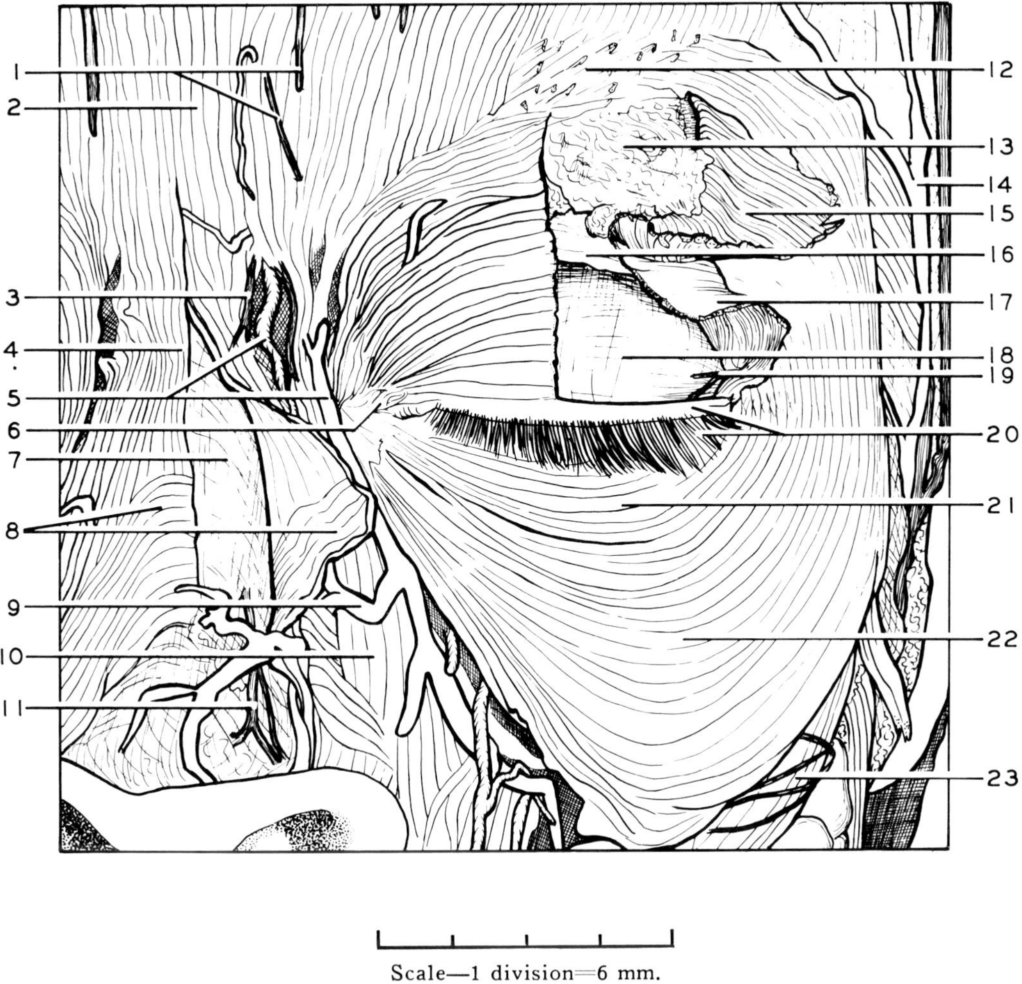

Scale—1 division=6 mm.

ANTERIOR ORBITAL DISSECTION

Orbicularis muscles and brow muscles

1. Ascending branches of infratrochlear n.
2. Frontalis m.
3. Descending branch of infratrochlear n.
4. Procerus m.
5. Angular a. and nasal branch of angular v.
6. Medial canthal tendon
7. Junction of nasal bone with lateral nasal cartilage
8. Transverse nasalis m. (left half reflected)
9. Nasal branch of angular a.
10. Levator anguli oris m.
11. External nasal ramus of anterior ethmoidal n.
12. Corrugator superciliaris m.
13. Layer of fat between orbicularis oculi muscle and orbital septum (not yet exposed)
14. Superficial temporal a.
15. Orbicularis oculi m. (orbital portion) (retracted)
16. Levator aponeurosis
17. Orbicularis oculi m. (palpebral part) (retracted)
18. Superior tarsus
19. Lacrimal n.
20. Margin of skin supporting cilia
21. Orbicularis oculi m. (palpebral part)
22. Orbicularis oculi m. (orbital part)
23. Zygomatic n.

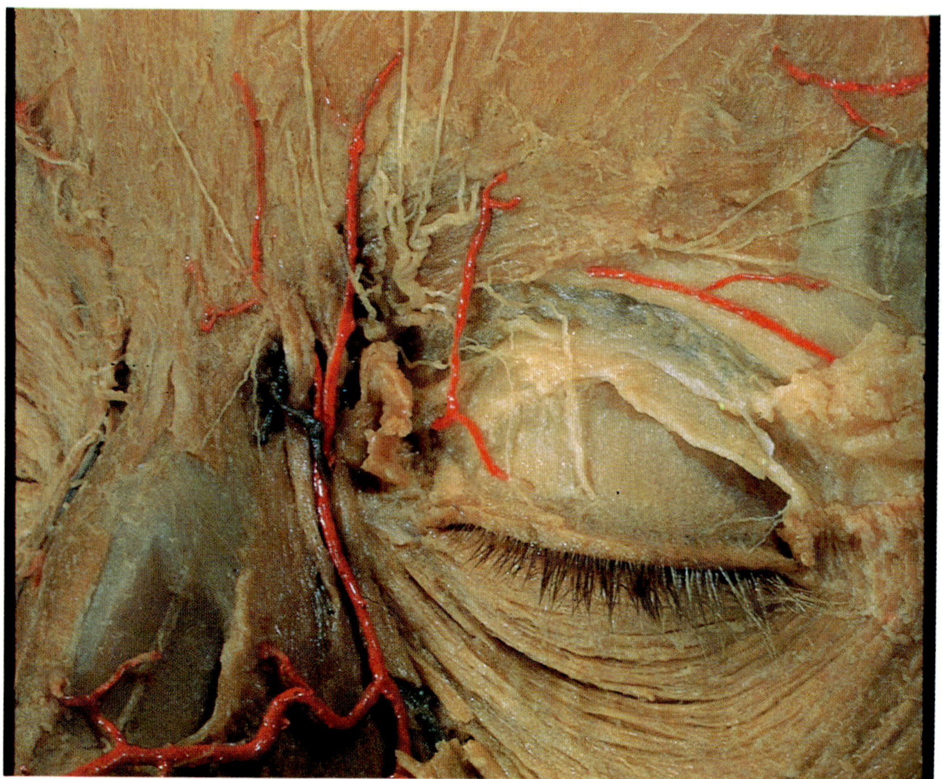

Figure 10.14

(From David L. Bassett, M.D., *Stereoscopic Atlas of Human Anatomy*, published by Sawyer's Inc., Portland, Oregon © 1954, color photographs by Wm. B. Gruber.)

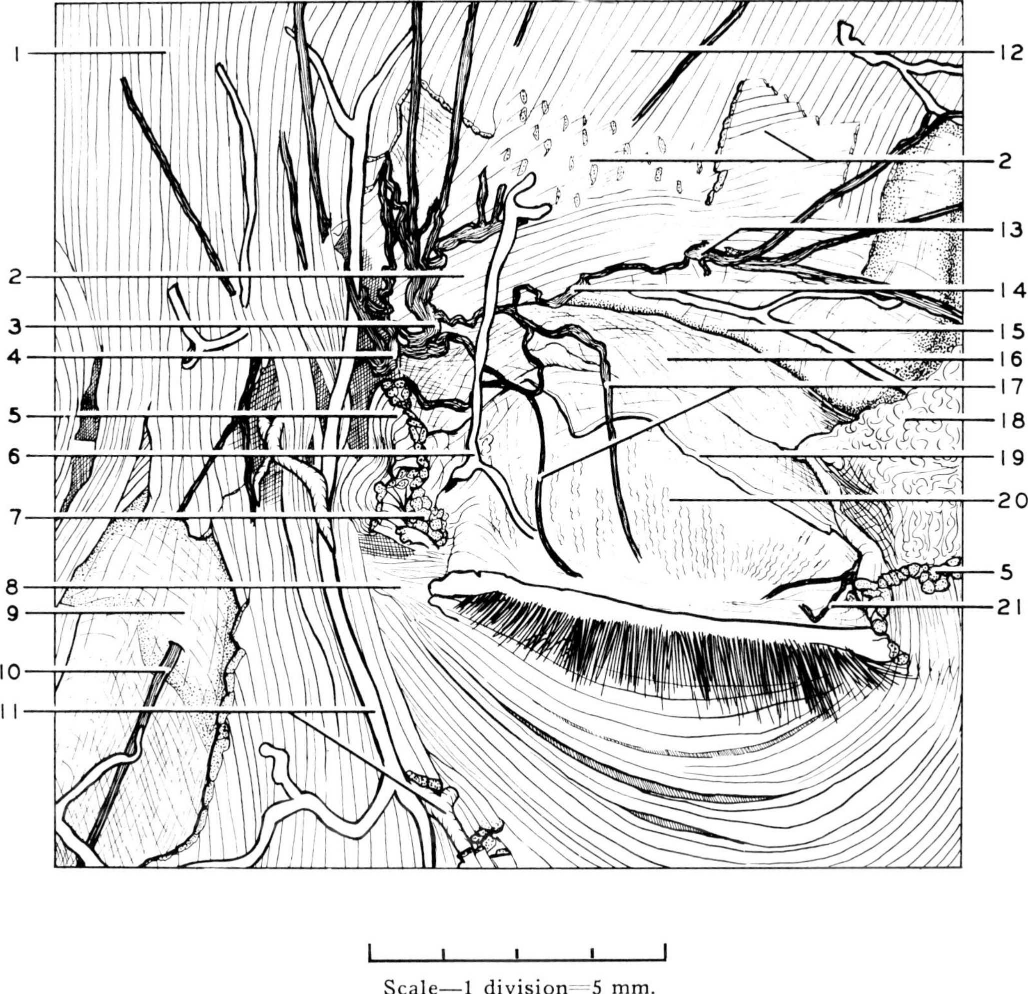

Scale—1 division=5 mm.

ANTERIOR ORBITAL DISSECTION

Insertions of levator and orbicularis muscles

1. Frontalis m.
2. Corrugator superciliaris m.
3. Supratrochlear n.
4. Infratrochlear n.
5. Orbicularis oculi m. (divided and partially removed)
6. Medial palpebral a. (note branch which passes superiorly to communicate with supraorbital artery)
7. Cut end of superior portion of Horner's muscle
8. Medial canthal tendon
9. Nasal bone (covered by periosteum)
10. External nasal ramus of anterior ethmoidal n.
11. Angular a. and v.

12. Frontalis m.
13. Supraorbital n. (communicates in part with temporal branch of facial nerve)
14. Frontal n. (note contribution to palpebral nerves)
15. Supraorbital margin (periosteum intact)
16. Orbital septum
17. Superior palpebral branches of frontalis supratrochlear and infratrochlear n.
18. Layer of fat beneath orbicularis oculi muscle
19. Aponeurosis of levator palpebrae superioris m.
20. Tarsus superior
21. Lacrimal n.

CLINICAL ORBITAL ANATOMY

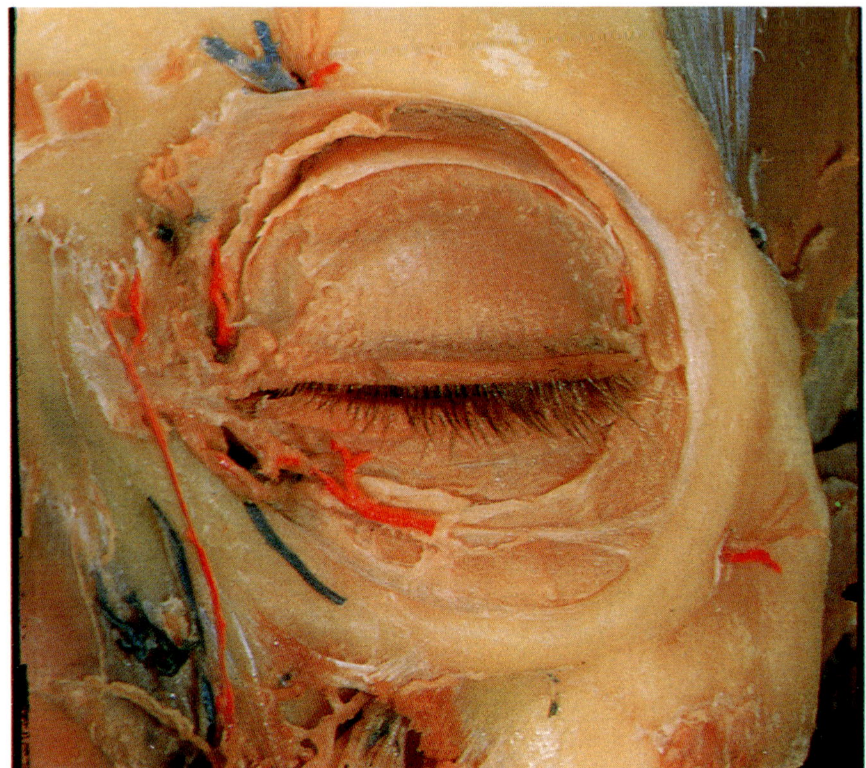

Figure 10.15

(From David L. Bassett, M.D., *Sterescopic Atlas of Human Anatomy*, published by Sawyer's Inc., Portland, Oregon © 1954, color photographs by Wm. B. Gruber.)

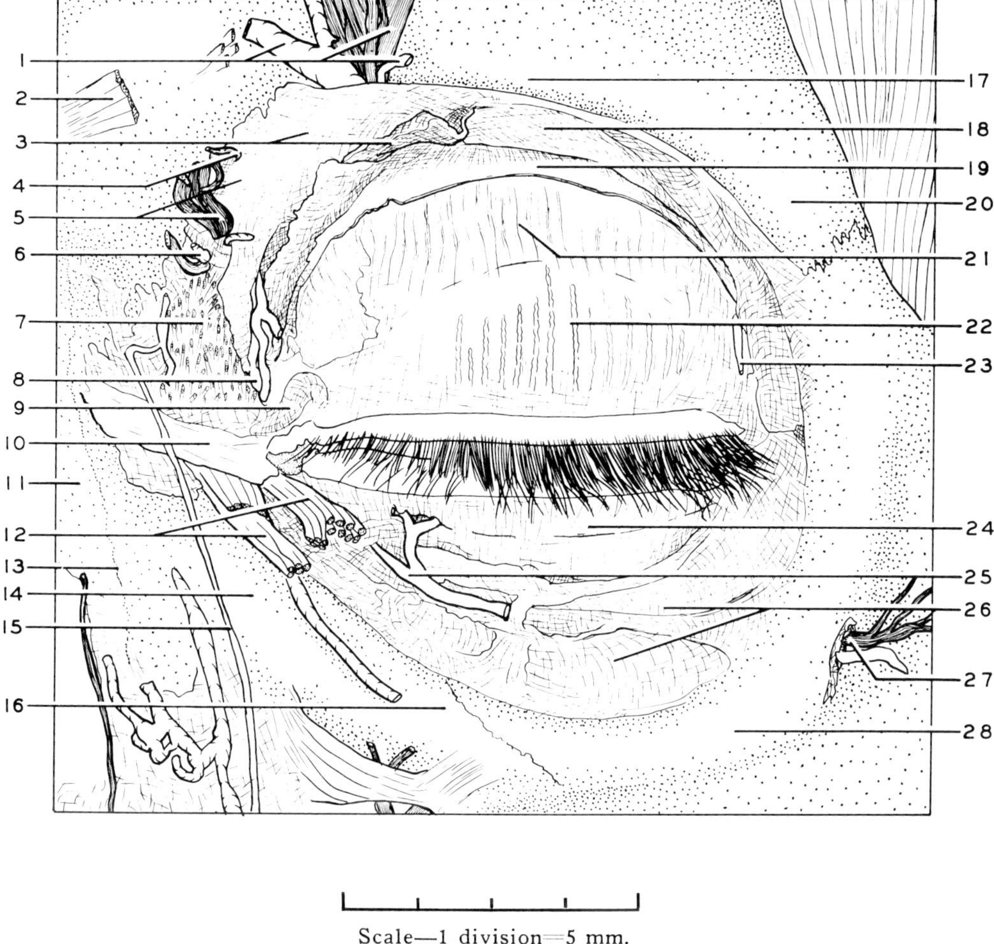

Scale—1 division=5 mm.

ANTERIOR ORBITAL DISSECTION

Orbital septum insertions of levator muscle

1. Supraorbital n., a., and v.
2. Remnant of origin of corrugator superciliaris m.
3. Orbital septum (detached from aponeurosis of levator palpebrae muscle)
4. Supratrochlear n.
5. Upper pointer: Position of trochlea
   Lower pointer: Infratrochlear n.
6. Nasofrontal v. emerging from orbit
7. Area of origin of upper part of orbicularis oculi m.
8. Medial palpebral a.
9. Superior canaliculus
10. Medial canthal tendon
11. Nasal bone
12. Fibers of orbicularis oculi m. (cut across near origin)
13. Nasomaxillary suture
14. Frontal process of maxilla
15. Small artery along periosteum
16. Zygomaticomaxillary suture
17. Superior orbital margin
18. Fascia of levator palpebrae superioris m.
19. Aponeurosis of levator palpebrae superioris m.
20. Zygomatic process of frontal bone
21. Superior tarsal m. (Müller's muscle)
22. Superior tarsus and Meibomian glands
23. Lateral palpebral a.
24. Inferior tarsus
25. Orbital branch (aberrant) from external maxillary a.
26. Orbital septum
27. Zygomaticofacial branch of zygomatic n.
28. Zygomatic bone

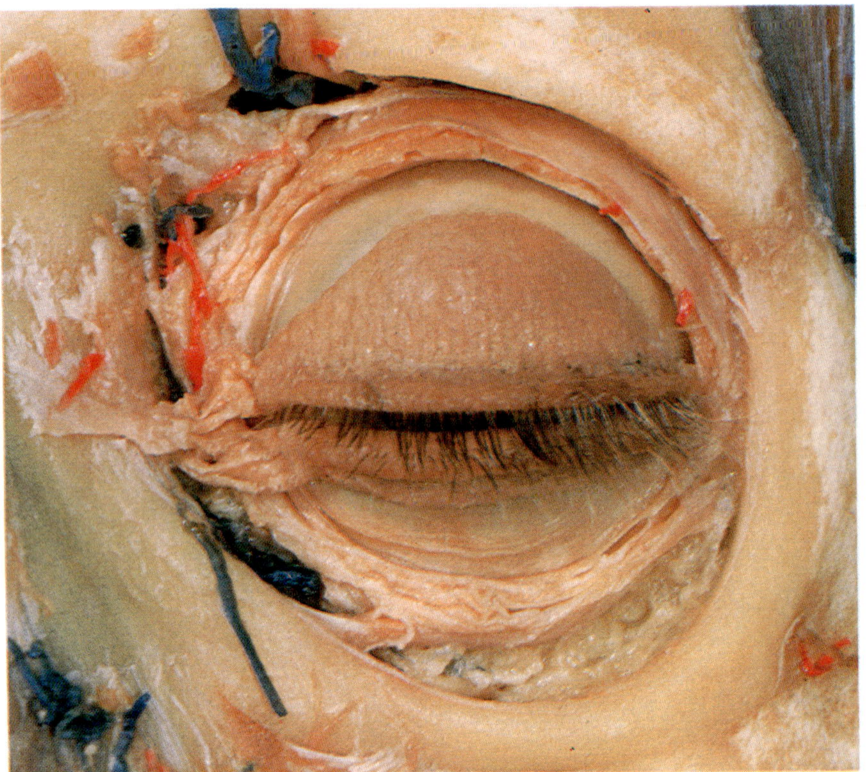

Figure 10.16

(From David L. Bassett, M.D., *Stereoscopic Atlas of Human Anatomy*, published by Sawyer's Inc., Portland, Oregon © 1954, color photographs by Wm. B. Gruber.)

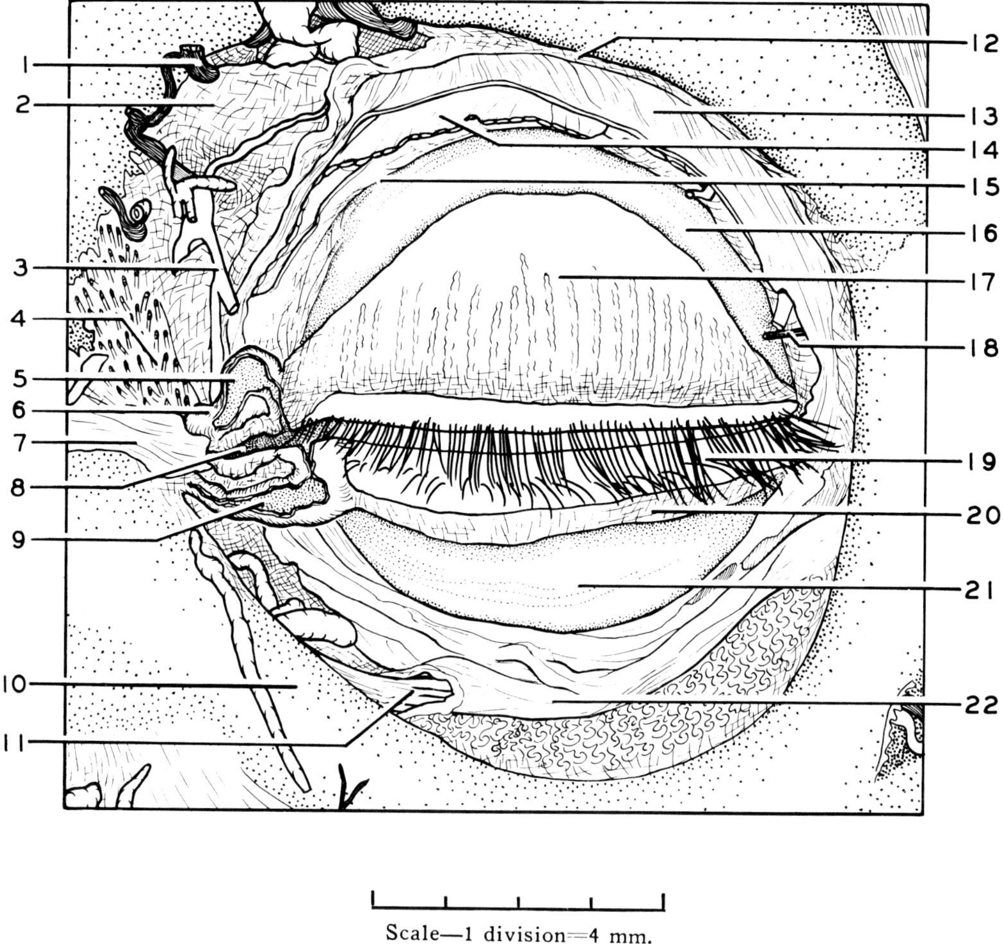

Scale—1 division=4 mm.

ANTERIOR ORBITAL DISSECTION

Suspensory ligament of Lockwood and canalicular system

1. Supratrochlear n.
2. Trochlea: Superior oblique m.
3. Medial palpebral a.
4. Lacrimal fascia overlying lacrimal sac (some fibers of origin of the orbicularis oculi muscle remain)
5. Ampulla superior canaliculus (opened)
6. Superior canaliculus
7. Medial canthal tendon
8. Medial canthal angle
9. Ampulla inferior lacrimal canaliculus (opened)
10. Infraorbital margin
11. Inferior oblique m.
12. Fascia of levator palpebrae superioris m.
13. Aponeurosis of levator palpebrae superioris m. (cut off)
14. Superior tarsal m. (cut across)
15. Superior conjunctival fornix
16. Conjunctiva
17. Superior tarsus
18. Lacrimal n.
19. Skin supporting cilia of lower lid
20. Inferior tarsus
21. Conjunctiva
22. Fascia forming suspensory ligament of eye

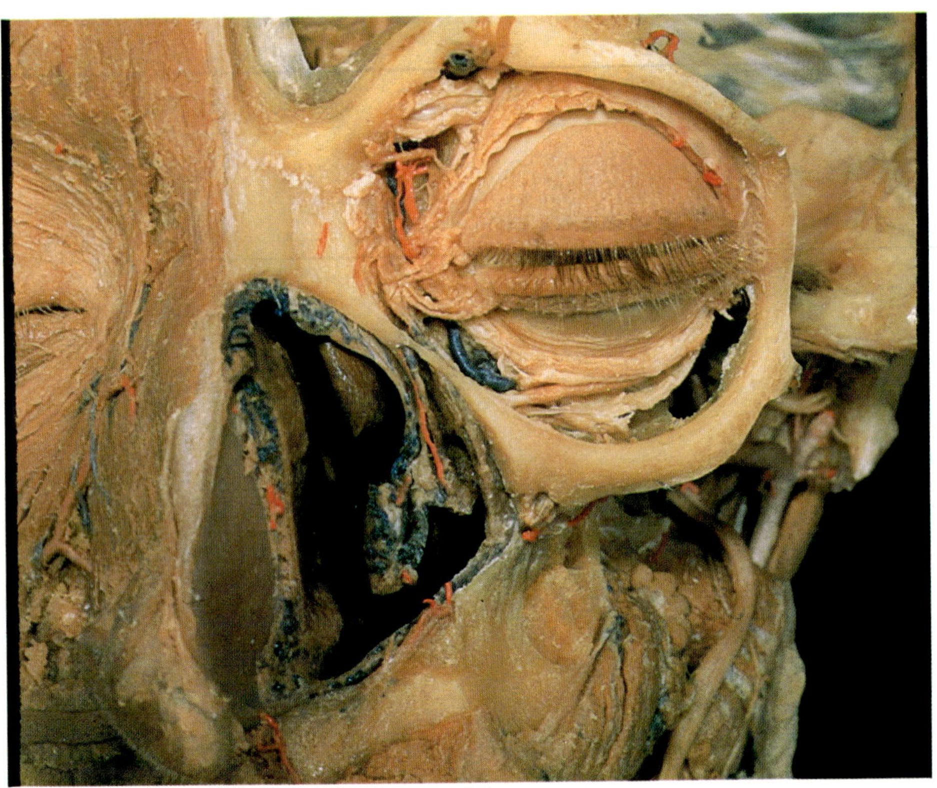

Figure 10.17

(From David L. Bassett, M.D., *Stereoscopic Atlas of Human Anatomy*, published by Sawyer's Inc., Portland, Oregon © 1954, color photographs by Wm. B. Gruber.)

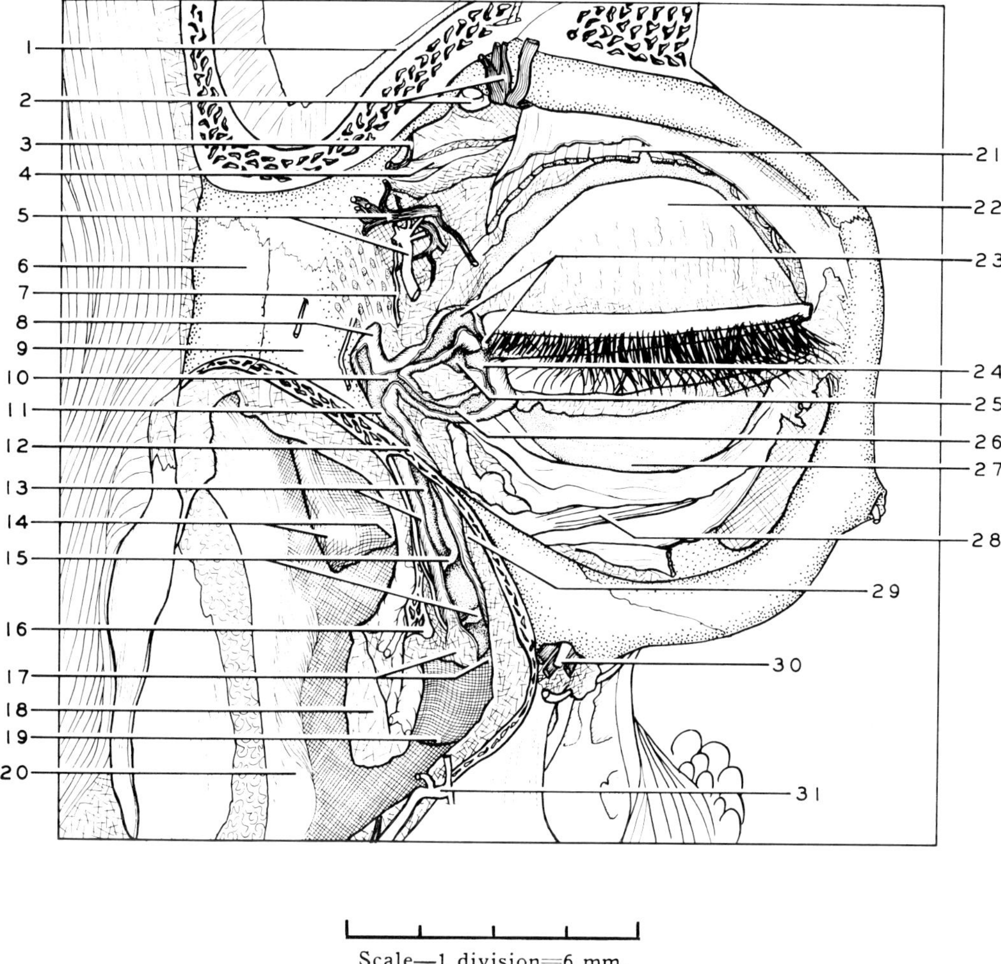

Scale—1 division=6 mm.

ANTERIOR ORBITAL DISSECTION

Suspensory ligament of Lockwood and dissection of the lacrimal excretory system

1. Dura mater
2. Supraorbital n. and v.
3. Supratrochlear n.
4. Tendon of superior oblique m.
5. Infratrochlear and dorsal nasal a.
6. Nasal bone
7. Inferior medial palpebral a.
8. Lacrimal sac (fornix)
9. Frontal process of maxilla
10. Site of entry of lacrimal canaliculi into lacrimal sac (sinus of Maier)
11. Lacrimal sac (opened)
12. Infraorbital margin
13. Nasolacrimal duct (opened; upper pointer, cavity; lower pointer, outer surface)
14. Middle turbinate and middle meatus
15. Mucosal folds within nasolacrimal duct
16. Inferior turbinate (cut across to illustrate bony wall of nasolacrimal canal)
17. Left pointer: Valve of Hasner
    Right pointer: Opening of nasolacrimal duct
18. Mucosa of inferior turbinate (note plexus venosus cavernosus conchae)
19. Inferior meatus
20. Nasal septum
21. Superior tarsal m.
22. Superior tarsus
23. Lacrimal punctum and ampulla of superior canaliculus
24. Inferior punctum
25. Medial canthal angle
26. Inferior canaliculus
27. Inferior conjunctival fornix
28. Inferior oblique m.
29. Periosteal lining of nasolacrimal canal
30. Infraorbital foramen
31. Anterior superior alveolar a.

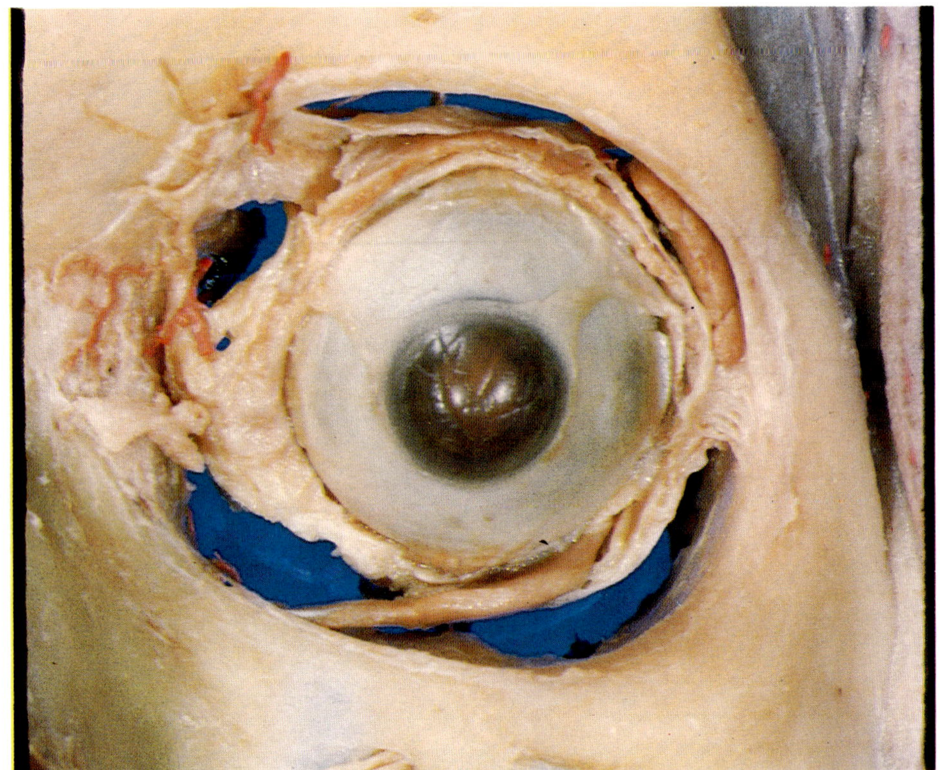

Figure 10.18

(From David L. Bassett, M.D., *Sterescopic Atlas of Human Anatomy*, published by Sawyer's Inc., Portland, Oregon © 1954, color photographs by Wm. B. Gruber.)

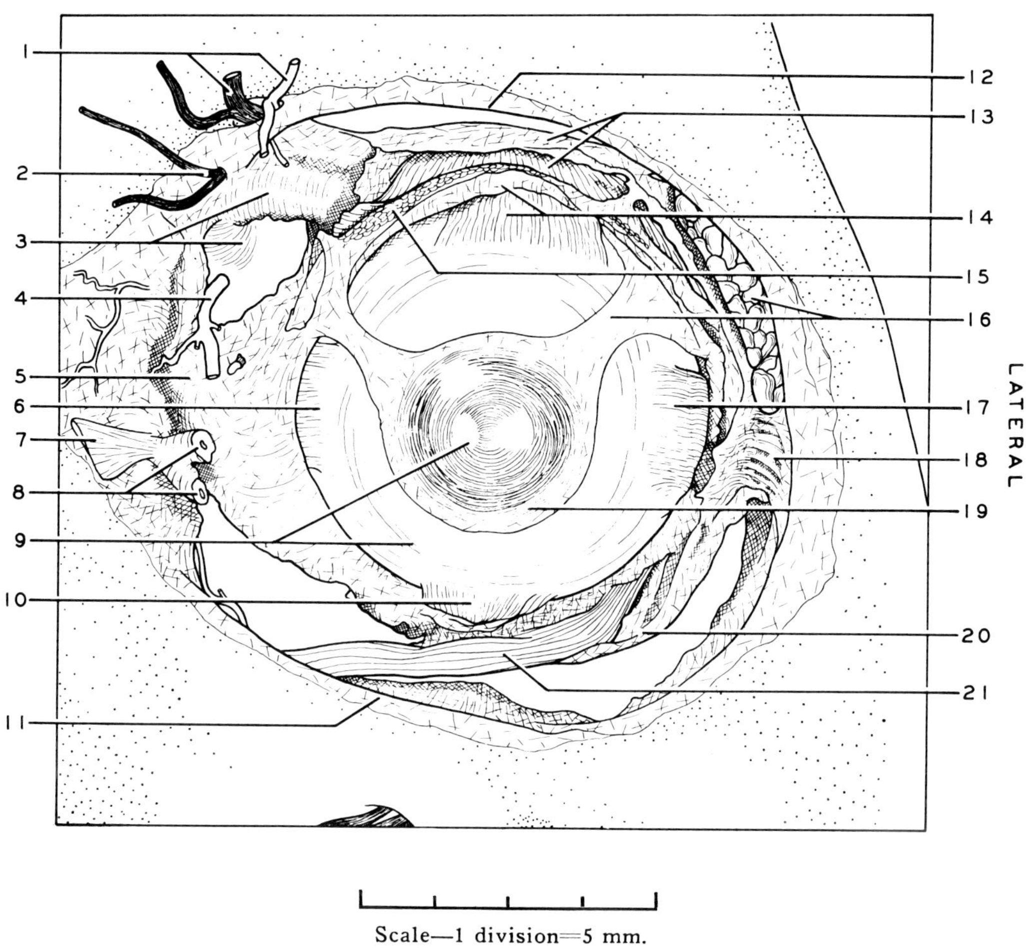

ANTERIOR ORBITAL DISSECTION

Inferior oblique muscle

1. Right pointer: Frontal a.
   Left pointer: Supratrochlear n.
2. Infratrochlear n.
3. Upper pointer: Trochlea superior oblique m. (covered by fascia)
   Lower pointer: Superior oblique m. (in background)
4. Medial palpebral a. (cut off)
5. Medial check ligament
6. Tendon of medial rectus m.
7. Medial canthal tendon
8. Lacrimal canaliculus
9. Upper pointer: Cornea
   Lower pointer: Sclera
10. Tendon of inferior rectus m.
11. Infraorbital margin
12. Supraorbital margin
13. Upper pointer: Fascia above levator palpebrae superioris m.
    Lower pointer: Aponeurosis of levator palpebrae superioris m. (cut off)
14. Upper pointer: Fascia between levator palpebrae superioris and superior rectus m.
    Lower pointer: Tendon of superior rectus m.
15. Superior tarsal m. (Müller) (cut across)
16. Upper pointer: Lacrimal gland
    Lower pointer: Conjunctiva and Tenon's capsule
17. Tendon of lateral rectus m.
18. Lateral check ligament (attached to orbital tubercle of zygomatic bone)
19. Annulus conjunctiva of overlapping corneal limbus
20. Inferior part of muscle fascia which forms suspensory ligament of eye
21. Inferior oblique m.

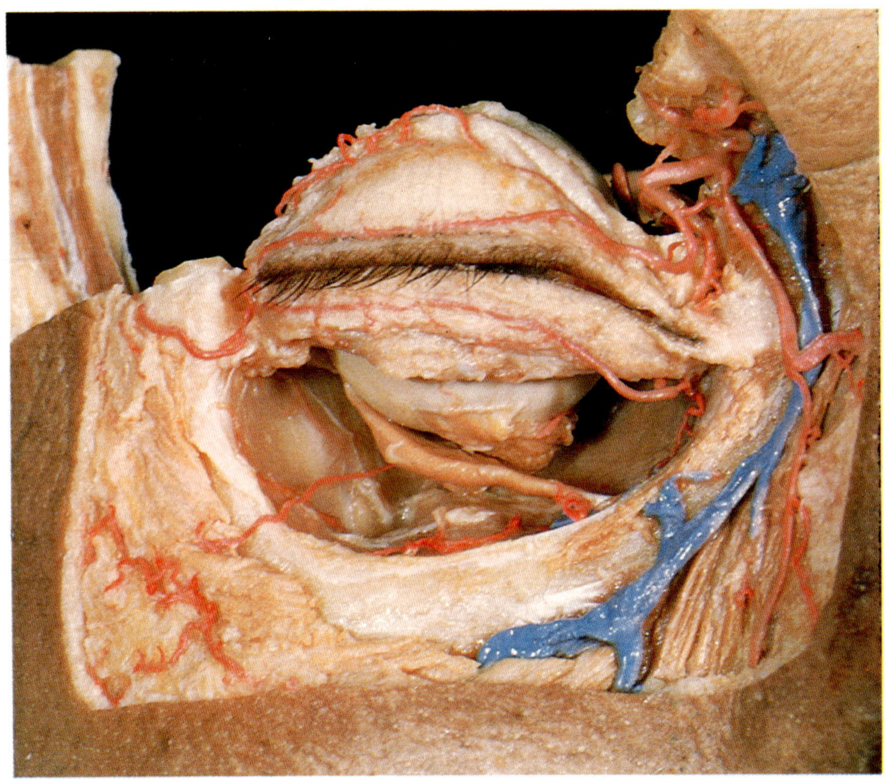

Figure 10.19

(From David L. Bassett, M.D., *Sterescopic Atlas of Human Anatomy*, published by Sawyer's Inc., Portland, Oregon © 1954, color photographs by Wm. B. Gruber.)

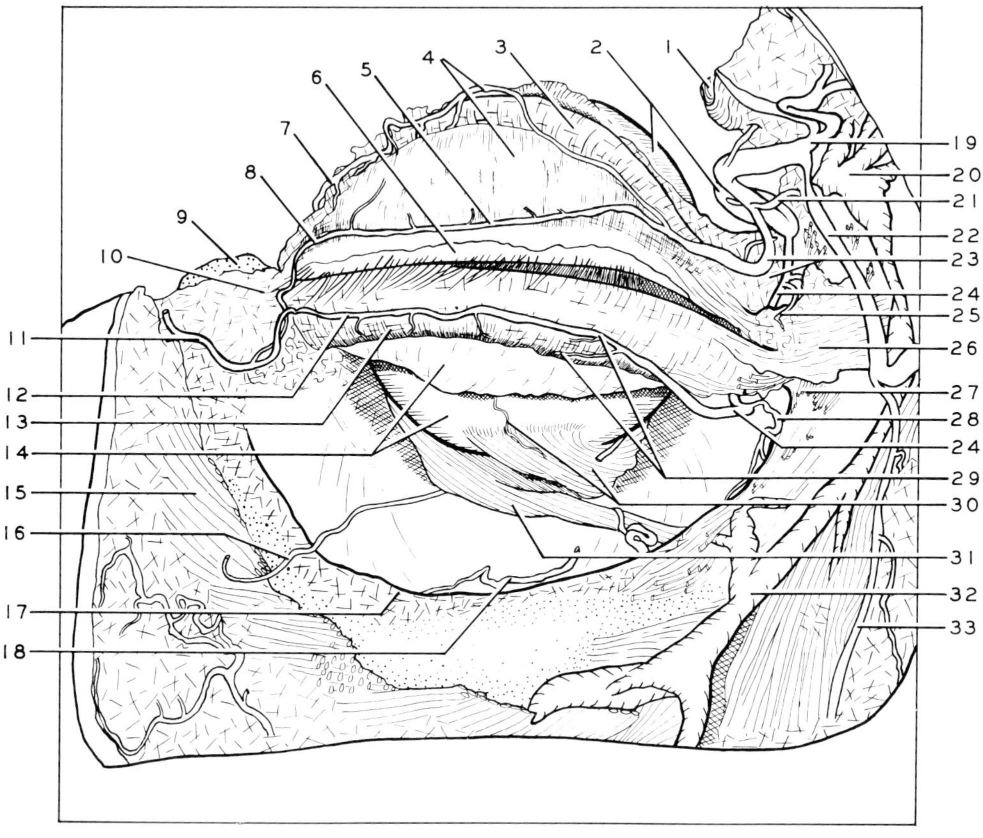

Scale—1 division=5 mm.

ANTERIOR ORBITAL DISSECTION

Palpebral arteries

1. Trochlea: Superior oblique m. (tendon removed)
2. Right pointer: Ophthalmic a. (in depths of dissection)
   Left pointer: Sclera
3. Conjunctiva (outer surface)
4. Upper pointer: Superior tarsal arcade
   Lower pointer: Superior tarsus
5. Superior tarsal arcade (marginal)
6. Skin of lid margin
7. Lateral palpebral branch of lacrimal a.
8. Superior tarsal branch of zygomaticoorbital a. (this vessel has only minute branches communicating with lacrimal a. (7))
9. Orbital margin (cut across)
10. Lateral palpebral raphe
11. Lateral palpebral branch of zygomaticoorbital a.
12. Inferior tarsal branch of zygomaticoorbital a.
13. Inferior tarsus
14. Upper pointer: Inferior tarsus conjunctiva palpebral (outer surface)
    Lower pointer: Sclera
15. Orbicularis oculi m.
16. Orbital branch of transverse facial a.
17. Infraorbital margin
18. Orbital branch of infraorbital a.
19. Frontal a.
20. Supraorbital v.
21. Small artery which enters osseous canal in frontal process of maxilla
22. Dorsal nasal a.
23. Upper pointer: Superior medial palpebral a.
    Lower pointer: Medial check ligament
24. Inferior medial palpebral a.
25. Arterial branch to lacrimal caruncle
26. Medial canthal tendon
27. Area of origin of orbicularis oculi m.
28. Arterial branch to lacrimal sac
29. Inferior tarsal arcade (marginal and peripheral arcades are present but not as well defined as in upper eyelid)
30. Left pointer: Anterior ciliary a.
    Right pointer: Inferior rectus m.
31. Inferior oblique m.
32. Angular v.
33. Angular a.

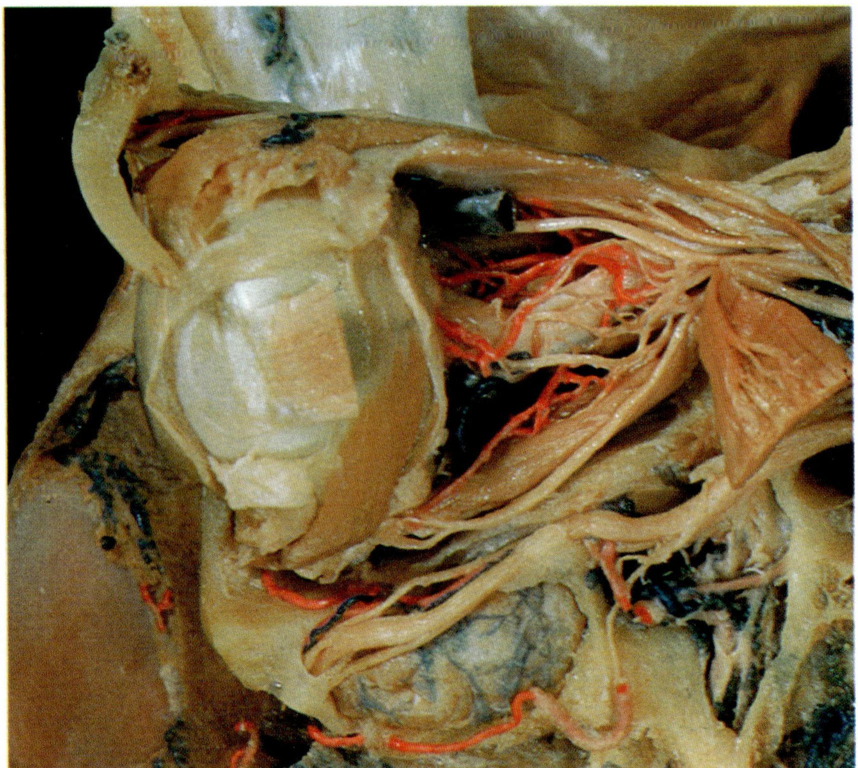

Figure 10.23

(From David L. Bassett, M.D., *Sterescopic Atlas of Human Anatomy*, published by Sawyer's Inc., Portland, Oregon © 1954, color photographs by Wm. B. Gruber.)

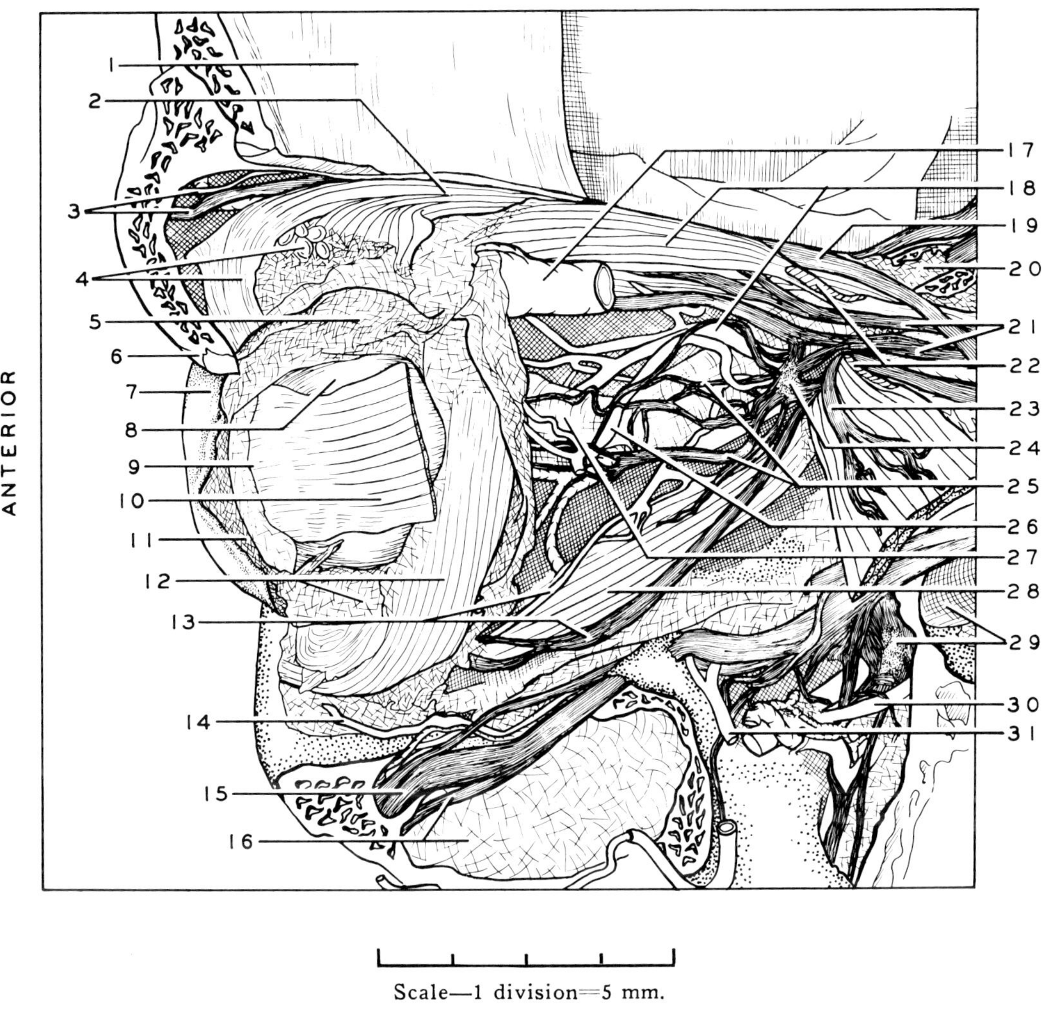

LATERAL ORBITAL DISSECTION

Orbital structures within the muscle cone

1. Falx cerebri
2. Levator palpebrae superioris m.
3. Supraorbital a. and n.
4. Upper pointer: Lacrimal gland
   Lower pointer: Lateral horn of aponeurosis of levator palpebrae superioris m.
5. Area of blending of fascia of lateral rectus m. with that of the bulb
6. Supraorbital margin
7. Conjunctiva
8. Sclera
9. Insertion of lateral rectus m.
10. Lateral rectus m.
11. Conjunctiva (elevated together with anterior extension of fascia bulbi)
12. Inferior oblique m. and related fascia
13. Artery and nerve to inferior oblique m.
14. Orbital branch of infraorbital n.
15. Infraorbital n. within infraorbital canal (cut open)
16. Upper pointer: Anterior superior alveolar n.
    Lower pointer: Mucoperiosteum of maxillary sinus

17. Superior ophthalmic v.
18. Upper pointer: Superior rectus m.
    Lower pointer: Ophthalmic a.
19. Frontal n.
20. Optic n. (optic canal opened)
21. Upper pointer: Branch of oculomotor n. to superior rectus muscle
    Lower pointer: Nasociliary n.
22. Annulus of Zinn (cut open to show contents of "oculomotor foramen")
23. Abducens n. (VI) entering lateral rectus m.
24. Ciliary ganglion
25. Short ciliary n. (no distinct long ciliary branches of nasociliary n. were present)
26. Long posterior ciliary a.
27. Optic n.
28. Inferior rectus m.
29. Sphenoid sinus and sphenopalatine ganglion
30. Vidian n. (nerve of pterygoid canal)
31. Infraorbital a. (cut off)

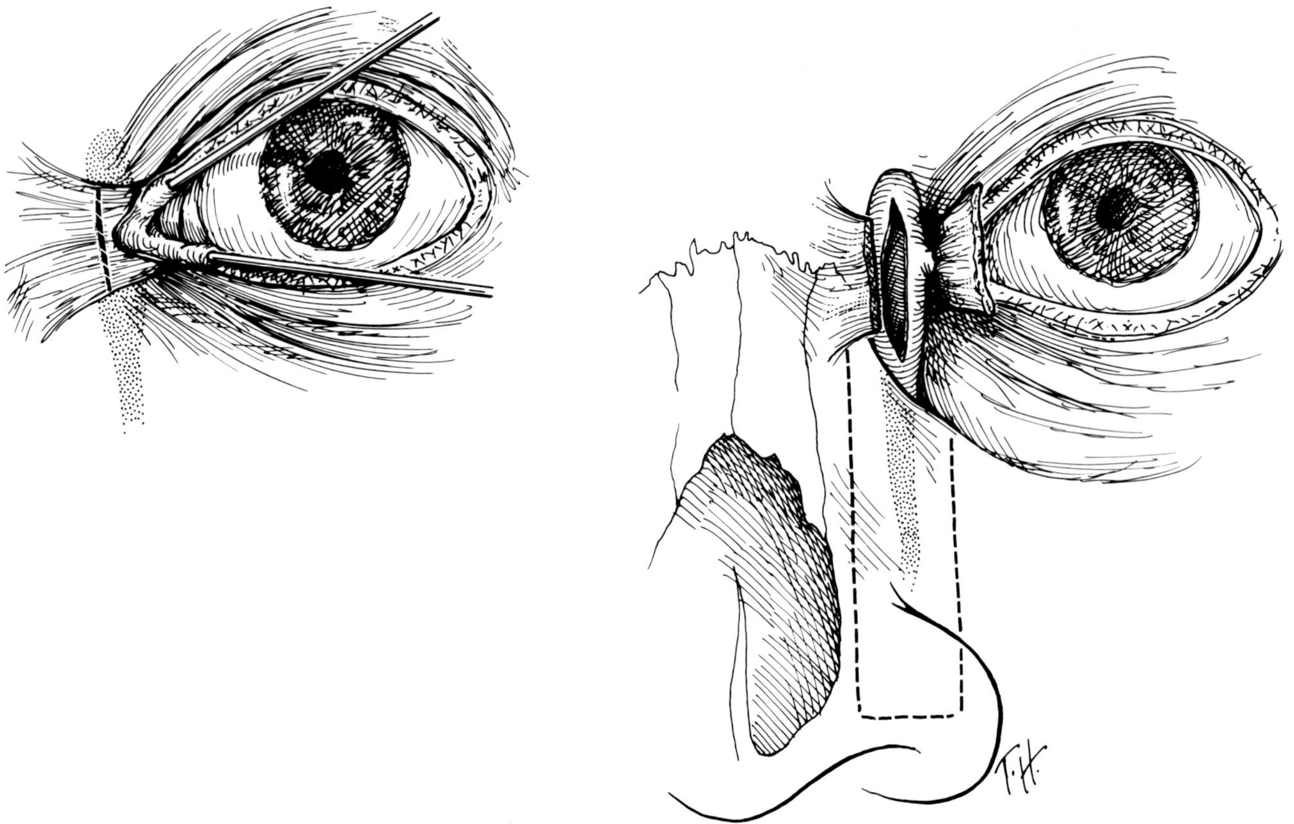

Figure 10.24. Lacrimal system. *Left*, probing of canalicular system; *right*, *dotted line* indicates area of bone removal to expose the nasolacrimal duct.

tering the medial wall of the lacrimal fossa. The lacrimal probe is held firmly against the medial wall of the lacrimal fossa and rotated superiorly so that it will pass inferiorly and slightly laterally to permit probe entrance into the nasolacrimal duct (Fig. 10.17). Once within the nasolacrimal duct, the probe is passed to enter the nose in the inferior meatus. Forced passage of the probe should be avoided in clinical conditions as it may create a false passage into the nose, sinus, or orbit. Appreciation of the lacrimal excretory system anatomy is essential for atraumatic nasolacrimal duct probing and silicone intubation.

One may become better acquainted with the normal course of the lacrimal system through repeated probing. After acquiring a familiarity with the system, the thin mucocutaneous lining of the punctum and canaliculus is incised with a probe in place. The canalicular system extends beneath the superficial component of the medial canthal tendon. To facilitate visualization of the nasal aspect of the canalicular system and the lacrimal sac, the medial canthal tendon is incised at its insertion on the frontal process of the maxilla and is then reflected laterally (Fig. 10.24). Prior to entering the lacrimal sac, the upper and lower canalicular systems unite to form the common canaliculus. Prior to entering the lacrimal sac, the common canaliculus dilates, forming the sinus of Maier. A flap of mucosa within the lacrimal sac acts as a valve at the internal common punctum. This is called the valve of Rosenmuller. It prevents reflux from the lacrimal sac into the canalicular system.

Deeper dissection exposes the bony anatomy of the medial canthal region. The lacrimal sac lies within the fossa between the anterior and posterior lacrimal crests and is surrounded by periorbita. The periorbita splits into two layers; the posterior medial layer lines the bone, while the anterior lateral layer, covering the lacrimal sac, forms the lacrimal diaphragm. The sac extends superiorly, several millimeters above the medial canthal tendon, and has a vertical dimension of 13–15 mm. Incision of the periosteum on the frontal process of the maxilla near the anterior lacrimal crest facilitates the elevation of the fused periosteum and the lacrimal sac from its fossa. The periosteum is adherent at the anterior lacrimal crest, prior to its entering the lacrimal fossa. A hemostat or a chisel can be used to break through the bone of the lacrimal fossa. A rongeurs may then be used to remove the anterior lacrimal crest surrounding frontal process of the maxilla and the lacrimal bone in the lacrimal fossa as in a dacryocystorhinostomy. The lacrimal sac should be opened vertically. The internal common punctum is identified by placing a probe in the canaliculus. The inferior continuation of the sac into the nasolacrimal duct is identified.

The anterior wall of the bony nasolacrimal duct should be removed with a hammer and chisel (Fig. 10.24). The inferior meatus of the nose is exposed upon bone removal. A probe placed within the nasolacrimal duct confirms its inferior and slightly lateral course to the inferior meatus. The mucosa of the nasolacrimal duct is incised. The mucosal flap noted at the opening of the nasolacrimal duct into the nose is the valve of Hasner. The nose should then be split near its midline with an oscillating saw and the turbinates and intranasal anatomy studied. The inferior, middle, and superior turbinates, inferior, middle and superior meatus, nasal septum, relationship of the roof of the nose to the cribriform plate, and relationship of the nose to the sinuses should be noted.

After completing the standard orbital dissection much can be learned by sacrificing orbital tissues to study areas otherwise inaccessible to examination. Tenon's capsule and the intermuscular septa can be studied posteriorly (Fig. 10.25). The eye is then removed and the gross anatomy of the globe and extraocular muscle insertions studied (Figs. 10.26 and 10.27). After the globe has been removed, the origins of various muscles and relationships of the bony orbit to surrounding structures can better be appreciated (Fig. 10.28). Following the removal of orbital tissues, the medial and inferior walls of the orbit should be studied. Remove the floor of the orbit and identify the infraorbital nerve and study the extent of the maxillary sinus. Identify the anterior and posterior ethmoidal foramina and remove the medial orbital wall and study the ethmoid sinus. Remove the posterior wall of the ethmoid sinus and explore the sphenoid sinus and its relationship to surrounding structures. If time is available remove the facial skin and try to dissect the five main branches of the seventh nerve (temporal, zygomatic, buccal, mandibular, and cervical) (Fig. 10.29). Their intimate relationship to the parotid gland should be noted. Many other pertinent superficial structures of the face and neck can be dissected if time permits.

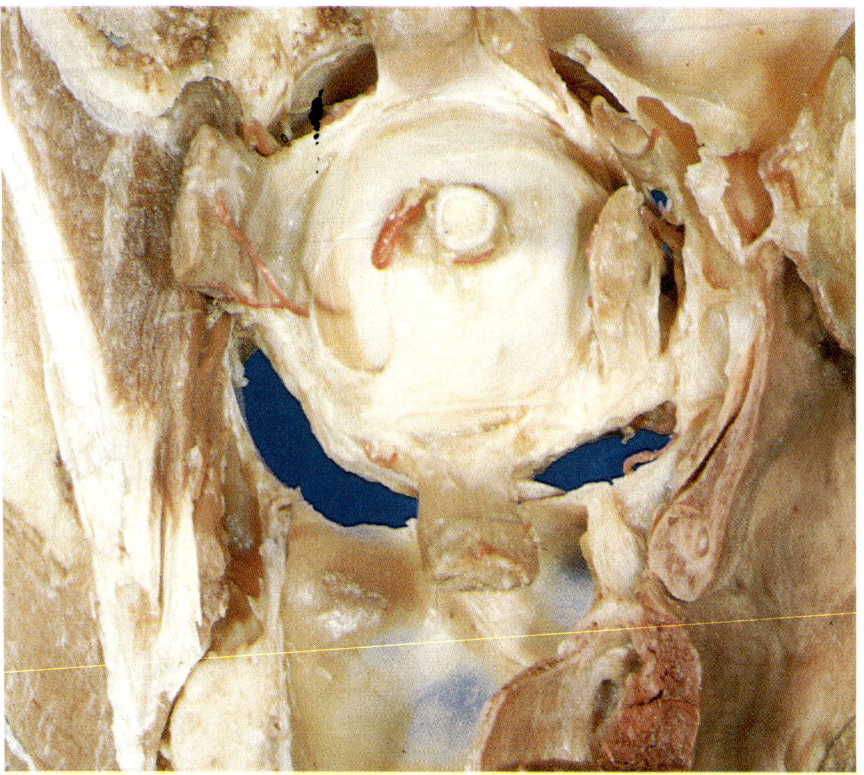

Figure 10.25

(From David L. Bassett, M.D., *Sterescopic Atlas of Human Anatomy*, published by Sawyer's Inc., Portland, Oregon © 1954, color photographs by Wm. B. Gruber.)

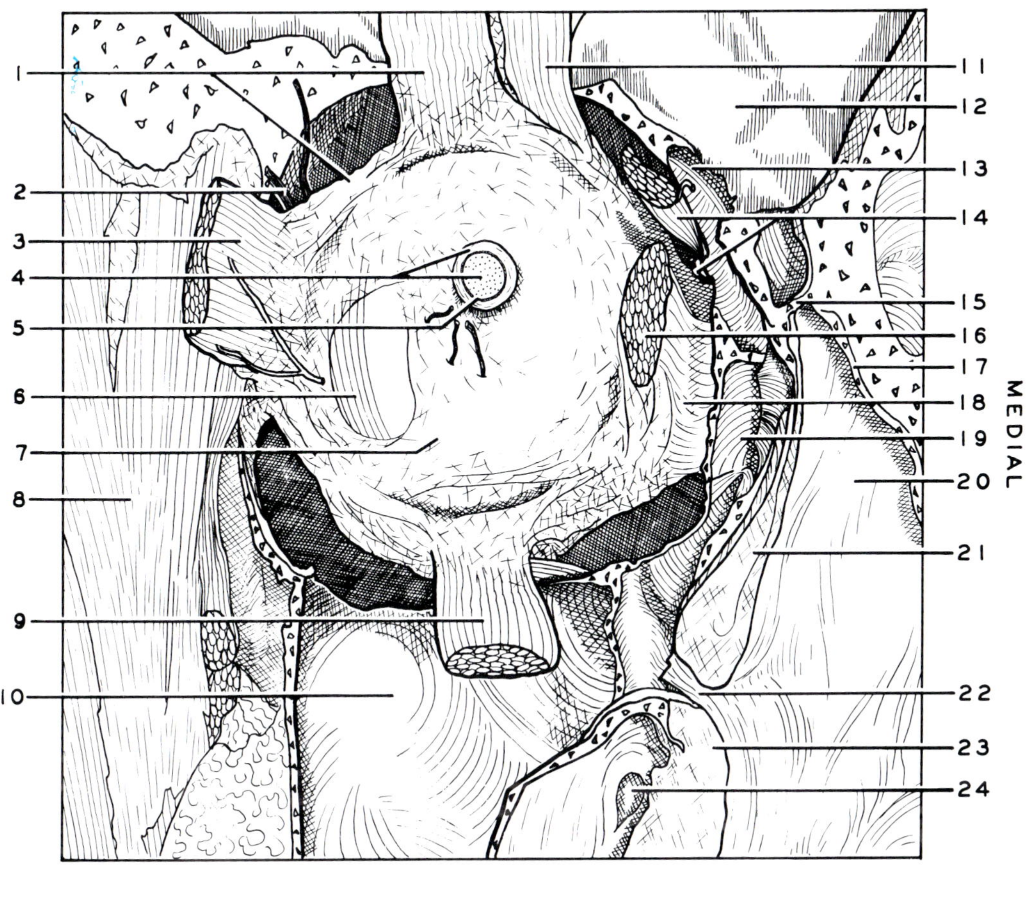

Scale—1 division=5 mm.

POSTERIOR ORBIT

Tenon's capsule

1. Upper pointer: Superior rectus m. (retracted superiorly)
   Lower pointer: Weblike fold of fascia between superior and lateral rectus muscles
2. Lacrimal n.
3. Lateral rectus m. (retracted laterally)
4. Upper pointer: Dura of optic n.
   Lower pointer: Optic n.
5. Cavum subarachnoidale (spatium intervaginalium)
6. Inferior oblique m.
7. Tenon's capsule
8. Temporalis m.
9. Inferior rectus m. (retracted inferiorly)
10. Maxillary sinus
11. Levator palpebrae superioris m. (retracted superiorly)
12. Anterior cranial fossa
13. Frontal sinus
14. Upper pointer: Superior oblique m.
    Lower pointer: Infratrochlear n.
15. Lamina cribrosa of ethmoid
16. Medial rectus m.
17. Nasal septum (cut through)
18. Medial check ligament
19. Ethmoid air cells (fossa for lacrimal sac lies anterior to this air cell)
20. Nasal mucosa
21. Middle turbinate
22. Middle meatus
23. Inferior turbinate
24. Opening of nasolacrimal duct into inferior meatus

Figure 10.26

(From David L. Bassett, M.D., *Sterescopic Atlas of Human Anatomy*, published by Sawyer's Inc., Portland, Oregon © 1954, color photographs by Wm. B. Gruber.)

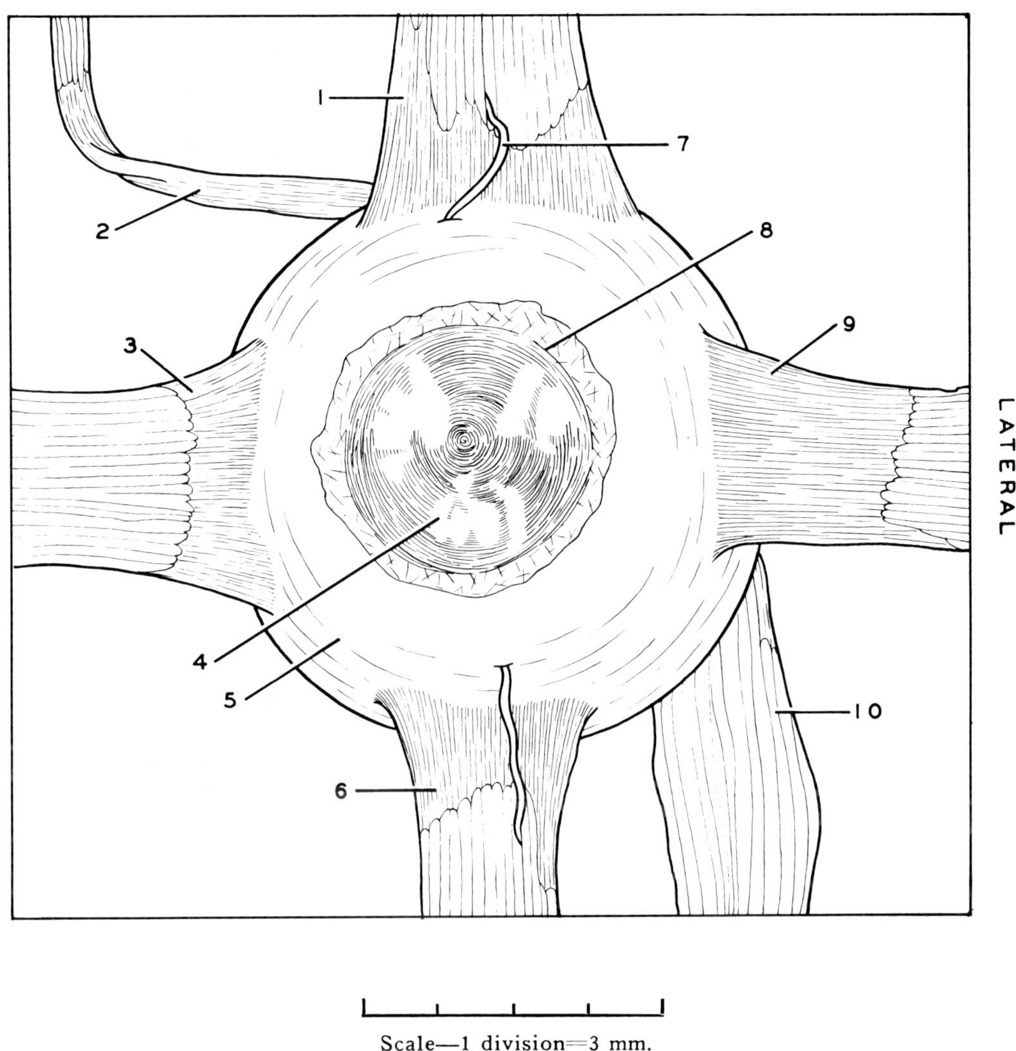

Scale—1 division=3 mm.

THE EXTRAOCULAR MUSCLES

1. Tendon of superior rectus m.
2. Tendon of superior oblique m. (portion of tendon which originally passed through trochlea turned upward)
3. Tendon of medial rectus m.
4. Cornea
5. Sclera (conjunctiva and fascia bulbi removed)
6. Tendon of inferior rectus m.
7. Anterior ciliary a.
8. Conjunctiva overlying corneal limbus
9. Tendon of lateral rectus m.
10. Inferior oblique m.

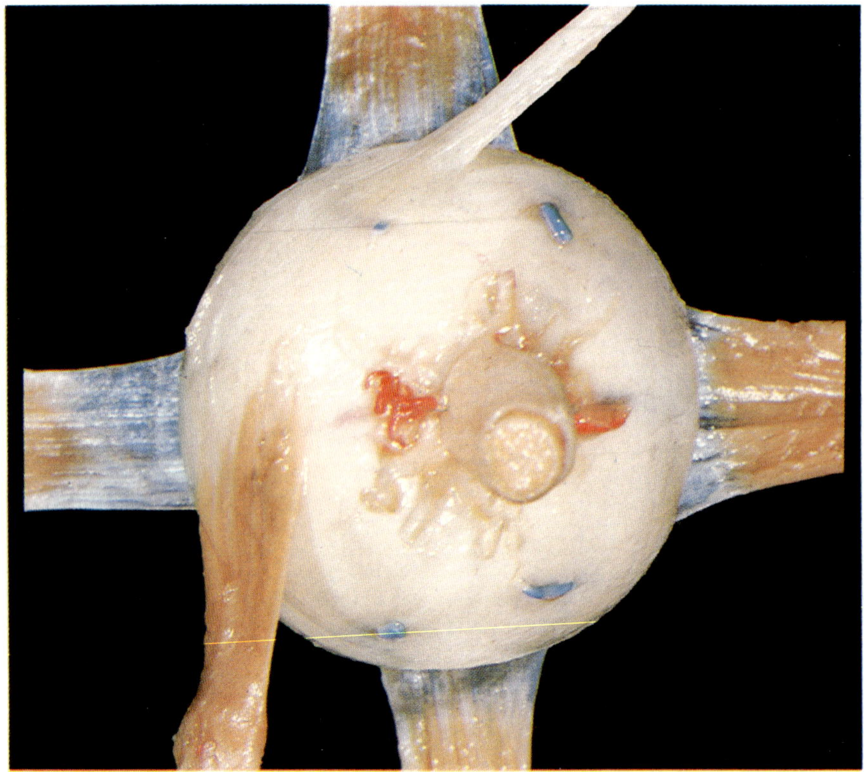

Figure 10.27

(From David L. Bassett, M.D., *Sterescopic Atlas of Human Anatomy*, published by Sawyer's Inc., Portland, Oregon © 1954, color photographs by Wm. B. Gruber.)

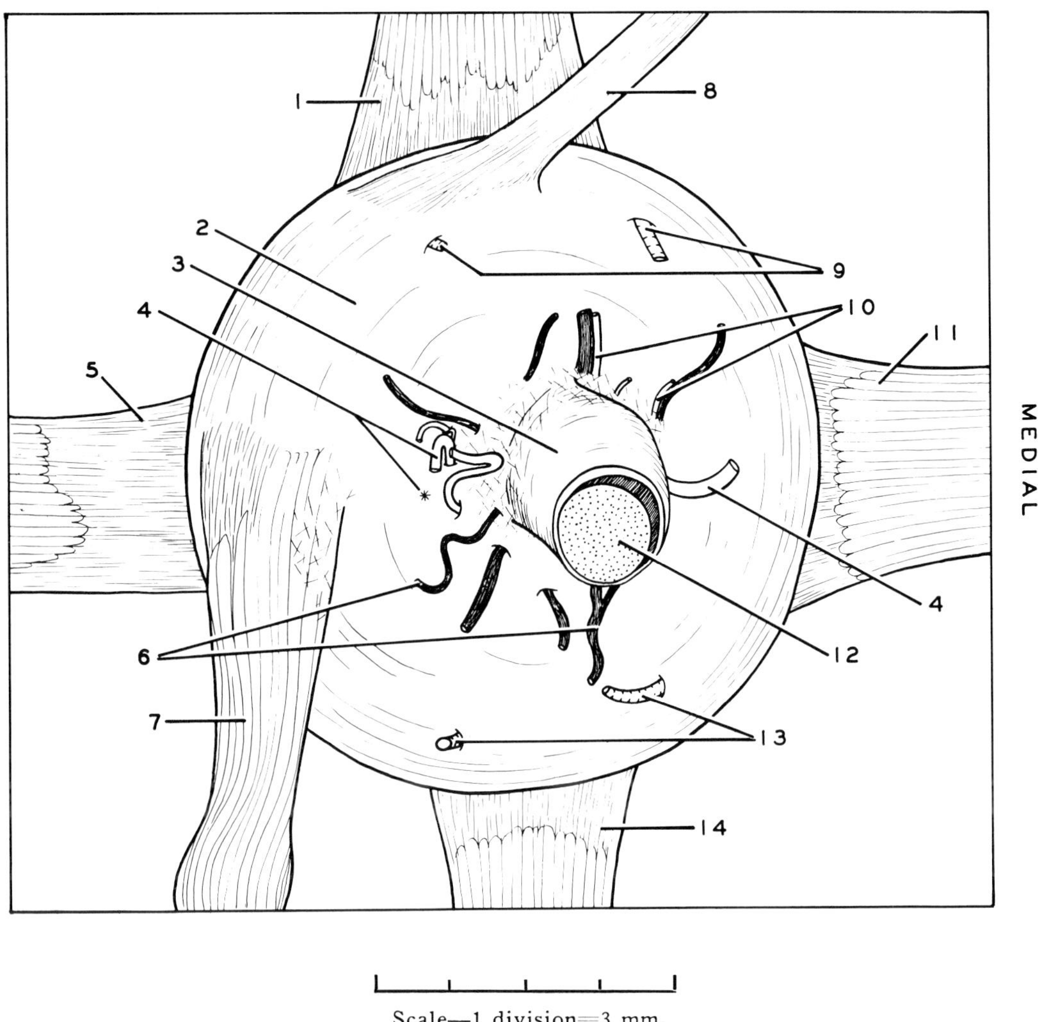

Scale—1 division=3 mm.

THE EXTRAOCULAR MUSCLES

1. Tendon of superior rectus m.
2. Sclera
3. Optic n.
4. Upper pointer: Long posterior ciliary a.
   Lower pointer: Position of macula lutea
     (marked by *)
5. Tendon of lateral rectus m.
6. Short ciliary m.
7. Inferior oblique m.
8. Tendon of superior oblique m.
9. Superior vortex vv.
10. Short posterior ciliaris aa.
11. Medial rectus m.
12. Optic n. (II)
13. Inferior vortex vv.
14. Tendon of inferior rectus m.

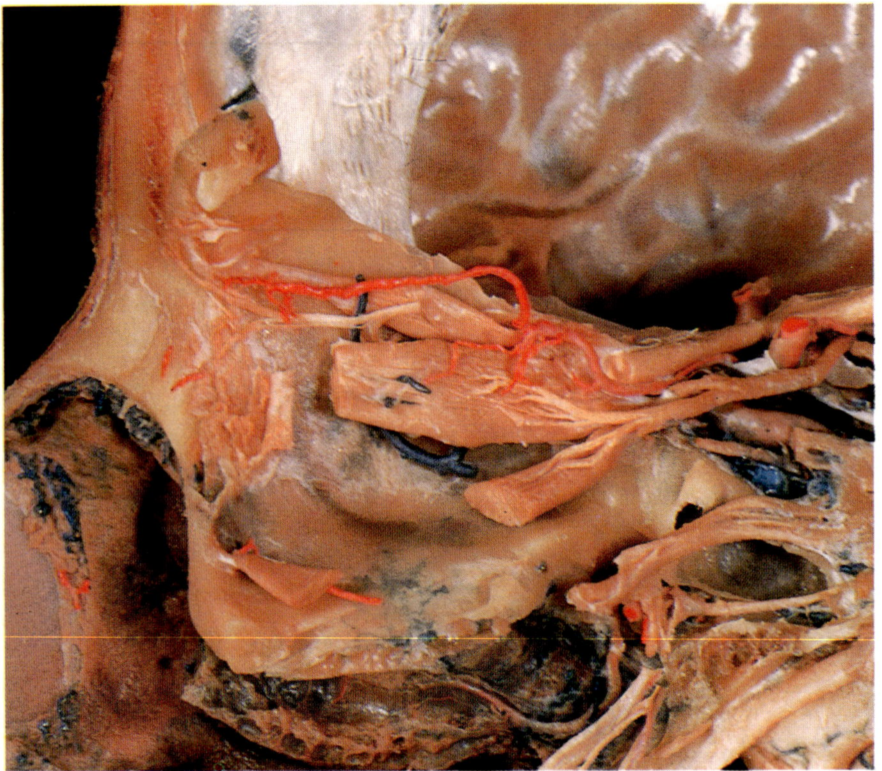

Figure 10.28

(From David L. Bassett, M.D., *Sterescopic Atlas of Human Anatomy*, published by Sawyer's Inc., Portland, Oregon © 1954, color photographs by Wm. B. Gruber.)

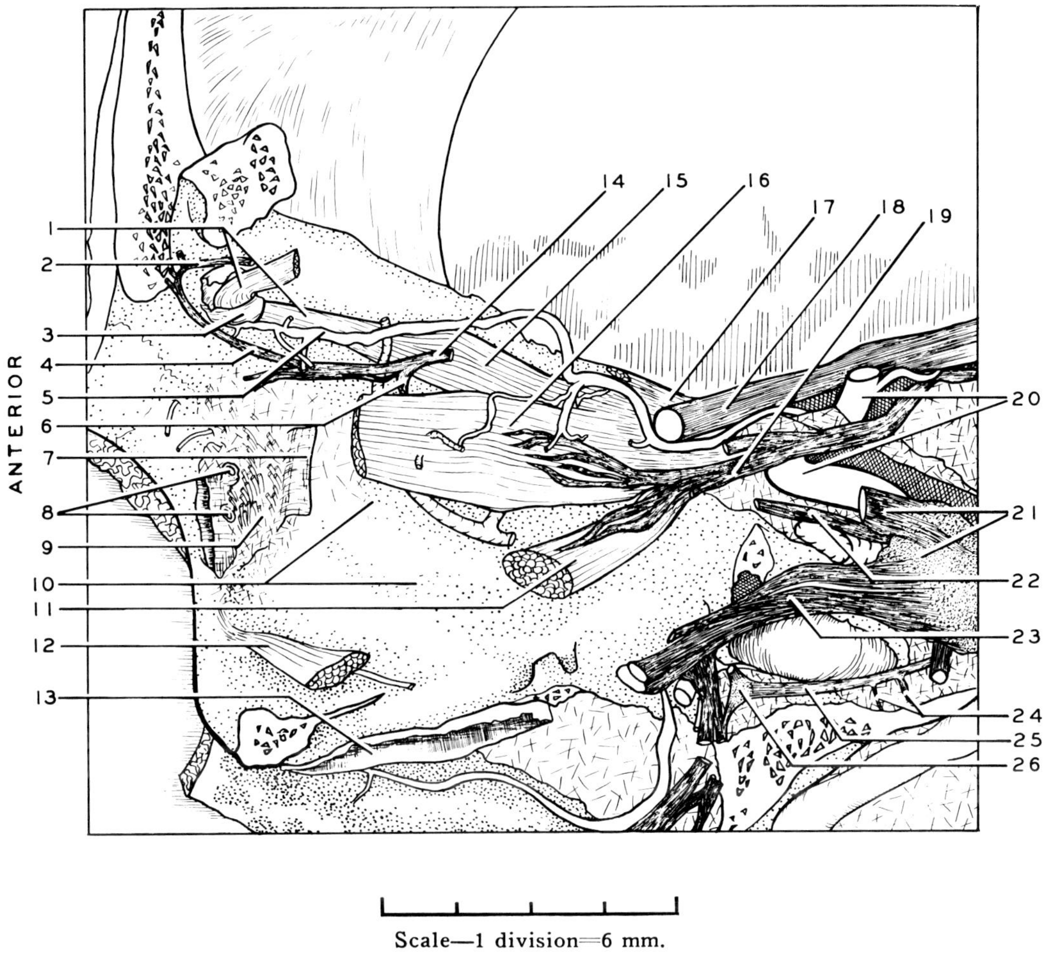

RELATIONSHIP OF ORBIT TO SURROUNDING STRUCTURES

1. Tendon of superior oblique m.
2. Supratrochlear n. (cut off within orbit but shown here in relation to trochlea and its anastomosis with the infratrochlear nerve)
3. Trochlea: Superior oblique m.
4. Infratrochlear n. (palpebral branches visible posterior to pointer)
5. Dorsal nasal a.
6. Anterior ethmoidal n.
7. Medial check ligament (attached to posterior lacrimal crest)
8. Lacrimal canaliculus (cut off)
9. Pars lacrimalis (Horner's) m. (attached to lacrimal fascia)
10. Upper pointer: Medial orbital wall
    Lower pointer: Inferior orbital wall
11. Inferior rectus m.
12. Inferior oblique m.

13. Infraorbital sulcus
14. Nasociliary n.
15. Superior oblique m.
16. Medial rectus m.
17. Annulus of Zinn (cut through)
18. Optic n. (II)
19. Oculomotor n. (III) (branches of inferior division of nerve enter medial and inferior rectus muscles)
20. Internal carotid a.
21. Upper pointer: Ophthalmic n. (V_1)
    Lower pointer: Gasserian ganglion
22. Abducens n. (VI)
23. Maxillary n. (V_2)
24. Major petrosal n. (usually described as a single nerve; here two filaments join the greater superficial petrosal nerve)
25. Vidian n. (nerve of pterygoid canal)
26. Sphenopalatine ganglion

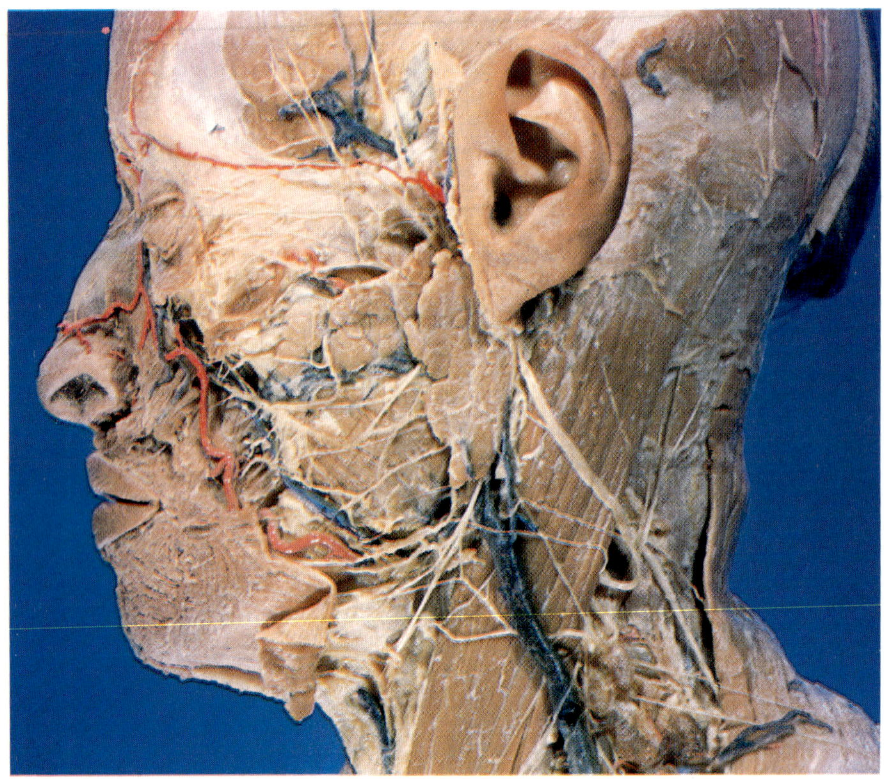

Figure 10.29

(From David L. Bassett, M.D., *Sterescopic Atlas of Human Anatomy*, published by Sawyer's Inc., Portland, Oregon © 1954, color photographs by Wm. B. Gruber.)

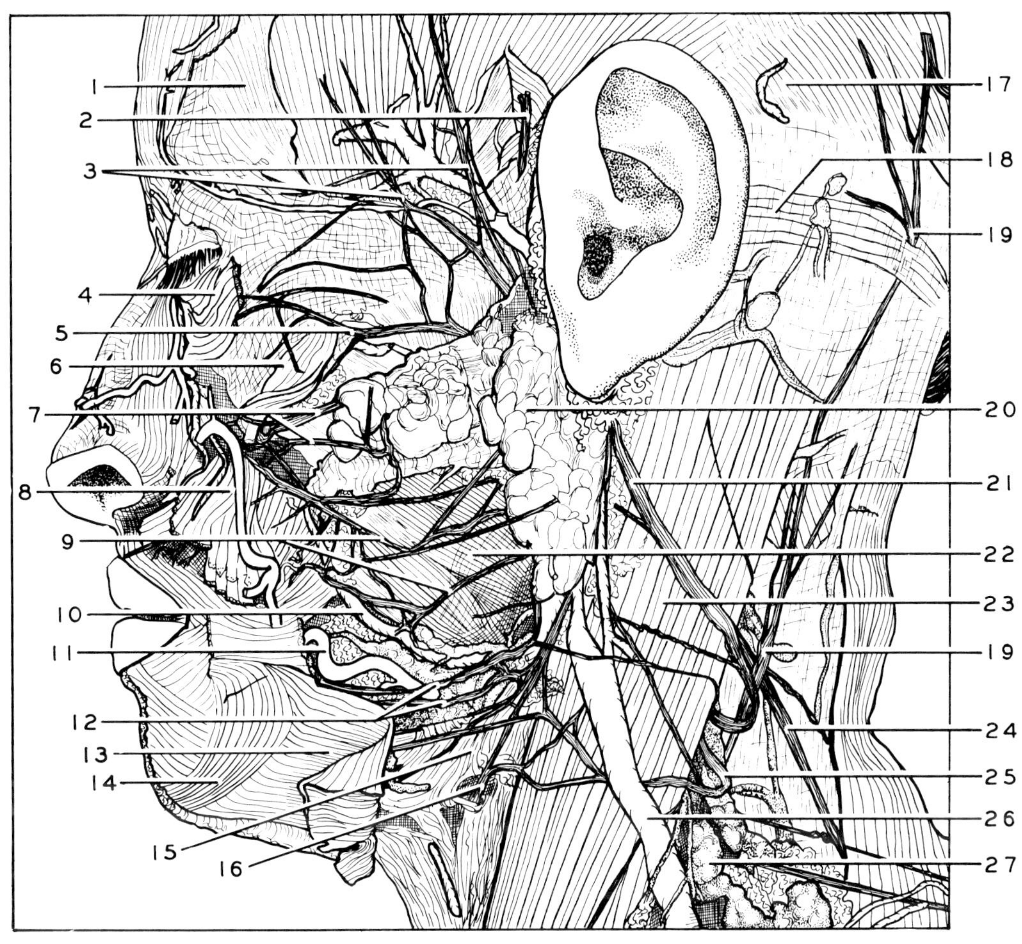

Scale—1 division=12 mm.

THE FACIAL NERVE

1. Temporalis fascia
2. Auriculotemporal n.
3. Temporal branch of facial n.
4. Orbicularis oculi m. (partially removed)
5. Zygomatic branch of facial n.
6. Zygomatic n. (cut across and reflected superiorly)
7. Infraorbital branches of facial n.
8. Levator labii superioris m.
9. Buccal branch of facial n.
10. Anterior facial v.
11. External maxillary a.
12. Mandibular branch of facial n.
13. Platysma (cut across and reflected)
14. Transversus menti
15. Fascia enclosing submaxillary gland
16. Cervical branch of facial n.
17. Occipitalis m.
18. Posterior auricular m.
19. Smaller occipital n.
20. Parotid gland
21. Greater auricular n.
22. Masseter m.
23. Sternocleidomastoideus m.
24. Accessory m. (branch to m. trapezius)
25. Cutaneous colli n.
26. External jugular v.
27. Superficial cervical node

Index